Dahlem Workshop Reports
Physical and Chemical Sciences Research Report 4
Atmospheric Chemistry

The goal of this Dahlem Workshop is:
to investigate the chemistry
of the lower atmosphere and the influences
of the biosphere, hydrosphere, and lithosphere

Physical and Chemical Sciences Research Reports
Editor: Silke Bernhard

Held and published on behalf of the
Stifterverband für die Deutsche Wissenschaft

Sponsored by:
Senat der Stadt Berlin
Stifterverband für die Deutsche Wissenschaft

Atmospheric Chemistry

E. D. Goldberg, Editor

Report of the Dahlem Workshop on
Atmospheric Chemistry
Berlin 1982, May 2-7

Rapporteurs:
M. O. Andreae · R. J. Cicerone · T. E. Graedel · J. H. Hahn

Program Advisory Committee:
E. D. Goldberg, Chairman · P. J. Crutzen · R. M. Garrels
J. H. Hahn · R. O. Hallberg · J. E. Lovelock · F. S. Rowland

Springer-Verlag Berlin Heidelberg New York 1982

Copy Editors: M. A. Cervantes-Waldmann, K. McWhirter
Photographs: E. P. Thonke

With 4 photographs, 53 figures, and 30 tables

ISBN 3-540-11651-6 Springer-Verlag Berlin Heidelberg New York
ISBN 0-387-11651-6 Springer-Verlag New York Heidelberg Berlin

CIP-Kurztitelaufnahme der Deutschen Bibliothek:

Atmospheric chemistry: report of the Dahlem Workshop on Atmospher. Chemistry, Berlin 1982, May 2-7/E. D. Goldberg, ed. Rapporteurs: M. O. Andreae . . . [Dahlem Konferenzen. Held and publ. on behalf of the Stifterverb. für d. Dt. Wiss. Sponsored by: Senat d. Stadt Berlin; Stifterverb. für d. Dt. Wiss.]. – Berlin; Heidelberg; New York: Springer, 1982.

(Physical and chemical sciences research reports; 4)
(Dahlem Workshop reports)

NE: Goldberg, Edward D. [Hrsg.]; Andreae, Meinrat O. [Mitverf.]; Workshop on Atmospheric Chemistry <1982, Berlin, West>; Dahlem Konferenzen; 1. GT

Printing: Proff GmbH & Co. KG, D-5340 Bad Honnef
Bookbinding: Graphischer Betrieb Konrad Triltsch, D-8700 Würzburg
2131/3014 – 5 4 3 2 1 0

Table of Contents

The Dahlem Konferenzen

DIRECTOR:

Silke Bernhard, M.D.

FOUNDATION:

Dahlem Konferenzen was founded in 1974 and is supported by the Stifterverband für die Deutsche Wissenschaft*, in cooperation with the Deutsche Forschungsgemeinschaft** and the Senat of the City of Berlin.

OBJECTIVES:

The task of Dahlem Konferenzen is:

- to promote the interdisciplinary exchange of scientific information and ideas,
- to stimulate international cooperation in research, and
- to develop and test different models conducive to more effective scientific meetings.

AIM:

Each Dahlem Workshop is designed to provide a survey of the present state of the art of the topic at hand as seen by the various disciplines concerned, to review new concepts and techniques, and to recommend directions for future research.

TOPICS:

The workshop topics (in the Life Sciences and the field of Physicochemistry) should be:

- of contemporary international interest,
- timely,
- interdisciplinary in nature, and
- problem-oriented.

PROCEDURE:

Dahlem Konferenzen approaches internationally recognized scientists to suggest topics fulfilling these criteria and to propose members for a Program Advisory Committee, which is responsible for the workshop's scientific program. Once a year, the topic suggestions are submitted to a scientific board for approval.

---

* The Donors Association for the Promotion of Sciences and Humanities

**German Science Foundation

PARTICIPANTS:

The number of participants is limited to 48 for each workshop. They are selected exclusively by a Program Advisory Committee. Selection is based on international scientific reputation alone and is independent of national considerations, although a balance between Europeans and Americans is desirable. Exception is made for younger German scientists for whom 10% of the places are reserved.

THE DAHLEM WORKSHOP MODEL:

A special workshop model has been developed by Dahlem Konferenzen, the *Dahlem Workshop Model*. The main work of the workshop is done in four small, interdisciplinary discussion groups, each with 12 members. Lectures are not given.

Some participants are asked to write background papers providing a review of the field rather than a report on individual work. These are circulated to all participants 4 weeks before the meeting with the request that the papers be read and questions on them formulated *before* the workshop, thus providing the basis for discussions.

During the workshop, each group prepares a report reflecting the essential points of its discussions, including suggestions for future research needs. These reports are distributed to all participants at the end of the workshop and are discussed in plenum.

PUBLICATION:

The Dahlem Workshop Reports contain:

- the Chairperson's introduction,
- the Background Papers, and
- the Group Reports.

The Dahlem Workshop Reports are available in two series:

1) Life Sciences Research Reports (LS) and
2) Physical and Chemical Sciences Research Reports (PC).

DAHLEM KONFERENZEN, Wallotstrasse 19, D-1000 Berlin 33, Federal Republic of Germany

Atmospheric Chemistry, ed. E.D. Goldberg, pp. 1-2. Dahlem Konferenzen 1982.
Berlin, Heidelberg, New York: Springer-Verlag.

Preface

E. D. Goldberg
Geological Research Division, Scripps Institution of Oceanography
La Jolla, CA 92093, USA

Atmospheric chemistry may be considered an old or a new discipline. Clearly, it was a part of the development of the mother field itself during the scientific revolution of the eighteenth century. Lavoisier, the father of modern chemistry, was well aware of the involvement of atmospheric gases in respiration. Oxygen was the intake gas and carbon dioxide the evolved gas in processes occurring at the surface of the earth. Still, up to the last several decades there has been an implicit mood among natural scientists that most of the gases of the atmosphere are inert.

Now atmospheric chemistry has developed into a field which encompasses nearly all branches of chemistry itself - atomic chemistry, photochemistry, organic chemistry, inorganic chemistry, and natural products chemistry among others. It is deeply involved in problems of climate, pollution, the development of our planet, and the origin of life. Research in this area is increasingly reported in such prestigious international journals as Science and Nature as well as in the renowned specialty publication, Journal of Geophysical Research. Although its practitioners have not organized themselves into a society and have not as yet initiated the publication of a journal, such activities can be anticipated.

The organizers of this Dahlem Workshop recognized that the entire spectrum of atmospheric chemical investigations could not be covered in a week by the participants. Thus, the focus was placed upon four problem areas: aqueous chemistry in the atmosphere, historical records of atmospheric composition, microbiological sources and sinks of non-conservative atmospheric gases and aerosols, and photochemical reactions in the troposphere - production from gas phases. The participants were chosen by the organizing committee, who were able to identify participants of distinction. In fact, they identified enough scholars for two or three workshops to consider the problem areas. I suspect that this gives some sense of the population of active workers in atmospheric chemistry.

Not unexpectedly, each of the four groups considering the problem areas operated in different formats. All have produced in their reports an assessment of present knowledge - concepts and technologies. All have in varying degrees of detail indicated the high priority research needs for the future. Sometimes these recommendations are incorporated in the report itself; sometimes they are summarized at the end.

In the various presentations there is some redundancy. This situation is not unexpected. The knowledge base of the nature and reactions of atmospheric constituents is as yet small. Still, it must be applied to a variety of problems such as the persistence of methylated metals in the troposphere, the deciphering of the past atmospheric compositions recorded in glaciers, and the chemical makeup of precipitation. Thus, I submit that we can tolerate the repetition of concepts and technologies that pervade some of the articles as a consequence of the quite heterogeneous interests of the involved scholars.

Finally, in a field such as atmospheric chemistry in which novel and revolutionary concepts continue to dramatize the lives of its practitioners, a volume such as this will have but a temporal interest. Still, it will provide a steppingstone in the next years for significant research by its scientists and as entry to the field for the younger converts.

Atmospheric Chemistry, ed. E.D. Goldberg, pp. 3-16. Dahlem Konferenzen 1982.
Berlin, Heidelberg, New York: Springer-Verlag.

Introduction: Chemistry of the Troposphere – Some Problems and Their Temporal Frameworks

R. M. Garrels
Dept. of Marine Science, University of South Florida
St. Petersburg, FL 33701, USA

Abstract. Attention of scientists from many scientific disciplines has been focusing more and more sharply on the chemistry of the lower atmosphere. The lower atmosphere is in contact with plants and animals, with forests, crops, and soils, with the ocean and its biota, as well as with the stratosphere above. Major environmental problems, such as carbon dioxide increase, acidic rain, smog, and radioactive gases and particles, involve the troposphere. Understanding these problems, as well as the broader one of the evolution of atmospheric composition through geologic time, requires knowledge of the time scales at which materials circulate through the troposphere into and out of the lithosphere, biosphere, hydrosphere, and stratosphere, as well as detailed understanding of the complex chemical reactions involving these materials.

INTRODUCTION

Over the past two decades, regional and global environmental problems have been recognized and brought to the attention of an interested and concerned public. It is not surprising that atmospheric scientists have led the way in assessing worldwide aspects of various types of pollution; the atmosphere mixes rapidly and circumnavigates the globe in less than two years. The atmospheric scientists found that to understand and perhaps solve the global problems they had discovered they needed to know the exchange of atmospheric constituents with plants and animals, with rocks and soils, and with the waters of the earth.

The lower part of the atmosphere in contact with the biosphere, the hydrosphere, and the lithosphere, is called the troposphere. The troposphere extends upward to about 10 kilometers and contains most of the mass of the atmosphere. Because it is heated from below, it is physically unstable and continuously mixes. Compared to the layered stratosphere above, the troposphere is less affected by photolysis and subsequent reactions of substances that are engendered by high energy solar radiation at great heights. It is, however, a much more important medium in terms of the magnitude of fluxes of gases and particulates into and out of it, and also because of the many compounds that enter it from the biosphere.

Understanding the exchanges between the troposphere and the other spheres requires experts in many disciplines - chemists, geologists, biologists, soil scientists, marine scientists - to aid in synthesizing global biogeochemical cycles of the elements.

Progress toward full understanding of such cycles has been slow. The early situation was rather like the Tower of Babel. Meetings involving mixtures of scientists revealed massive ignorance of the important literature in each other's fields and an inability to talk to each other. At first most progress was made by a handful of individuals who assiduously tried to cross disciplinary boundaries; at present more progress is being made by organizing teams of scientists who cooperate in the solutions of problems. Funding for this interdisciplinary research has been a problem everywhere; the usual sources of funds are wedded to a particular brand of science, and there is considerable resistance to giving money to "outsiders." But progress is being made as the "generalists" in each field find each other. It thus seemed appropriate to have this Dahlem Workshop that brings together discussants from many disciplines to focus on the chemistry of the troposphere.

CURRENT PROBLEMS

Some of the areas of research that have focused attention on the problems created by man's intrusion into natural chemical cycles of the elements are outlined below. The list is by no means comprehensive.

Carbon Dioxide

One of the truisms of science is that genuinely new discoveries are indeed rare. Many of the exciting, fast-moving fields of today are revivals of work done many generations ago. The probable effects of a carbon dioxide increase on global climate were predicted in 1896 by Arrhenius (1), better known as the originator of the theory of electrolyte solutions. His predictions of the climatic effects of an increase of atmospheric carbon dioxide differ little from the complex computer simulations of today. Relations between atmospheric carbon dioxide levels and climates of the geologic past were brillinately interpreted in 1898 by Chamberlin (4), again anticipating some current views.

The interdisciplinary nature of the carbon dioxide problem is illustrated by the work of Keeling, a chemist working at a marine institution. His careful measurements of atmospheric CO_2 values since 1958 have provided documentation of a continued increase in CO_2 content of approximately one part per million by volume each year (Keeling and Bacaston (6)). His results, combined with those of a handful of other scientists who recognized long ago the likelihood of an important increase in atmospheric carbon dioxide as a result of man's influence, have inspired major efforts toward modeling the effects of carbon dioxide increase on climate and have engendered the conclusion that within the next century mean global temperatures will rise 1 to $3^{\circ}C$, with small increases in the tropics and large (8° to $10^{\circ}C$) increases at high latitudes.

Until a few years ago the carbon dioxide increase was attributed entirely to the burning of fossil fuels and production of cement, with about 60% of the added CO_2 remaining airborne. The "missing" CO_2 was supposed to have dissolved into the upper few hundred meters of the ocean, except for a small fraction perhaps taken up by the fertilization of the biosphere by the increased CO_2. Currently there is controversy; estimates of the role of the biosphere range from its being a source comparable to fossil fuels in magnitude, because of clearing of forests, especially in the tropics, to no source at all because of mid-latitude reforestation and the rapid secondary growth in clearings. Many workshops around the world have been convened, but the problem is unresolved.

The burning of fossil fuels has also given rise to much interest in the environmental effects of combustion products other than carbon dioxide.

Acidic Rain

Nitrogen and sulfur released to the atmosphere react complexly there to produce sulfuric and nitric acids which return to the earth's surface in precipitation. The entire eastern half of the United States and much of eastern Canada now receive rain with pH values as low as 4; western and northern Europe are similarly afflicted. The impact of such rains on lakes can be drastic; at pH values less than 5, fish ofttimes cannot survive. The effects on vegetation are less clear; changes in plant associations certainly will occur, as well as changes in the microflora and -fauna, but predictions of the effects on overall productivity are not yet possible because of the mixture of toxic and nutritional effects attributable to sulfur and nitrogen compounds. There has been a strong tendency to ascribe all acidic rains to man's influences; we need more information on the acidity of natural rains.

Ozone Jeopardy

Studies of the importance of fixed nitrogen compounds in the atmosphere (e.g., Svensson and Söderlund (9)) were strongly stimulated by the recognition that the chlorinated hydrocarbons used as carriers for the contents of spray cans are so stable in the troposphere that they could reach the ozone layer of the stratosphere and reduce significantly the concentration of ozone there. Such studies were triggered by earlier work on the possible effects of the emissions of fleets of high stratospheric airplanes. A decrease in ozone would result in an increase of ultraviolet light reaching the earth's surface. One highly published result would be an increase in human skin cancer. Today there is a grand scale inventory of the terrestrial and oceanic nitrogen cycles, with emphasis on N_2O.

Smog

One of the environmental issues that has been studied long and hard is the pollution of urban atmospheres by automobile, power plant, and smelter emissions involving nitrogen oxides, carbon monoxide, oxidized sulfur gases, ozone, and a wide variety of particulates. The smog problem is experienced by most cities of the industrialized world with populations over several hundred thousand.

Radioactive Substances in the Troposphere

Since the first nuclear bomb explosion in 1948, there has been great concern and much research on the fate of the radionuclides, both particulate and gaseous, that have been put into the atmosphere by nuclear explosions. The amount of toxic material from atmospheric tests is so great that explosions of nuclear devices, with a few exceptions, have been carried out underground for the last decade. Prior to 1968, 308 nuclear devices were fired on or above the earth's surface (7).

Similarly, there are intermittent emissions to the atmosphere from some of the 200 nuclear reactors in operation today scattered over the northern hemisphere, and there is a potential for massive emissions. The nuclides in the gaseous wastes are chiefly ^{13}N and ^{41}Ar, as well as isotopes of Kr and Xe, and halogens.

More than 60 radionuclides occur naturally in the earth's environment, so that the earth and the atmosphere have always had substantial background radiation. At present, natural radiation levels are higher than those created by nuclear weapons testing, leaks from nuclear power plants, or other additions by man.

There is much current scientific interest in the use of ^{14}C and ^{3}H, generated chiefly by atmospheric nuclear tests in the 1960s, as tracers of the rate of mixing of the atmosphere with the ocean. The two nuclides can be measured as a function of their depths of penetration in the ocean and are used to estimate the rate of uptake of other substances, such as atmospheric CO_2, that do not have such precisely measurable signals. Purposefully introduced man-made substances may become important as tracers in the future (Liss, personal communication).

The toxic effects of radioactive nuclides in the atmosphere are not now important relative to those of natural sources, but it goes without saying that research on the fate of the many atmospheric radioactive nuclides that would be produced during a nuclear war or by various possible nuclear accidents is being carried out in many countries. The subject is so vast and so complex that mention here is restricted to emphasis on the extreme importance of continuing research.

Pesticides and Industrial Chemicals

One publication that helped to spur the environmental renaissance, if I may so term the return of interest in problems such as those that occupied Arrhenius and Chamberlin, was the book Silent Spring, by Rachel Carson, published in 1962 (3). Her interest was in the lethal effects of such pesticides as DDT on animals, especially birds. The book was a major factor in causing a ban in the United States on DDT as an insecticide. Research stimulated by her book demonstrated the mobility of DDT and related compounds, which have found their way through the atmosphere to the poles of the earth. This rather unexpectedly important mobility of the insecticides spurred research on other mobile organic compounds in the atmosphere. There are at least a dozen important mobile organic compounds that are carried in the atmosphere in sufficient concentrations to be toxic. For example, the level of the industrially used PCBs (polychlorinated biphenyls) in fish of Lake Michigan is above acceptable values. Eighty-five to ninety percent of the PCBs in fish come from the atmosphere.

A somewhat similar influence on research was exerted as a result of the mercury poisonings at Minimata, Japan, between 1953 and 1960. There it was found that inorganic mercury was converted to methyl mercury by microorganisms. Methyl mercury acts as a nerve poison and is far more toxic than inorganic mercury compounds. These researches led to further studies of the role of methylated compounds in controlling the concentrations of trace metals in the atmosphere. Methylated compounds of mercury, arsenic, and selenium may be important in their atmospheric behavior.

Extensive work has gone into determining the trace metal contents of atmospheric particulates and the importance of man's inputs. The ratio of metals in particulates of urban air to trace metals in particulates of remote areas is commonly 10/1 and may range up to 100/1 (cobalt). Some investigators would increase these ratios by an order of magnitude.

Ocean Influence

The difficulties of assessing, or even finding, the important element inputs to and outputs from the atmosphere are illustrated by the importance to atmospheric research of the recent discovery of hot springs on the floor of the deep Pacific Ocean. The springs range in temperature up to 375°C, and they are part of a convectional system in which seawater circulates through hot rocks where lava emerges from the sea floor. The entire ocean circulates through these ridges in ten million years or less. It is still too early to generate reliable quantitative data, but certainly this circulation can significantly change ocean chemistry over long periods of time, and hence also affect tropospheric chemistry, for the gases of the two systems are coupled. For example, dissolved oxygen is stripped from seawater in its passage through the basalts, creating a significant sink. The oxygen removed eventually must be replaced in the seawater from the troposphere. Reduction of sulfate by hot basalt may be an even more important oxygen sink.

Long-term Evolution of the Atmosphere

One of the fast moving areas of recent research concerns the evolution of the atmosphere through geologic time. Until recently it was subject to speculation, without means to choose among endless hypotheses. The details of the chemistry of the present atmosphere are now being worked out, permitting more reliable models of reactions in early atmospheres for various assumed compositions. Biologists and geologists are rapidly describing the evolutionary patterns of microorganisms and finding both fossil and stable isotopic evidence of their existence at various stages in early earth history. Because microorganisms are so important in the global carbon, sulfur, and nitrogen cycles, this new information puts us closer to a firm understanding of the major aspects of the atmospheric compositions and climates of the geologic past (e.g., (10)).

SUMMARY

This brief discussion of some of the major problems that have increased interest in the chemistry of the troposphere, and their opening of many new topics of research, show why this Dahlem Workshop is appropriate now.

How now to organize a program of research?

TROPOSPHERIC CHEMISTRY - WHEELS WITHIN WHEELS

The natural global cycles of the elements range over many orders of cycling rates. These metabolic cycles of the earth are like the geared wheels of a clock that measures seconds, minutes, months, years, and centuries with an array of interlocking wheels. Some spin quickly; others cannot be seen to move, so slow is their rotation, but all are interconnected.

The fast wheels in natural cycles are mostly those of the atmosphere. Trace gases that enter it tend to be removed or converted to other gases in a few days, weeks, or months. An increase in an entering flux is usually mirrored by an increase in the exit flux, so that the increase in the entering flux is not simply stored in the atmosphere. The oceanic biota are similarly fast wheels; the residence time of carbon in the biota is only a month or two.

The terrestrial biota and the waters of the oceans are the next fastest wheels. The oceans mix in a thousand years or so; they can rise 100 meters or more in a few thousand years; forests migrate hundreds of kilometers in a few thousands of years in response to climatic changes.

The fast atmospheric responses to perturbation and the secondary oceanic and biotic responses are both within the current sphere of human concern and attempts to control the environment. Opinions on the lifetime of reasonable environmental legislation range up to as many as 100 years in the future, but not much farther.

Fortunately for the finite earth, with its almost entirely self-enclosed chemical systems of the surface environment, the slowly moving cycles have the inertia to brake the fast cycles that have been speeded even more by human interference. It may be possible for the fast wheels to be turned so fast that their connection to the basic metabolism of the earth is broken.

The relations between the fast wheels and the slow wheels provide a general framework for assessing research on the chemistry of the troposphere. Solutions to some of the innumerable questions about the future of the earth's surface environment rest on determination of response times of the positive and negative feedbacks to man's perturbations of the atmosphere. Some examples of response times follow.

Carbon Dioxide

The basic reason for the carbon dioxide problem, in which atmospheric carbon dioxide has increased 15% or so since the beginning of industrialization in the middle of the 19th century, is that the increase of the input has not been mirrored by a nearly equal increase of the output. The input wheels are spun by the glut of CO_2 from automobiles and power plants. The output wheel has many fewer gears, dominated by the rate of removal of the input by the slow rotation of the ocean mixing wheel. The input wheel of fossil fuel burning will slow down in, at most, a few centuries, but the output wheel of ocean mixing will continue its slow pace. Eventually, centuries hence, the current disturbance will disappear. One legacy, of course, is that future human beings will not have the opportunity to perturb the atmosphere with much fossil fuel carbon dioxide (2).

Fixed Nitrogen

The fixed nitrogen problem defies efforts to compare increase of the fixation rate, driven by fossil fuel combustion and industrial fixation of nitrogen, with changes in the rate of denitrification which sends N_2 and N_2O back into the atmosphere.

The input wheel, representing nitrogen compounds fixed annually with valences other than the zero of atmospheric nitrogen, has now been increased by about 25% over total natural global fixation by prokaryotic organisms - bacteria and algae. Furthermore, the industrial fixation rate has increased by about 10% per year and the fossil fuel burning fixation rate by about 4% per year for the last ten years, and both might well continue at this rate of increase for at least the next decade (8).

The speed of rotation of the output wheel returning fixed nitrogen to N_2 is not known. Atmospheric nitrogen is fixed naturally into ammonia, ammonium, organic nitrogen, nitrous oxide, nitric oxide, nitrogen dioxide, and nitric acid; it is returned to atmospheric nitrogen through a complex system of bacterially controlled "denitrification" reactions resulting in a general reversal of the fixation reactions, but not necessarily through the same pathways. Most fixed nitrogen is returned to the atmosphere as N_2. Figure 1 is a diagram of the nitrogen cycle.

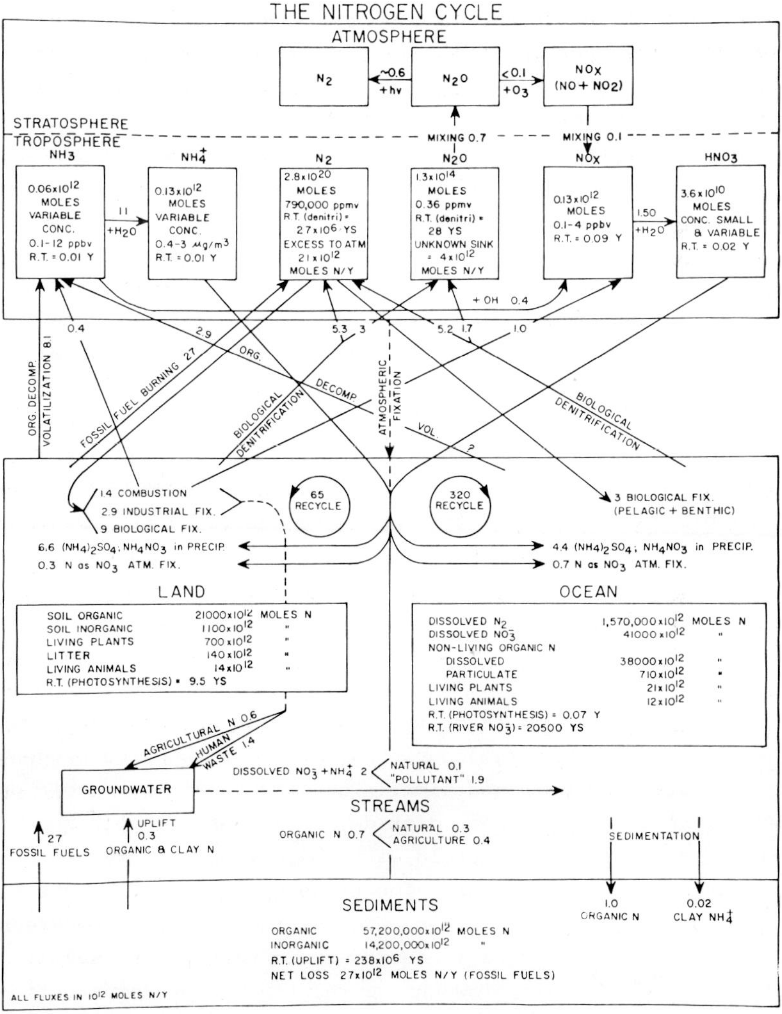

FIG. 1 - A model of the global nitrogen cycle (from (5)). Revised 1981.

The numbers are not very trustworthy; the figure illustrates chiefly the complexity of this cycle which represents a typical cycle for a nutrient element.

It is commonly assumed that the pre-man nitrogen fixation and denitrification system was in steady state (not a safe assumption!). The question arises as to whether denitrification has kept pace with fixation, so that the reservoirs of ammonia, nitrous oxide, nitrite, and nitrate have remained a constant source of supply for the biota. It has already been shown that one facet of the increase in nitrogen fixation over natural rates is a contribution to the acidity of rain. If natural denitrification is lagging seriously behind the increased fluxes of nitrogen fixation caused by man, the consequences will be important in global ecology. We may have the nitrogen analogy to carbon dioxide, in which outputs run on a cycling wheel slower than man-induced inputs. Whether the resultant increases in the fixed nitrogen inventory will increase primary productivity and the existing biomass, or will conceivably result in toxic amounts of fixed nitrogen in the environment, is unknown. It is clearly important to investigate carefully the rate at which the denitrification wheel turns. Walker (personal communication) points out that the response of organisms to increased anthropogenic fixation of nitrogen may be a reduction in the natural rate of fixation.

A most important example of the interaction of fast and slow wheels is related to the concentration of atmospheric oxygen as a function of geologic time. The atmospheric reservoir of oxygen, relative to carbon dioxide, is so large that carbon dioxide could change manyfold without influencing oxygen concentration in the atmosphere importantly. So the flux wheels rotate at the same rate during photosynthesis, but the oxygen flux is geared to such a large reservoir that the reservoir hardly changes. This slow change relative to photosynthetic rates, if compared to the changes in the oxygen reservoirs of sedimentary rocks, is fast. Most of the oxygen that has been generated by the conversion of CO_2 to buried organic matter in sediments is stored in the oxygen of sulfate in the ocean or in rocks, and in the ferric iron of silicate and oxide minerals. The slowest wheels in the terrestrial machine of fluxes and reservoirs are those associated with the storage of elements in the reservoirs of sedimentary rocks. Of all the oxygen released by photosynthesis of carbon dioxide and storage of the resultant organic material in sedimentary rocks (some 1300×10^{18} mols), 95% is stored in sulfates and ferric compounds, and only 5% resides in the atmosphere.

SUMMARY

In summary, among the problems that must be solved to understand the chemistry of the troposphere are the delineations of the rates at which the many wheels of the coordinated earth metabolic system turn. We measure large annual inputs - do the annual outputs match, and if not, how important is the net increase? Will the increase in reservoirs result in an increase of output at a new steady state, and how high will the new reservoir steady-state concentration be? How long will it take for the basic metabolism of the earth to remove man-made perturbations?

The task ahead is monumental. The articles that follow help to show where we stand today in understanding the wheels within wheels that have carried us through more than 3 billion years of life, and the degree to which we may yet shatter this wonderful machine, the earth.

Acknowledgements. I should like to thank P.J. Crutzen and J.C.G. Walker for their helpful comments on an earlier draft of this paper.

REFERENCES

(1) Arrhenius, S. 1896. On the influence of carbonic acid in the air upon the temperature of the ground. Phil. Mag. J. Sci. (Fifth series) 41: 237-276.

(2) Bolin, B.; Degens, E.T.; Kempe, S.; and Ketner, P., eds. 1979. The Global Carbon Cycle. SCOPE Report 13, ICSU. Chichister, New York, Brisbane, Toronto: John Wiley & Sons.

(3) Carson, R.L. 1962. Silent Spring. Boston: Houghton Mifflin.

(4) Chamberlin, T.C. 1898. The influence of great epochs of limestone formation upon the constitution of the atmosphere. J. Geol. 6: 609-621.

(5) Garrels, R.M.; Mackenzie, F.T.; and Hunt, C. 1975. Chemical Cycles and the Global Environment. Los Altos, CA: William Kaufmann, Inc.

(6) Keeling, C.D., and Bacaston, R.B. 1977. Impact on industrial gases on climate. In Energy and Climate. Studies Geophys., pp. 72-95. Washington, DC: U.S. National Academy of Science.

(7) Panel on Radioactivity in the Marine Environment, Committee on Oceanography, National Research Council. 1971. Radioactivity in the Marine Environment. Washington, DC: U.S. National Academy of Science.

(8) Simpson, H.J., et al. 1977. Man and the global nitrogen cycle. In Global Chemical Cycles and their Alterations by Man, ed. W. Stumm, pp. 253-274. Berlin: Dahlem Konferenzen.

(9) Svensson, B.H., and Söderlund, R., eds. 1976. Nitrogen, Phosphorus, and Sulfur - Global Cycles. SCOPE Report 7. Ecol. Bull. (Stockholm) 22: 23-73.

(10) Walker, J.C.G. 1977. Evolution of the Atmosphere. New York: MacMillan Publishing Co., Inc.

Atmospheric Chemistry, ed. E.D. Goldberg, pp. 17-40. Dahlem Konferenzen 1982.
Berlin, Heidelberg, New York: Springer-Verlag.

Factors Governing the pH, Availability of H^+, and Oxidation Capacity of Rain

J. J. Morgan
Environmental Engineering Science
California Institute of Technology, Pasadena, CA 91125, USA

Abstract. The acidity of rain is coupled to redox reactions in air and in atmospheric water. The pH, an intensive quantity, needs to be distinguished from the base neutralizing capacity. For acidic rain observed at most locations, H_2SO_4, HNO_3, NH_3, and $CaCO_3$ are dominant components. Their local availability or production rates govern net acidity. pH is thus almost entirely determined by these major "strong" components imposed on a CO_2 background, with some influence by SO_2(aq), smaller concentrations of HNO_2 and weak organic acids and minor bases, e.g., Fe_2O_3, yielding acid aquo metal ions. Total global emissions to the atmosphere of H_2SO_4 precursors outweigh those of HNO_3 by a factor of 2-3 on an equivalent basis. In specific settings HNO_3 may be comparable to H_2SO_4 in rain. Total atmospheric acidity appears to be a useful quantity for estimating potential acidity of rain at different locations. There are indications in photochemical models of HNO_3 and H_2SO_4 that feedback among S and N species may be important. Heterogeneous oxidations of SO_2 in cloud, fog, and rain play important roles in the acidification process. "Background" acidities of rain appear to be highly variable; pH values are expected to range from below 5 to above 6. Present-day SO_2 and NO_x fluxes account for a pH lowering of $\sim$ 0.5 to 1.5, depending upon source location and transport-conversion rates.

INTRODUCTION

Acidity (base-neutralizing capacity) and alkalinity (acid-neutralizing capacity) are complementary concepts. The acidity versus alkalinity balance of rain (and drizzle, snow, fog, and clouds) is intrinsically coupled to oxidation-reduction reactions in the air and in atmospheric water. Sillén (17) likened the geochemical evolution of the earth's atmospheric-ocean system to a set of gigantic, somewhat coupled acid-base and oxidant-reductant titrations in which the "volatiles," acids from the earth's interior (H_2O, HCl, SO_2, CO_2, and others), were titrated by bases from the rocks and in which reduced components (H_2, NH_3, Fe_3O_4, C) were titrated by the O_2 of the evolving atmosphere-biosphere system. As a result of such global titrations, we arrive at a present-day atmosphere which is 20% O_2, 79% N_2, and 0.03% CO_2, and a world ocean with a pH of 8 and an alkalinity of 2.3 μeq/ℓ. The system is not a stationary equilibrium, of course. It is open, with high fluxes. Still, the titration metaphor may be useful in thinking about the state of the atmosphere-hydrosphere-biosphere system today, on global, regional, and local scales. Evidence over the last century shows increased emissions of strongly acidic precursors such as sulfur dioxide and nitric oxide to the atmosphere. Fossil fuel burning has also led to an increase of global atmospheric CO_2 by some 15%. These emissions, through photochemical and chemical reactions in the atmosphere, lead to an increase in the total acidity of the atmosphere and consequent increases in local and regional fluxes of acidity to the earth's surface through rainfall of increased acid content and lower pH (wet deposition) and through absorption and fallout of gases and aerosols (dry deposition). Present fluxes of acidity and possible future increases have important implications for aquatic and terrestrial ecosystems, water resource systems, weathering of natural and man-made environments, and for the longer-term chemical state of atmosphere-water systems locally and regionally, if not globally.

Concepts of H^+ Availability and Acidity

Acids are proton donors; bases are proton acceptors. In water as a solvent the "strong" acids, e.g., HNO_3, donate protons to water essentially completely.

$$HNO_3 + H_2O \rightleftarrows H_3O^+ + NO_3^-;$$

the equilibrium acidity constant, K_a, is the measure of acid strength. Here, $K_{a,HNO_3} = 1$ for the standard state conventions adopted. The aquated proton, H_3O^+, is the strongest acid that can exist in water. Aqueous CO_2 is a weak acid. The reaction

$$CO_2(aq) + H_2O \rightleftarrows H_3O^+ + HCO_3^-$$

has an effective acidity constant $K_{a,CO_2} = 10^{-6.4}$ at 25°C. Bases, e.g., OH^- ion, CO_3^{2-} ion, or NH_3, are similarly strong or weak. For example, the strengths of OH^-, CO_3^{2-}, and NH_3 as bases can be ranked by their acceptance of protons (19)

$$OH^- + H_3O^+ \rightleftarrows 2H_2O,$$

$$CO_3^{2-} + H_3O^+ \rightleftarrows HCO_3^- + H_2O, \text{ and}$$

$$NH_3 + H_3O^+ \rightleftarrows NH_4^+ + H_2O,$$

with respective equilibrium constants $10^{14} > 10^{10.3} > 10^{9.2}$ at 25°C.

The concentrations or activities of a given acid are readily linked to those of related components in the gas phase or in solid phases by appropriate two-phase constants, e.g., Henry's Law and solubility equilibrium constants. Similarly, acids and bases can be formed from precursors by overall oxidation-reduction processes, e.g., $N_2 + \frac{5}{2} O_2 + H_2O = 2HNO_3$.

Four key concepts which define the resultant acid-base characteristics of rain are: (a) energetics, i.e., equilibrium properties of component acids and bases and acid-base precursors; (b) the total acid or base neutralizing capacity with respect to some proton reference level, i.e., a specified acid-base solution; (c) the aqueous proton concentration, $[H_3O^+] \equiv [H^+]$, or activity, $\{H^+\}$, related intensity factors representing free aqueous protons or immediately available protons, (Bates (1) refers to proton activity, proton availability, and degree of acidity synonymously); and (d) kinetics of reactions and transfer processes leading to production of acids or bases in rain.

It is essential to distinguish between $H_3O^+(aq)$ and a base neutralizing capacity, or reservoir of ultimate H^+ availability. These notions can be illustrated by means of simple examples for various component acids at concentration levels such as those in rain. A group of acids of different strengths (see Fig. 3, below) and concentrations was selected. Initial pH values (neglecting minor activity coefficient corrections) and base neutralizing capacities have been calculated. No redox changes have been assumed. The solution is closed to the atmosphere. The results are shown in Table 1.

Some points can be noted. Comparing equimolar (10 µM) solutions one at a time, we see that pH values range from 4.7 (H_2SO_4) to neutral (for the weak acid, NH_4NO_3), whereas the total acidity (TOTH) is the same on an equivalent basis (H_2SO_4, hydrated SO_2, and hydrated CO_2 are diprotic). As remarked by Odén (14), for strong acid solutions (e.g., H_2SO_4 and HNO_3), the total concentrations of these acids fix the pH value, hence the H^+ activity or availability. For a weak acid

TABLE 1 - Available H^+ (pH) and base neutralizing capacities of dilute acid solutions.

Acid	Conc. µM	pH (25°C)	Base Neutralizing Capacity µeqℓ^{-1}	
			to pH	BNC*
H_2SO_4	10 50	4.7 4.0	7.0 7.0	20 100
HNO_3	10 50	5.0 4.3	7.0 7.0	10 50
$SO_2(aq)$	10	5.0	6.1 8.1	10 20†
CH_3COOH	10 50	5.1 4.7	7.1 7.3	10 50
NH_4NO_3	10 50	7.0 6.8	8.9 9.4	10 50
$CO_2(aq)$	10	5.7	7.6 9.0	10 20†

*BNC with respect to least-protonated level is also known as TOTH. †Least-protonated level of weak diprotic acid.

such as CH_3COOH, with $pK_a = 4.7$, a dilute solution leads to extensive dissociation and pH 5.1. If CO_2 at 10 μM is titrated to pH ∿ 9, 20 $\mu eq\ell^{-1}$ of acidity will be measured. We notice that SO_2 with $pK_a = 1.8$ is effectively a "strong" acid at these concentrations.

At fivefold higher concentrations of H_2SO_4, HNO_3, or CH_3COOH, pH is lowered considerably. If H_2SO_4, HNO_3, CH_3COOH, and NH_4NO_3 were mixed at 50 μM concentration levels, the resulting pH would be approximately 3.8. The CH_3COOH dissociation would contribute only about 6 $\mu eq\ell^{-1}$ to the free acidity (available protons). The NH_4NO_3 contributes nothing. The base neutralizing capacity with respect to a CO_2/H_2O two-phase system (with aqueous CO_2 = 10 μM) would be 195 $\mu eq\ell^{-1}$ (because 10% of the 50 μM weak acid, CH_3COOH, remains undissociated at pH 5.7).

Were NH_3 rather than NH_4NO_3 introduced at 50 μM, the basicity of NH_3 would lead to the reaction $NH_3 + H^+ = NH_4^+$, and the acidity would decrease by 50 $\mu eq\ell^{-1}$, the pH rising to 4.0. If 50 μM NH_3 were oxidized to HNO_3 (in a receiving ecosystem), $NH_3 + 2O_2 \rightarrow H^+ + NO_3^- + H_2O$, the base neutralizing capacity would increase by 100 $\mu eq\ell^{-1}$, or about 50% of the original 195 $\mu eq\ell^{-1}$. If H_2SO_4 would be reduced (in an anoxic sediment environment) to H_2S ($pK_a = 7$), the base neutralizing capacity towards CO_2/H_2O would decrease by almost 100 $\mu eq\ell^{-1}$. Oxidation of all the CH_3COOH to CO_2 would decrease BNC by 45 $\mu eq\ell^{-1}$ with respect to the original CO_2/H_2O reference level. Finally, if SO_2 were present at 10 μM, oxidation of SO_2, $SO_2 + \frac{1}{2} O_2 \rightarrow H_2SO_4$, would increase the BNC towards CO_2 by 10 $\mu eq\ell^{-1}$, lowering pH slightly but leaving TOTH unchanged. It is clear that the acid-base and oxidant-reductant states of precipitation are strongly related.

ACIDITY, ALKALINITY, AND pH

The earliest known quantitative description of the chemistry of rain and its acidity was given for the United Kingdom by Smith in his book Air and Rain in 1872 (18). His analysis included acidity (as H_2SO_4), ammonia, nitrate, sulfate, chloride, and organic matter. Arrhenius' theory of electrolytic dissociation came 15 years later; Smith's acidity roughly corresponded to $[H^+]$ of 200 µM. Sörensen introduced the pH scale in 1909; Smith's rain had a pH of perhaps 3.7. Thinking in terms of the components required to establish the rain composition which Smith analyzed at Manchester (5), we could choose H_2O, H_2SO_4, HNO_3, HCl, NH_3, NaCl, and Na_2SO_4 (the latter two for the marine component). The approximate charge-balance relationship in the rainwater would then be $[H^+] + [NH_4^+] + [Na^+] = 2[SO_4^{2-}] + [NO_3^-] + [Cl^-]$. We now recognize that this chemical balance is incomplete. Later rainfall analyses (e.g., Gorham (4), Junge (8), and Granat (6,7)) reveal that the following ions are nearly always found in acidic rainfall in significant concentrations, the relative importance depending on provenance: SO_4^{2-}, NO_3^-, Cl^-, NH_4^+, Na^+, K^+, Ca^{2+}, Mg^{2+}, and H^+. For weakly acidic and alkaline precipitation HCO_3^- ion is an important species. Table 2 shows ion concentration data and pH for acidic rainfall at three locations.

TABLE 2 - Chemical analyses of acidic rain at three locations.

Species	Concentration µeq ℓ^{-1}		
	Sjöängen, S. Sweden[a]	Hubbard Brook, New Hampshire[b]	Pasadena, California[c]
SO_4^{2-}	69	110	39
NO_3^-	31	50	31
Cl^-	18	12	28
NH_4^+	31	22	21
Na^+	15	6	24
K^+	3	2	2
Ca^{2+}	13	10	7
Mg^{2+}	7	32	7
H^+	52	114	39
pH	4.30	3.94	4.41

[a]1973-75 (7); [b]1973-74 (10); [c]1978-79 (12).

Observations on major ions of rain at many locations for the past several decades suggest a universal set of input components which account for the net acidity, pH, and resulting chemical species in most rain. Recent observations on rainfall chemistry at different locations in the Los Angeles basin of southern California (Fig. 1) provide a convenient illustration of variations in components. The ratio of $[SO_4^{2-}]$ to $[NO_3^-]$ on an equivalent basis ranges from 1.5 at Westwood (6 km from the Pacific coast) to 1.1 at Pasadena (40 km inland) to 0.9 at Riverside (70 km inland). The majority of fossil-fuel burning stations are sited near the coast. Qualitatively similar variations in composition are seen over far greater distances in northern Europe and the eastern United States.

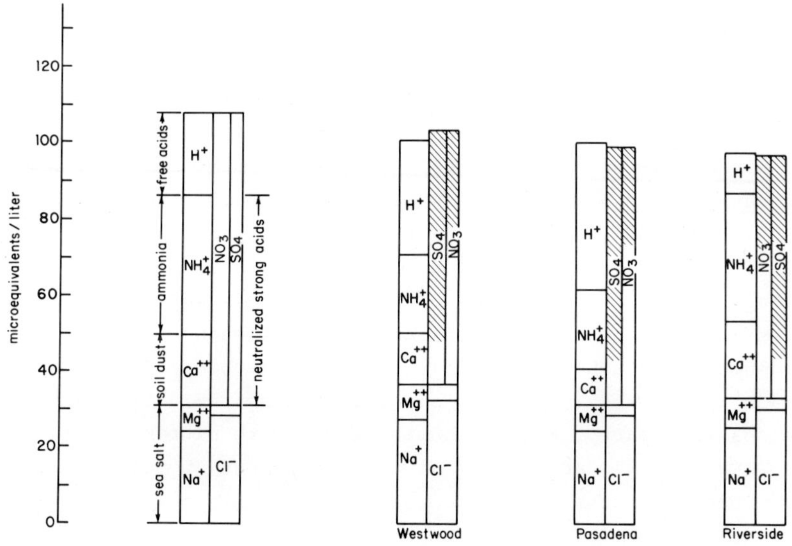

FIG. 1 - Composition of rain in southern California, 1978-79, interpreted in terms of input components and source type. (Coast → inland, left → right; widths of SO_4^{2-} and NO_3^- bars indicate relative amounts; crosshatchings indicate stationary sources, and non-hatching indicates mobile sources) (12).

A set of components that accounts for observed net acidity or alkalinity of rainfall at European stations was proposed by Granat (6). Figure 1 illustrates application of the concept in terms of the "components" strong acids, sea salt, soil dust, and ammonia. The corresponding chemical components could then be (consistent with Table 2 and Fig. 1): NaCl, KCl, $MgSO_4$, $CaCO_3$, NH_3, H_2SO_4, and HNO_3. This choice of acid-base-salt components accounts for commonly-reported chemical analyses. Other components can be chosen to describe ultimate sources of acidity or alkalinity in rain and to relate sources to gas, aerosol, or aqueous phase species involved in transport and transformation processes leading to acidification. In the absence of significant anthropogenic emissions, organic sulfur compounds and H_2S are major precursors for H_2SO_4 formation in the atmosphere. For nitrogen the principal natural and anthropogenic precursor of atmospheric and rainfall acidity is nitric oxide, NO (which, when grouped together with NO_2, is referred to as NO_x). The dominant man-made precursor emission of sulfur is sulfur dioxide, SO_2. Interplay of acid-base (proton transfer) and oxidant-reductant (electron transfer) reactions in air and water is better accounted for by the emission components: NaCl, KCl, $MgCl_2$, $MgSO_4$, $NaHCO_3$, $CaCO_3$, NH_3, HCl, H_2S, RSR', SO_2, H_2SO_4, and NO. In the oxidizing atmosphere, H_2S, RSR', and SO_2 are transformed to H_2SO_4 (with characteristic times ranging from less than a day to weeks, depending on chemical and physical conditions in the atmosphere). Both homogeneous photochemical oxidations and heterogeneous aqueous-phase (and possibly particulate solid phase) oxidation processes take part. Emitted NO is oxidized to nitrous acid, HNO_2, and nitric acid, HNO_3, over homogeneous photochemical and chemical pathways in the gas phase.

The general picture for acid and basic components entering rain is summarized schematically in Fig. 2.

Acidity and Base Neutralizing Capacity

The commonly-used definitions of acidity and alkalinity in water chemistry are based on the choice for a reference level

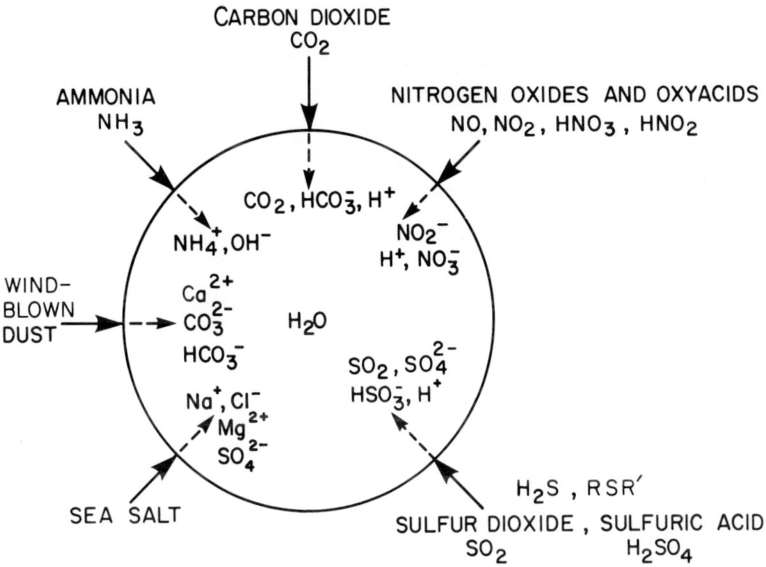

FIG. 2 - Source and chemical components known to take part in rain acid-base chemistry.

of CO_2 aqueous solutions resulting from equilibrium between water and CO_2 at its partial pressure (fugacity) in the atmosphere. A temperature of interest is specified, thus fixing the pH, which depends on two acidity constants for CO_2(aq) and Henry's Law constant for CO_2. At an atmospheric CO_2 concentration of 335 ppmv (molecules of CO_2 per million molecules of air), the reference pH is 5.65. Solutions of pH > 5.65 and at equilibrium with the atmosphere have acid neutralizing capacity. This capacity factor is defined as the alkalinity. Solutions of pH < 5.65 at equilibrium with the atmosphere have base neutralizing capacity. This capacity factor is referred to as "mineral acidity"; it results from the presence of acids stronger than CO_2 in water.

The "strong acids" (HCl, HNO_3, H_2SO_4) are, of course, all stronger than CO_2. Chemical compositions in Table 2 and Fig. 1 show that base neutralizing capacity of rain with respect to the CO_2 reference level, $[BNC]_{CO_2}$, is attributable entirely to the strong acids H_2SO_4 and HNO_3. The precipitation-weighted mean Pasadena rain (Table 2) was formed (per liter of rain) from 18 µmol H_2SO_4 (deducting the SO_4^{2-} contributed by seawater), 31 µmol HNO_3, 21 µmol NH_3, and 3.3 µmol $CaCO_3$. The value of $[BNC]_{CO_2}$ for rain with only these components is $[BNC]_{CO_2} = 2[H_2SO_4]_o + [HNO_3]_o - 2[CaCO_3]_o - [NH_3]_o$, or 39 $\mu eq\ell^{-1}$. At Pasadena, the potential acidity or input acidity was 67 $\mu eq\ell^{-1}$. Component bases NH_3 and $CaCO_3$ neutralized 42% of the potential acidity. At Hubbard Brook (10) approximately 30% of the potential acidity was neutralized, and at Sjöängen (7), approximately 45%.

The charge balance for rain with HNO_3, H_2SO_4, and HCl as the net acid components and with $CaCO_3$ and NH_4 as the only net basic components is $[H^+] + 2[Ca^{2+}] + [NH_4^+] = 2[SO_4^{2-}] + [NO_3^-] + [Cl^-] + [HCO_3^-] + 2[CO_3^{2-}] + [OH^-]$. The base neutralizing capacity in terms of actual solution species is $[BNC]_{CO_2} = [H^+] - [HCO_3^-] - 2[CO_3^{2-}] - [NH_3] - [OH^-]$. For rain at equilibrium with atmospheric CO_2, knowledge of $[BNC]_{CO_2}$ and the equilibrium constants for CO_2 solubility, CO_2 and HCO_3^- acidity and water ionization, plus the constants for NH_3 solubility and NH_4^+ acidity is sufficient to find $[H^+]$ or $\{H^+\}$ and all other species activities. (If other acid-base gases, e.g., SO_2, HNO_2, NO_2, etc., are involved, their equilibria enter into the calculation via the charge balance.) $[BNC]_{CO_2}$ can be obtained experimentally by titration to the CO_2 reference pH or calculated for known or assumed input components.

Alkaline rain existed earlier at locations now acidic and exists today in regions with strong sources of alkaline wind-blown dust and ammonia in excess of sources of SO_2, H_2S, organic sulfur, and NO_x. The US NADP network shows extensive regions of western North America with rain of pH > 5.7, unlike the eastern US and Canada, and unlike northern Europe. Using 1955-56 rain analyses for the western US, it is possible to estimate pH values by application of charge-balance, CO_2-equilibrium relationships. The pH values obtained are considerably higher than those either observed today in the northeast US or calculated for 1955-56 (11) from charge balances. Geographical relation to sources of acid and basic emissions and the transport-transformation times from source locations are generally recognized to be among the factors governing pH of rain.

Alternative Proton Reference Levels

It should be kept in mind that the reference level chosen for defining a neutralizing capacity is an arbitrary datum taken for convenience (analytical or computational) and may not necessarily have direct environmental significance without further interpretation. Chemical speciation, i.e., pH, metal ion, ligand, and complexes in both rain and aquatic systems depends on absolute input concentrations or acid-base components and redox components. A reference level has no intrinsic significance for describing changes in H^+ availability and acidity. The concept of a proper background level for neutralizing capacity or proton activity should be distinguished from that of a reference level. For example, if rain at a location has been characterized at an earlier time by, e.g., pH 7.5 and ALK = 100 $\mu eq \ell^{-1}$, and at a later time by pH 4.3 and $[BNC]_{CO_2}$ = 50, it is clear that the overall change in the acidity-alkalinity balance is $\Delta[BNC]_{CO_2} = +150\ \mu eq \ell^{-1}$. Total acid (proton) input is $\sum \Delta[BNC]$ and can be compared with changes in emissions or changes in resultant $[BNC]_{CO_2}$ of aquatic systems (waters, sediments, soils) (14).

ACIDITY AND OXIDATION (pH AND pε)

Proton transfer and electron transfer are analogous notions. The pH corresponds to activity of aqueous protons. A solution of low pH has a relatively strong proton-donating tendency; one of high pH has relatively strong proton-accepting tendencies. Electron transfers formally describe an oxidation-reduction process. For example, the half-reactions for peroxide reduction,

$H_2O_2(aq) + 2H^+ + 2e = 2H_2O$,

and SO_2 oxidation,

$SO_2(aq) + 2H_2O = SO_4^{2-} + 4H^+ + 2e,$

combine to yield the overall reaction of $SO_2(aq)$ with H_2O_2:

$H_2O_2 + SO_2 = SO_4^{2-} + 2H^+,$

generating increased H^+ and $[BNC]_{CO_2}$. Figure 3 summarizes the energetics of proton transfer reactions (a) and electron transfer reactions (b) in terms of the standard equilibrium constant for proton donation, K_a, and the standard equilibrium constant for unit electron acceptance, K_e, which is simply a measure of the free energy per electron transferred. For H_2O_2 reduction to H_2O, log K_e is 30; for SO_2 oxidation to H_2SO_4, log K_e^{-1} is -3; for SO_2 oxidation by H_2O_2 (two electrons), log K = 54, which corresponds to the standard potential difference of 1.6 volts. The pε and E_H scales are linearly related by pε = $E_H/2.3\ RTF^{-1}$ (19). Figures 3(a) and 3(b) emphasize a useful analogy between acid-base and oxidant-reductant reactions. The analog of pH = -log{H^+} is pε = -log{e^-}. Systems of low

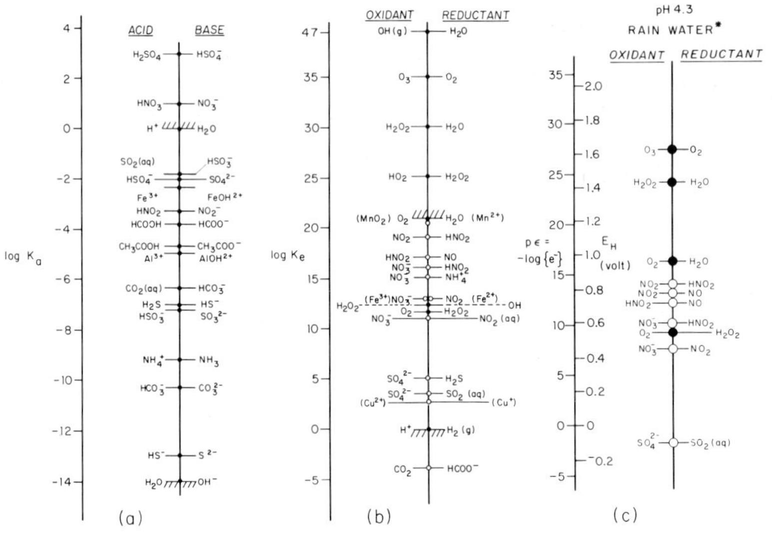

FIG. - 3 - Energetic levels for (a) proton-transfer (HA = A^- + H^+, K_a) and (b) electron-transfer (OX + e = Red, K_e) reactions (19). (c) shows non-equilibrium pε values estimated for rain at Pasadena (13).

pε are strong electron-donating (reducing) systems. As pH = pK_a + log (base/acid), so pε = $p\varepsilon^o$ + log (oxidant/reductant) for each half-reaction system. At complete equilibrium, the pH and pε values reflect the respective acid/base and oxidant/reductant compositions for all pairs.

In Fig. 3(a), acids higher on the scale can transfer protons to the bases lower on the scale. In Fig. 3(b), higher oxidants and lower reductants have a similar relationship. For acids the "energy ladder" leads to the concepts of $[BNC]_{CO_2}$ and TOTH.

Any acidity capacity implies a defined redox reference level as well. Thus NH_3 oxidation to HNO_3, H_2SO_4 reduction to H_2S, CH_3COOH oxidation to CO_2, etc., change acidity capacity by creating or destroying acidity. This simply means that the complete equilibrium state requires specification of both pH and pε, i.e., the acid-base and oxidant-reductant states.

An oxidation capacity for rain can be readily identified in terms of Fig. 3(b). Any electron reference level can be selected, e.g., $K_{e,ref}$. Then, the oxidation capacity, [OX], is $\sum$[equivalent oxidants, $K_e > K_{e,ref}$] - [equivalent reductants, $K_e < K_{e,ref}$]. If the $SO_4^{2-}/SO_2(aq)$ electron level is chosen, only Cu^+, H_2, $HCOO^-$, and SO_2 among the reductants would enter into the oxidation capacity. Not all half-reactions included are compatible in an equilibrium solution. In practice, as for BNC, only a limited number of species, e.g., SO_2, HSO_3^-, and SO_3^{2-}, are related by proton equilibria.

Figure 3(c) is an example of calculated apparent (nonequilibrium) pε values for rain at Pasadena (13), based on measured aqueous concentrations or gas-phase concentrations of O_3, H_2O_2, O_2, SO_2, HNO_2, NO_3^-, NO, and NO_2. (Ozone and hydrogen peroxide would be unstable in the rain, being much stronger oxidants than O_2. NO_2, NO, and NO_3^- are out of equilibrium because of slow two-phase kinetics.) From the concentrations of the oxidants and reductants an apparent oxidation capacity, [OX], can be calculated with respect to the SO_4^{2-}/SO_2 reference level. For $[O_2] = 3x10^{-4}M$, $[NO_3^-] = 8x10^{-5}M$, and $[SO_2] + [HSO_3^-] = 3x10^{-6}M$, $[OX] = 1.38x10^{-3}$ $eq\ell^{-1}$. O_2 and NO_3^- dominate. The median levels of the known powerful and kinetically active oxidants, O_3 and H_2O_2, $3x10^{-10}M$ and $2x10^{-8}M$, respectively, are less than equivalent to the median S(IV) reductant level, $3x10^{-6}M$, at air-rain equilibrium for pH 4.3. However, true levels of H_2O_2 are not well-known. The equilibrium scavenging level could be as high as $10^{-5}M$, or $2x10^{-5}$ $eq\ell^{-1}$ of aqueous H_2O_2 for an initial air concentration of 1 ppbv, based on Henry's Law constant for H_2O_2 at 25°C.

TOTAL ATMOSPHERIC ACIDITY

Acid-base and redox processes may take place in the gas phase, in aerosol phases, and in cloud and fog. The acidity of rain is a reflection of the components and species entering into the various atmospheric phases, e.g., HNO_3, H_2SO_4, H_2S, SO_2, NH_3, $(CH_3)S$, NO_2, etc. It can be useful to define the total acidity (over several phases) of a volume of atmosphere, by stoichiometrically relating actual components and species to acid and base species that would donate or accept protons. For example, H_2SO_4 releases two protons; upon oxidation, NO_2 releases one proton, etc. (An extension to total oxidation capacity is evident.) As for acidity of rain, alternative pH levels and redox states can be chosen, e.g., $[BNC]_{CO_2}$ and most-oxidized or TOTH and partially-oxidized, etc. Assuming some set of levels,

$$\begin{Bmatrix}\text{Total}\\ \text{Acidity}\end{Bmatrix} = \begin{Bmatrix}\text{Aerosol}\\ \text{Acidity}\end{Bmatrix} + \begin{Bmatrix}\text{Gas Phase}\\ \text{Acidity}\end{Bmatrix} + \begin{Bmatrix}\text{Cloud}\\ \text{Acidity}\end{Bmatrix} + \dots .$$

In terms of known gas partial pressure, p_i, concentrations in water, $[X_i]$, aerosol-phase concentrations, (X_i), and a corresponding set of stoichiometric acid, (>0), or base, (<0), coefficients b_i, g_i, and a_i, which depend on H^+ and e^- reference levels chosen,

$$\begin{Bmatrix}\text{Total}\\ \text{Acidity}\end{Bmatrix} = \frac{1}{RT} \sum b_i p_i + L\sum g_i [X_i] + \sum a_i (X_i),$$

where the acidity is referred to unit volume of air and L is liquid water volume per volume of air. Total acidity represents a way of defining potential acidification by rain, fallout, etc., and a basis for constructing an acidity balance in terms of a conserved property, subject to precise definition of H^+ and redox reference levels. Different levels are appropriate for different environmental situations, e.g., short-term vs. long-term or oxidizing vs. reducing conditions.

PRODUCTION OF RAIN ACIDITY

Availability of H^+ and the base neutralizing capacity of rain are determined by: (a) transfer or scavenging of preexisting acidic and basic aerosol and gaseous species, and (b) generation of acids in cloud and rain. Mechanisms and rates of the scavenging processes determine the initial composition of clouds and rain and thus set the aqueous chemical conditions (pH, metals, buffers, ligands, aqueous SO_2 and NO_x, and oxidants such as O_3 and H_2O_2 for heterogeneous SO_2 oxidation) which in turn affect the H^+ and acidity of rain. Gaseous and aerosol species scavenged by atmospheric water are formed by homogeneous photochemical reactions and multiphase gas-to-particle conversion processes. Emitted reduced sulfur compounds, H_2S, RSR', and SO_2, are partially oxidized in the troposphere to H_2SO_4 via photochemical, free-radical pathways involving O_3, OH, H_2O_2, HO_2, RO_2, and $O(^3P)$. Emitted NO and NO_2 are partially oxidized to HNO_2 and HNO_3 by similar photochemical pathways (in which NO and NO_2 themselves play key roles in O_3 formation) with OH, O_3, RO, and H_2O_2 as reactants and products.

Unoxidized SO_2 in the air dissolves in clouds or aerosol water where O_2, O_3, or H_2O_2 oxidation produces H_2SO_4. Unoxidized SO_2 is also removed from the troposphere by dry deposition. Reaction of H_2SO_4 with NH_3 leads to ammonium sulfate aerosols, e.g., NH_4HSO_4 or $(NH_4)_2SO_4$. The aerosols may be scavenged during cloud formation (nucleation scavenging) and by aerosol-droplet collisions in the clouds (diffusion). Both SO_2(gas) and sulfate aerosol are scavenged by raindrops. Efficiency of raindrop aerosol scavenging depends strongly upon the aerosol size distribution (smaller particles are poorly captured).

SO_2 scavenging by clouds and rain can be treated as a reversible gas-aqueous distribution, since mass transfer is generally rapid. Dry deposition is slow for the submicron aerosol fraction, the predominant fraction for H_2SO_4.

Nitric acid reacts rapidly with ammonia to produce NH_4NO_3 aerosol which is scavenged by clouds and rain or removed by dry deposition. NH_4NO_3 has a relatively high vapor pressure for $NH_4NO_3 \rightarrow NH_3 + HNO_3$, and the variation with temperature is great. This may have implications for scavenging efficiency. Gas-phase HNO_3 can be scavenged by clouds and rain (as can HNO_2) or be absorbed at the earth's surface. At higher relative humidity HNO_3 will tend to condense on aerosol particles.

A potential liquid water sink for NO and NO_2 is reaction via the two-phase equilibria

$$2NO_2(g) + H_2O(\ell) \rightleftarrows 2H^+ + NO_3^- + NO_2^-, \text{ and}$$

$$NO(g) + NO_2(g) + H_2O(\ell) \rightleftarrows 2H^+ \; 2NO_2^-.$$

However, the overall two-phase rates are not rapid, showing characteristic times much longer than typical cloud life or rainfall time (9). Possibilities of other, more rapid two-phase mechanisms involving NO and NO_2 need exploration (e.g., catalysis in aerosols or in cloud water). Other products of photochemical oxidations of NO and NO_2 (N_2O_5, organic nitrates, peroxynitric acid) are either dry-deposited or taken up in aerosols or precipitation. N_2O_5 yields nitric acid directly in an aqueous phase.

Factors which should govern rain acidity from SO_2 and H_2SO_4 components include: (a) concentrations of SO_2 and sulfuric acid aerosol in the air and their vertical distributions; (b) aerosol particle size distributions and associated chemical distributions of H_2SO_4 and $(NH_4)_xH_y(SO_4)_{(x+y)/2}$ species; (c) vertical distributions of O_3 and H_2O_2 concentrations in the air, with their possible roles in aqueous phase H_2SO_4 formation from SO_2; (d) initial air concentration of neutralizing bases, e.g., NH_3 and $CaCO_3$, which set initial pH values together with CO_2, SO_2, H_2SO_4, and HNO_3; (e) microphysical characteristics (sizes and numbers of cloud droplets as well as other characteristics, e.g., evaporation); (f) raindrop size distributions; and (g) rate of precipitation ("dilution effect").

For acidification by the NO- and NO_2-derived acids, HNO_3 and HNO_2 (and N_2O_5 intermediate), governing factors are roughly similar. The tropospheric concentrations of gas-phase HNO_3, N_2O_5, and HNO_2, and of species associated with the aerosol HNO_3 component are of central importance. Size distributions (submicron vs. above) and temperature effects on NH_4NO_3 volatility are of particular interest.

It is clear that the acidic species in the air available to be scavenged by clouds and rain at a particular location depend on source magnitudes of S and N compounds, geography of sources, air trajectories, and regional photochemical setting (and thus upon <u>all</u> relevant emissions, e.g., hydrocarbons, SO_2, NO, CO, etc., and upon season of the year), in addition to scavenging factors. Theoretical considerations suggest that cloud nucleation scavenging, interception and impaction of coarse aerosol particles, and SO_2 cloud scavenging followed by oxidation (O_3, H_2O_2), in that order, are dominant acidification mechanisms leading to H_2SO_4 in rain (3). For a given distribution of HNO_3(g) and nitrate particle concentrations, it appears that nucleation scavenging in clouds can play an important role, depending on the size distributions, particle properties, and temperature. If gaseous HNO_3 concentrations in air are comparable to those for aerosol NO_3^-, gas scavenging in clouds and by rain will be important. HNO_3 scavenging can be close to 100% because of the large (10^6M atm^{-1}) Henry's constant for HNO_3/H_2O. If larger aerosol particle sizes are important for HNO_3, impaction and interception by raindrops can contribute significantly to acidification as well. Particle size distribution data are not abundant for nitrates. It is difficult to draw conclusions about scavenging of HNO_3 in comparison to that of sulfur acids.

Concentrations from Scavenging

For aerosol scavenging of component X, the resulting cloud or rain concentrations can be related by $[X] = (X)\varepsilon f/L$ (8), in which $[X]$ is aqueous concentration, (X) is aerosol concentration (e.g., mole per volume of space), ε is efficiency, f accounts for net cumulative evaporation or dilution, and L is the liquid water content per volume of air. For equilibrium gas scavenging, aqueous concentration is given by (scavenged component being conserved per unit volume of space): $[X_i] = (P_{i,o}K_H\alpha_o^{-1})/(1 + LRTK_H\alpha_o^{-1})$, in which $P_{i,o}$ is initial gas partial pressure, K_H is Henry's Law constant ($M\ atm^{-1}$), and α_o is the fraction of undissociated dissolved gas in solution, a function of pH and acidity constants, hence temperature. For a "completely" scavenged and undissociated gas, the concentration is just $[X_i] = P_{i,o}/LRT$. At the pH of acidic rain, HNO_3 is completely scavenged, NH_3 is scavenged well in excess of 90%, and SO_2 is scavenged to just a few percent. H_2O_2 is completely scavenged, independent of pH. O_3 is scavenged to a very small extent.

Rain chemistry data at Pasadena (Table 2, Fig.1) and the sulfate and nitrate aerosol data at times of precipitations suggest that scavenging of aerosols may contribute substantially to observed rain concentrations, perhaps greater than 50% for each. However, there is considerable uncertainty as to scavenging efficiencies and washout ratios. At least 12% of excess sulfate at Pasadena can be accounted for by SO_2 scavenging and subsequent oxidation; the corresponding value at Riverside is in excess of 24% (12). With respect to annual emissions, for which the equivalent potential acid ratio of NO_x to SO_2 (1975-76) was ~ 2.3, HNO_3 is less efficiently scavenged by precipitation in the Los Angeles Basin than is H_2SO_4 ($[HNO_3]/2[SO_4^{2-}]$ ~ 0.8 for mean rain. This may represent kinetic differences in sulfur and nitrogen rainfall removal processes; it may also reflect, indirectly, the influences of dry deposition, fog and dew removal processes, and advection.

Importance of Heterogeneous Oxidations

The residence time of sulfur in the troposphere is believed to be on the order of one week or so. Photochemical reactions yield characteristic $SO_2 \rightarrow H_2SO_4$ conversion times on the order of a week, but not under winter conditions. Known homogeneous photochemical mechanisms for sulfur seem unable to account for observed year-round levels of SO_4^{2-} in rainfall. Oxidation of absorbed SO_2 by H_2O_2 and O_3 in clouds and fog appears to account for very high rates, where photochemical paths cannot (15). Figure 3(b) shows that MnO_2 and Fe^{3+} can oxidize SO_2 in acid solutions, suggesting that a catalytic process involving Mn or Fe can be initiated. In Los Angeles high sulfuric acid formation rates are correlated with high relative humidity, giving added support for an important aqueous phase pathway for SO_2 conversion. A kinetically-attractive alternative path to photochemical oxidation of NO_x has not been delineated; catalyzed oxidation in aqueous aerosol phases or clouds has been mentioned as a possibility.

Interactions and Feedback in the Sulfur-nitrogen System

The key role of the OH radical (and also H_2O_2) in the photochemical oxidation mechanisms of both NO_2 and SO_2 together with a faster rate for NO_2 by a factor of ~10 have suggested the possibility that the rate of homogeneous atmospheric oxidation of SO_2 might be slowed by high emissions of NO_2. Rodhe, Crutzen, and Vanderpol (16) have advanced a simple photochemical and transport model which indicates that long-range transport of HNO_3 can be less pronounced than that for H_2SO_4. In the model NO_2 emissions delay and decrease the oxidation of SO_2 to H_2SO_4 because of competition for OH. The tentative conclusion was advanced that a linear dependence of atmospheric concentrations on emission rates may not be assumed because of nonlinear photochemical mechanism interactions.

In heterogeneous aqueous phase conversion of scavenged SO_2 (SO_2(aq) and HSO_3^- species) certain rate laws, e.g., those with O_2 or O_3 as oxidant, predict slower conversions upon acidification, in part because total dissolved $[SO_2 + HSO_3^-]$ decreases with lowering of pH. Rapid HNO_3 scavenging might thus lead to inhibition of the aqueous SO_2 oxidation rate. Interestingly, the laboratory rate law for oxidation of $[SO_2 + HSO_3^-]$ by H_2O_2(aq) does not exhibit this negative feedback at lower pH. Interactions between the gas-phase and aqueous-phase oxidation of SO_2 and NO_2 deserve greater attention under chemical conditions actually encountered in the atmosphere.

Background Acidities and pH

The global background emissions of H_2S, RSR', SO_2, and NO_x might provide a rough estimate of background inputs of $[BNC]_{CO_2}$ and the resulting pH of rain (although the idea of a "global" background is probably untenable, with S residence times as short as a week or less). If we take the following estimated input fluxes for sulfur and odd nitrogen based on the SCOPE report (20), F_S = 30 Tg S y^{-1} and F_{NO_x} = 15 Tg N y^{-1}, the equivalent acid flux is 3 T eq y^{-1}. Perhaps two-thirds enters the rain. Dividing by the global annual rainfall, $4.5 \times 10^{17} \ell\ y^{-1}$, gives a $[BNC]_{CO_2}$ of 4 $\mu eq \ell^{-1}$ and pH of 5.3 (by solving the proton balance equation for 25°C). If a background ammonia flux is estimated, e.g., 30 Tg N y^{-1} (20), then, since NH_3 is extensively scavenged by rain, $[NH_4^+] \sim 5\ \mu eq \ell^{-1}$, $[BNC]_{CO_2} \sim -1\ \mu eq \ell^{-1}$ and pH $\sim$ 5.8. More base, e.g., sea salt or alkaline dust, would further elevate the pH; more acidic aerosol particles or gas would lower it. The buffer intensity, $d[BNC]_{CO_2}/d\,pH$, is small in this pH region. A background pH is difficult to characterize. Small variations in fluxes of acids and bases could easily cause pH variations between, e.g., 5.0 and 6.0.

Another approach is to examine the pH of cloud water for certain assumed aerosol, gas-phase, and water content conditions. If, for a remote location, (HNO_3) is 0.5 $\mu g/m^3$, (H_2SO_4) is 1.1 $\mu g/m^3$, p_{SO_2} is 10^{-10} atm, and p_{NH_3} is 10^{-10} atm, the pH of cloud can be computed for any degree of scavenging by solving the charge balance equation with p_{CO_2}, $p_{NH_{3,0}}$ (the initial value), and $p_{SO_{2,0}}$ as constants and assuming a cloud liquid water content, e.g., L = 10^{-6} (vol water/vol air). Alternatively, a titration curve can be constructed showing added net nonvolatile acids and bases (H_2SO_4, HNO_3, $CaCO_3$), or $[BNC]^*_{CO_2}$, vs. pH. Response to variations of total input BNC can then be examined directly. Figure 4 shows the result for assumed constraints (25°C). A value of fε ~ 1 has been taken for aerosol scavenging. The resulting scavenged acid concentrations are $[HNO_3]_0$ = 8 μM and $[H_2SO_4]_0$ = 11 μM, yielding $[BNC]^*_{CO_2}$ of 30 μeqℓ$^{-1}$. This gives pH = 4.6 (curve A). In the absence of NH_3 and SO_2 (curve B), the pH would be ~ 0.1 unit lower. The pH of rain would depend on the actual values of fε locally. At smaller net acid, dissolved NH_3 provides greater buffering. Lower temperatures would raise curve A and lower curve B in the pH range ~ 4.5-6.0. Around $[BNC]^*_{CO_2}$ ~ 0, NH_3, alkaline dust, and small acid inputs can alter background pH significantly. (The alkalinity contribution from sea salt is ~ 0.2 μeqℓ$^{-1}$ per mg/ℓ of Na^+.) Similar considerations have recently been put forward in an analysis by Charlson and Rodhe (2).

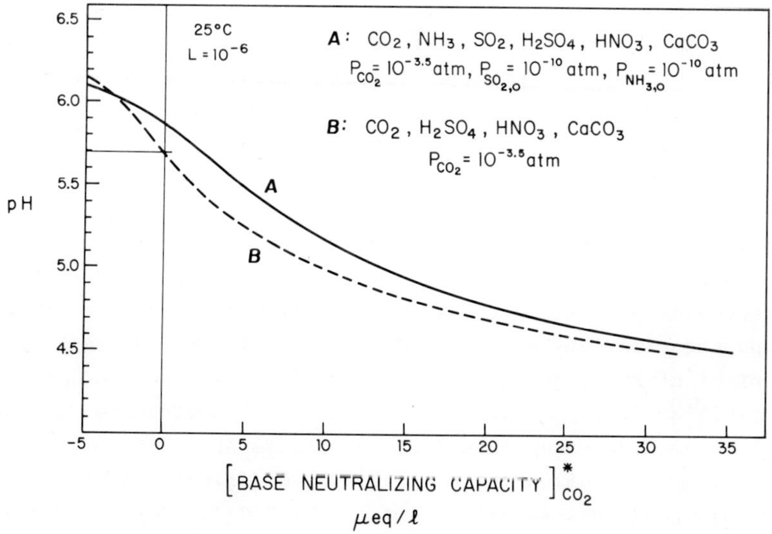

FIG. 4 - Titration curve for two-phase aqueous-gas systems with indicated components. $[BNC]^*_{CO_2}$ is the total added equivalent concentration of nonvolatile acids or bases. (The titrations combine mass-conservation in two phases with open-phase equilibria.)

Present-Day Global Acid Fluxes

The SCOPE report estimates (20) place recent (1970) man-made NO_x emissions at around 20 Tg N y^{-1} and man-made SO_2 emissions at around 70 Tg S y^{-1}. This corresponds to an equivalent flux of ~ 6 T eq y^{-1}. A rough estimate of increased acidity, $\Delta[BNC]^*_{CO_2}$, in rain would be about 9 $\mu eq \ell^{-1}$ on a "global" basis. If the inputs were spread uniformly, then ΔpH of rain would be about 0.3 to 0.6 "globally," depending on the background pH. Local and regional scales have been most affected. It is not unexpected to find pH values in the range of 5 to 4 and even lower in rain downwind from local and regional NO_x and SO_2 emissions which together yield equivalent acid wet deposition fluxes in excess of regional alkaline inputs by ~ $2x10^2$ to ~ 10^3 eq H^+ ha^{-1} y^{-1}. "Global" potential H^+ loadings (~ 10^2 eq ha^{-1} y^{-1}) and eastern North American loadings (~ 10^3) are in this range. The physical and chemical details of regional acid-base and redox "titrations" are highly complex in time and space and depend, in addition to the matrix of emissions, on the different climatologic, meteorologic, photochemical, and geologic features of regions. Present-day global and regional emission loadings predict acidic rain; experience confirms this in many regions of the world.

SOME QUESTIONS AND PROBLEMS

It seems important to ask whether photochemical mechanisms can predict HNO_3 inputs to rain spatially and temporally. If not, what processes can account for observed NO_3^- levels in rain? The suggested nonlinear character of emissions vs. atmospheric concentrations of HNO_3 and H_2SO_4 invites further examination. Control strategy questions may hinge on this aspect of atmospheric acidity. Better understanding of local scavenging of H_2SO_4- and HNO_3-bearing species by clouds and rain requires detailed information on particle size distributions of HNO_3 and H_2SO_4 aerosol acidities and on the $HNO_3(g)/NO_3^-$ (particle) distribution. Key questions related to heterogeneous SO_2 oxidation (cloud, aerosol) concern chemical speciation factors in the solution for reductants, oxidants, and catalysts.

Acknowledgements. I thank H.M. Liljestrand of the University of Texas for his help and advice. I also thank J.N. Galloway and G.E. Likens for interesting me in acidic rain. The support of the Air Resources Board of California is acknowledged.

REFERENCES

(1) Bates, R.B. 1959. Concept and determination of pH. In Acids-Bases in Analytical Chemistry, eds. I.M. Kolthoff and S. Bruckenstein, pp. 362-404. New York: Interscience.

(2) Charlson, R.J., and Rodhe, H. 1982. Factors controlling the acidity of natural rainwater. Nature 295: 683-685.

(3) Garland, J.A. 1978. Dry and wet removal of sulphur from the atmosphere. Atmos. Env. 12: 349-362.

(4) Gorham, E. 1955. On the acidity and salinity of rain. Geochim. Cosmochim. Acta 7: 231-239.

(5) Gorham, E. 1981. Scientific understanding of atmosphere-biosphere interactions: a historical overview. In Atmosphere-Biosphere Interactions. Washington, DC: National Academy Press.

(6) Granat, L. 1972. On the relation between pH and the chemical composition in atmospheric precipitation. Tellus 24: 551-560.

(7) Granat, L. 1978. Sulfate in precipitation as observed by the European atmospheric network. Atmos. Env. 12: 413-424.

(8) Junge, C.E. 1963. Air Chemistry and Radioactivity. New York: Academic Press.

(9) Lee, Y.-N., and Schwartz, S.E. 1981. Evaluation of the rate of uptake of nitrogen dioxide by atmospheric and surface liquid water. J. Geophys. Res. 86: 11,971 - 11,983.

(10) Likens, G.E.; Wright, R.F.; Galloway, J.N.; and Butler, T.J. 1979. Acid rain. Sci. Am. 241: 43-51.

(11) Liljestrand, H.M., and Morgan, J.J. 1979. Error analysis applied to indirect methods for precipitation acidity. Tellus 31: 421-431.

(12) Liljestrand, H.M., and Morgan, J.J. 1980. Spatial variations of acid precipitation in Southern California. Env. Sci. Technol. 15: 333-339.

(13) Liljestrand, H.M., and Morgan, J.J. 1982. Chemical source, equilibrium and kinetic models of acid precipitation. In Energy and Environmental Chemistry: Acid Rain, ed. L.H. Keith, vol. 2. Ann Arbor: Ann Arbor Science.

(14) Odén, S. 1976. The acidity problem - an outline of concepts. Water, Air, Soil Poll. 6: 137-166.

(15) Penkett, S.A.; Jones, B.M.R.; Brice, K.A.; and Eggleton, A.E.J. 1979. The importance of atmospheric ozone and hydrogen peroxide in oxidizing sulfur dioxide in cloud and rainwater. Atmos. Env. 13: 123-137.

(16) Rodhe, H.; Crutzen, P.; and Vanderpol, A. 1981. Formation of sulfuric acid and nitric acid during long-range transport. Tellus 33: 132-141.

(17) Sillén, L.G. 1965. Oxidation state of earth's ocean and atmosphere. Ark. Kemi 25: 159-175.

(18) Smith, R.A. 1872. Air and Rain. London: Longmans, Green.

(19) Stumm, W., and Morgan, J.J. 1981. Aquatic Chemistry. New York: Wiley-Interscience.

(20) Svenssen, B.H., and Söderlund, R., eds. 1976. Nitrogen, Phosphorus and Sulphur - Global Cycles. SCOPE Report 7. Ecol. Bull. (Stockholm) 22: 89-134.

Atmospheric Chemistry, ed. E.D. Goldberg, pp. 41-56. Dahlem Konferenzen 1982.
Berlin, Heidelberg, New York: Springer-Verlag.

The Chemical Composition of Precipitation: A Southern Hemisphere Perspective

G. P. Ayers
Division of Cloud Physics, CSIRO
Sydney, Australia

Abstract. Measurements made in the Australian region form the basis for a discussion on the chemical composition of precipitation. Three categories are considered: remote maritime, remote continental, and polluted conditions. In the first two cases the bulk of the dissolved material can be accounted for by a knowledge of the composition of the atmospheric aerosol. Trace gases play a smaller role than in the third case, where anthropogenic emissions of SO_2 and NO_x can lead to sulfuric and nitric acids being the dominant dissolved species.

INTRODUCTION

The redistribution of liquid water performed by clouds and rain, vital to most forms of life, leads also to a redistribution of trace elements and nutrients derived from atmospheric trace gases and aerosols. Studies of the composition of precipitation thus contribute both to our knowledge of the trace gas and aerosol chemistry of the atmosphere and to our understanding of the biogeochemical cycles of many elements. Current concern over the acidification of precipitation and the fate of anthropogenic effluents such as trace metals and organics has generated considerable support for such studies.

Where possible in what follows points will be illustrated by reference to studies carried out in the Australian region. In this way more of the scarce southern hemisphere data will be added to the body of information already available from the northern hemisphere.

SCAVENGING BY CLOUD (for background information see (21,22))

Nucleation Scavenging

Cloud droplets form when cooling of a saturated air parcel produces a slight water vapor supersaturation, S (S = relative humidity - 100%). In natural clouds S evolves according to

$$\frac{dS}{dt} = \alpha \frac{dz}{dt} - \beta \frac{dW}{dt} , \qquad (1)$$

where α and β are positive constants, z is height, and W is liquid water content. In a steady updraft the first term (cooling rate) is constant, so S passes through a maximum, S_{max}, as the second term (decrease in supersaturation due to condensation) increases from zero, equals, and exceeds the first term. S_{max} is on the order of 0.1-0.2% for stratiform clouds and 0.3-1% for cumuliform clouds.

Each cloud droplet forms on a soluble particle that requires a certain critical supersaturation, S_c, for activation as a nucleation site. For ammonium sulfate, a typical nucleus, particle radius and S_c are related by

$$r = 1.45 \times 10^{-2}\ S_c^{-2/3} , \qquad (2)$$

where r is in µm and S_c in %. The effect of composition on this relationship is not large. For example, at a given value of S_c a change in composition to sodium chloride would lower r by only 13% from the value predicted by Eq. 2.

Setting S_c equal to S_{max} implies that particles as small as 0.01 µm may be activated in vigorous cumulus clouds, while in stratiform clouds subject to gentle uplift, the minimum size activated may be much larger, on the order of 0.1 µm. In areas not too heavily influenced by anthropogenic contributions to the aerosol it is common for most of the total aerosol volume to be associated with particles larger than 0.1 µm (for example, see Fig. 1). In such cases, therefore, most of the aerosol volume would be incorporated into cloud, irrespective of cloud type. However, chemical species in the aerosol generally are not distributed uniformly as a function of particle size. Thus the extent to which any species concentrated mainly in the size range of the minor volume mode (near 0.07 µm in Fig. 1) would be scavenged by nucleation depends very much on cloud type.

<u>In-cloud Scavenging</u>

Particles are transported to cloud droplets principally by Brownian diffusion. A simple Maxwellian description of the process would give

$$n = n_O \exp(-4\pi \bar{R}NDt), \qquad (3)$$

where n is number concentration of particles having diffusion coefficient D, $\bar{R}$ is mean cloud droplet radius, N is droplet concentration, and t is time. Based on the number distribution in Fig. 1(a), Eq. 2 predicts droplet concentrations of 191 and 1163 cm^{-3} for S_{max} values of 0.1 and 0.3%. For a liquid water content of 0.3 g m^{-3} the respective mean drop radii

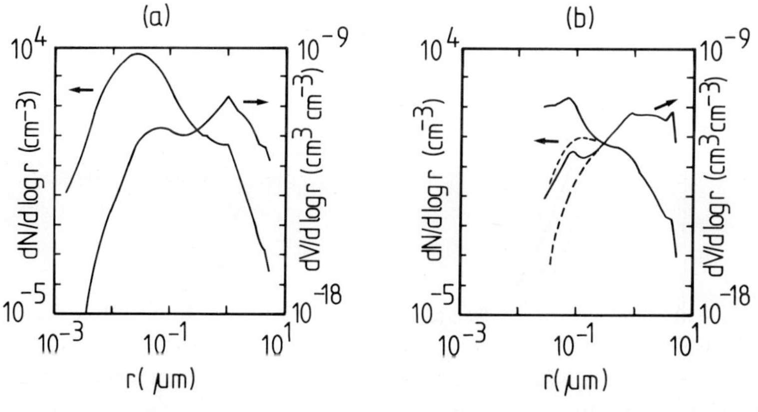

FIG. 1 - Aerosol number distributions and associated volume distributions measured at the Australian Baseline Station (3). (a) Urban-influenced conditions, mean of five distributions. (b) Clean conditions, mean of seven distributions. The broken lines show the underlying sea salt distribution.

would be 7.2 and 3.9 µm. Inserted in Eq. 3 these values yield the size-dependent scavenging curves shown in Fig. 2, where the size fraction removed by nucleation is also shown. Comparison with the volume distributions in Fig. 1 shows that a negligible fraction of total aerosol mass is actually removed by diffusion. However, since gases of interest have radii <0.001 µm ($D \approx 0.1\ cm^2\ s^{-1}$), Eq. 3 implies that diffusional transport of gases to cloud droplets would be very efficient, being practically complete in tens of seconds.

SCAVENGING BY RAIN

The important aerosol scavenging processes appear schematically in Fig. 3. Particles in the range <0.01 µm are most effectively removed by diffusive processes, while inertial impaction is efficient for particles ≥1 µm but ineffectual at smaller sizes. The so-called "Greenfield Gap" near 0.1 µm is the region in which neither diffusion nor impaction is an efficient removal process, so phoretic effects probably dominate at this size. Slinn (this volume) makes clear the level of uncertainty that for various reasons still surrounds the exact form of the curve depicted in Fig. 3. He also discusses the case of scavenging by ice particles, where again there are large gaps in our knowledge.

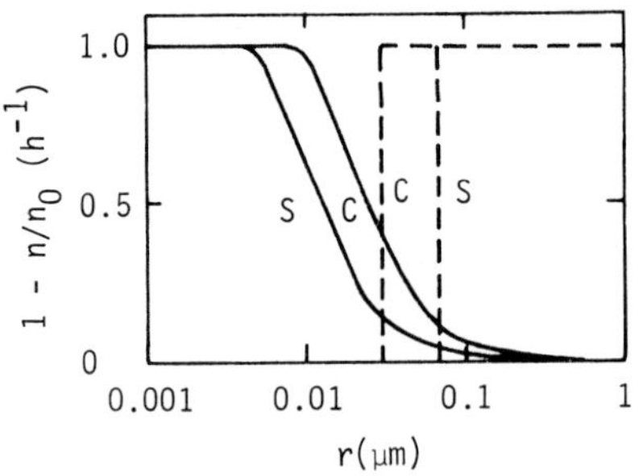

FIG. 2 - Fraction of particles removed per hour in stratiform (S) and cumuliform (C) clouds by diffusion (solid curves) and nucleation (broken lines).

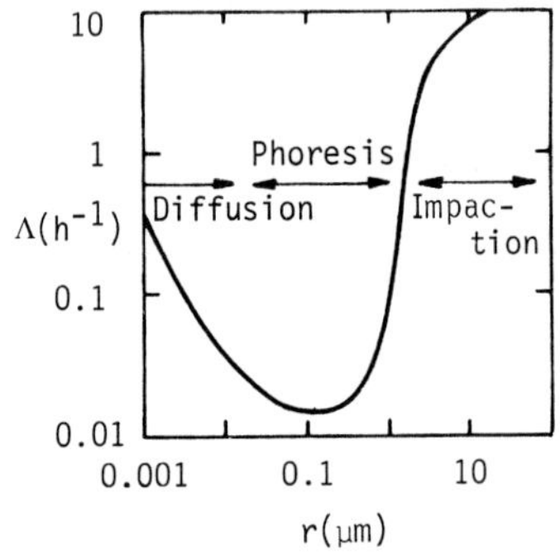

FIG. 3 - Schematic representation of the fractional depletion rate for aerosol scavenged by raindrops.

CHEMICAL COMPOSITION

Ions commonly determined in precipitation are H^+, NH_4^+, Na^+, K^+, Mg^{2+}, Ca^{2+}, Cl^-, NO_3^-, SO_4^{2-}, and HCO_3^-, as they comprise the bulk of the dissolved material. Less often analyses are undertaken for total organic carbon, or nitrogen, silicates, phosphates, heavy metals and organics. Most of this discussion concerns the common ionic species.

Remote Maritime Atmosphere

Far from land the inorganic fraction of the aerosol consists mainly of large sea-salt particles ($r \gtrsim 0.1$ µm) and smaller ammonium sulfate particles ($r \lesssim 0.1$ µm), with most of the total aerosol mass coming from the salt particles. Concentrations of trace gases such as NH_3, SO_2, and HNO_3 are low, usually on the order of 0.1 ppbv. As a result the sea-salt aerosol, which should be efficiently scavenged during cloud formation and below cloud by precipitation, is expected to be the major contributor to total dissolved ionic species in both cloud water and rainwater. However, the composition of the sea-salt aerosol can be modified from that of bulk seawater both during the aerosol formation process at the sea surface, and subsequently by reaction with trace gases (HNO_3, NH_3, O_3, SO_2, etc.).

One such modification is the enrichment of NO_3^- and SO_4^{2-} often observed in salt particles, sometimes in conjunction with a Cl^- deficiency. Enrichment is usually given the symbol E, and can be defined as

$$E_Y(X) = \frac{(X/Y)\ \text{air}}{(X/Y)\ \text{seawater}} . \qquad (4)$$

Y is a species characteristic of seawater that passes across the air-sea interface without enrichment. $E > 1$ implies enrichment of X in the atmosphere relative to seawater, while $E = 1$ implies no enrichment and a seawater source for X.

In the Australian region, data on the composition of rainwater that is essentially maritime in nature have been published for collection sites located at Cape Grim (41°S) and Maatsuyker

Island ($44^{o}S$) (3). Volume-weighted mean concentrations of a number of ions are listed in Table 1.

It is noticeable that concentrations of the sea-salt components (the metals, Cl^-, and SO_4^{2-}) are considerably higher than those of nitrate and ammonium ions, in agreement with our expectation based on the relative atmospheric concentrations of trace gases and the aerosol. It is also worth noting that the absolute concentrations of the sea-salt components are higher than those reported for many other places, Samoa, for example (8). The reason is probably that the Australian sites are located in the region of the "roaring forties" and are thus subject to stronger winds than is the Samoan site, leading to higher sea-salt aerosol loadings in the Australian cases. This instance underlines the importance of meteorological factors to rain-water chemistry in general.

Enrichment factors calculated from the data in Table 1 are given in Table 2. Most of the modest enrichments evident from

TABLE 1 - Monthly (approximately), wet-only precipitation samples. Volume-weighted means shown ($\mu mol.dm^{-3}$).

Place	No. of samples	pH	Na^+	K^+	Mg^{2+}	Ca^{2+}	NH_4^+	Cl^-	NO_3^-	SO_4^{2-}
Maatsuyker Is.[a]	12	5.67	596	21.6	69.6	20.2	2.1	689	5.8	44.9
Cape Grim[b]	56	5.99	1297	32.3	122	42.9	2.0	1349	5.0	79.1

[a]Samples collected between June 1977 and February 1979 (3).
[b]Samples collected between April 1977 and August 1981 (3).

TABLE 2 - Enrichment factors relative to Cl^-, calculated from the data in Table 1.

Place	Na^+	K^+	Mg^{2+}	Ca^{2+}	SO_4^{2-}	NH_4^+	NO_3^-
Maatsuyker Is.[a]	1.0	1.7	1.0	1.6	1.3	>>1	>>1
Cape Grim[b]	1.1	1.3	0.9	1.7	1.1	>>1	>>1

[a]Samples collected between June 1977 and February 1979 (3).
[b]Samples collected between April 1977 and August 1981 (3).

Table 2 are not reflected as significant in Table 3, where enrichment factors were obtained not from the mean concentrations, but by regression analysis of the raw data.

Since the intercepts in the regressions were positive, it seems likely that the apparent enrichments in Table 2 do not reflect enrichment at the air-sea interface but rather the addition of extra components to the bulk aerosol. Recent studies at Cape Grim suggest that one additional component is local soil material that should not be considered as part of the maritime aerosol ((1), and Ivey, personal communication). This case illustrates the type of difficulty that contaminants can introduce to the interpretation of data. Another difficulty often encountered is evident from the enrichment factors for sulfate shown in Table 3: an enrichment of sulfate postulated on the basis of other information is masked in the precipitation data by poor analytical precision. Clearly analytical techniques of high precision are necessary in this work.

From Table 3, the deficit of Mg^{2+} and enrichment of Ca^{2+} at Cape Grim appear to be significant. In the latter case the aerosol studies of Andreae (personal communication) support the idea that enrichment of Ca^{2+} does occur during aerosol formation at the ocean surface; however, there was no such evidence for Mg^{2+}. Thus the deficit in Mg^{2+} in rain at Cape Grim remains unexplained, although one suggestion is that Mg^{2+} may be lost by ion exchange processes involving the insoluble local soil contaminant that gives rise to the apparent enrichment factors in Table 2 (Ivey, personal communication).

The large enrichments of NH_4^+ and NO_3^- in rainwater (Table 2) compared to seawater can be ascribed to the addition of gaseous

TABLE 3 - Enrichment factors relative to Cl^-, with 95% confidence limits. Obtained by linear regression analysis of raw data.

Place	Na^+	K^+	Mg^{2+}	Ca^{2+}	SO_4^{2-}
Maatsuyker Is.	0.93±0.09	1±1	1.0±0.1	1.2±0.8	1.1±0.8
Cape Grim	1.0±0.2	1.2±0.2	0.85±0.06	1.5±0.3	1±17

ammonia and nitric acid to the sea-salt aerosol prior to cloud formation, and subsequently to cloud water and rainwater. Contributions to NH_4^+ (and SO_4^{2-}) would also be expected from the smaller size fraction of the aerosol ($r \lesssim 0.1$ µm). The size of these contributions can be estimated roughly by using what little information is available to predict the composition of cloud water at Cape Grim under "typical" conditions. The predictions appear in Table 4.

Note that the concentrations scale with the inverse of liquid water content, so that although the choice of 0.3 $g.m^{-3}$(STP) may be a "typical" figure for Cape Grim, values in the range 0.1-1 $g.m^{-3}$(STP) certainly occur, with consequent effects upon ion concentrations.

In the absence of published data on cloud water composition at Cape Grim (there is a dearth of such data everywhere!), the value of the predictions should not be overemphasized. However, the rainwater data in Table 1 give a useful perspective: given the uncertainty in gaseous and aerosol nitrate levels at Cape Grim, the cloud water concentrations predicted in Table 3 can

TABLE 4 - Aerosol and trace gas composition of air at Cape Grim ($nmol.m^{-3}$(STP)) and predicted cloud water composition ($\mu mol.dm^{-3}$), assuming 0.3 $g.m^{-3}$(STP) liquid water.

Component	Na^+	NH_4^+	Cl^-	NO_3^-	SO_4^{2-}
sea-salt aerosol[a]	144	0.0?	168	~1.6[c]	9.9
ammonium sulfate[b] aerosol		5.9			2.6
trace gases		3.5[d]		~3.2[c]	
total:	144	9.4	168	~4.8	12.5
cloud water	480	28.5	560	~16	41.6

[a]Based on sea-salt particle size distribution, Fig. 1(b); SO_4^{2-} enriched 15% relative to seawater (12).
[b]Based on ammonium sulfate particle size distribution, Fig. 1(b); smallest size included assumes S_{max} = 0.3%, see Eq. 2.
[c]Nitrate loadings for southern maritime atmosphere taken from GAMETAG results (14).
[d]Observed long-term mean gas concentration at Cape Grim (2).

be said to be of the same order as the nitrate levels in Table 1. This is similar for the sulfate levels; in fact, the total sulfate enrichment that would be deduced from the predictions is the same as that found by Andreae in aerosol samples at Cape Grim. On the other hand, the predicted ammonia concentration seems too high in comparison with those of Table 1. Two explanations can be offered. The first recognizes the ample worldwide evidence showing that ammonia concentrations are not stable in precipitation samples held for lengthy periods after sample collection. This problem must be unavoidable where monthly samples are collected, as at Cape Grim (and as recommended for the WMO network). Clearly many of the reported ammonia concentrations in the literature must be treated with caution. The second complication arises because the prediction assumes a conservation of total ammonia mass plus partitioning between gaseous and aqueous phases according to Henry's Law. The validity of this calculation is questionable while there is still uncertainty regarding the kinetics of ammonia dissolution and the appropriate value for the Henry's Law constant for ammonia (13).

In summary, the foregoing discussions reveal that even in the relatively uncomplicated situation of the "clean" maritime atmosphere there are many factors that may influence the composition of rainwater (or cloud water) as finally determined by the analytical chemist in the laboratory (and it should be emphasized that we have considered only the few most concentrated inorganic species). Examples of such factors are: local meteorology; atmosphere/cloud dynamics; cloud/rain microphysics; local site contamination; sample degradation during storage; etc. - no doubt many readers could add more.

Remote Continental Atmosphere

Even in remote continental regions the multiplicity of natural aerosol and gas sources ensures that variations about a "typical" composition for precipitation are greater than in the maritime case. An added complication is the ever present seasalt contribution to continental rain, albeit one that decreases inland. For Cl^- the data available (from (15,17,18)) are well fitted by an equation of the form

$$Cl^- = a\,d^{-0.25} - b, \qquad (5)$$

where d is distance from the coast, and a and b are empirical parameters that are determined by local meteorology. Although as much as 80% of the sea-salt contribution is lost in the first 10 km, significant contributions very often extend for some hundreds of kilometers inland. The resultant wide range of ionic concentrations that can be found in continental rain is summarized in Table 5.

Recast in terms of enrichment factors referenced to seawater, the Australian data reveal the pattern shown in Table 6. The enrichment factors in the bottom line of Table 5 were calculated from Hutton's data (16) on the mean ionic composition of a range of Australian vegetation types. When Hutton's data were taken into account with a knowledge of regional soil types, most authors interpreted their data as showing that the remote continental rains sampled contained a mixture of vegetation-derived and soil dust components, plus a small but variable sea-salt component.

Sulfate enrichments similar to maritime values reflect the fact that SO_2 levels over unpopulated regions of Australia are lower than those over many "clean" areas of the US or Europe. A similar conclusion appears valid for NO_3^-, continental concentrations (Table 6) being hardly above those in maritime rain. For NH_4^+, however, the known disparity between maritime and continental gas concentrations (2) is reflected in rainwater

TABLE 5 - Concentration ranges in Australian continental rain[a] (μmol dm^{-3}).

pH	Na^+	K^+	Mg^{2+}	Ca^{2+}	NH_4^+	Cl^-	NO_3^-	SO_4^{2-}	HCO_3^-	Phos-phates
5-8	1-1000	0.1-20	0.1-150	0.1-150	1-30	1-1000	1-10	5-50	1-200	0.1-1(?)

[a]From ((4,5,9,15,18,20,23,24), Reeve and Fergus (unpublished data), and Khanna (personal communication).

TABLE 6 - E_{Cl} values based on seawater composition. Continental rain.

Location	Ref.	Na^+	K^+	Mg^{2+}	Ca^{2+}	SO_4^{2-}	HCO_3^-
N.T.	(23)	1.1	8.1	1.66			
N.T.	(24)[a]		5.01	2.61	2.63		
N.S.W.	(9)	0.65	7.89	1.74	18.3		
A.C.T.	(9)	0.65	14.9	2.08	36.3		
W.A.	(4)	1.05			2.13		
Victoria	(5)	0.87	5.60	0.33	3.70	2.80	
Queensland	(20)	0.77	2.69	1.02	12.8	2.65	
W.A.	(15)	1.02	5.45	1.56	18.1	2.88	118
A.C.T.[b]		0.94	15.8	3.92	23.5	7.63	
	(11)	0.98	77.4	6.33	50.9		

[a] Ratios to Na^+.
[b] Khanna (personal communication).

composition: 1-10 μmol dm^{-3} for maritime rain, extending up to 30 μmol dm^{-3} for continental rain (although here again it is not clear that reported ammonia concentrations are free from losses during sample storage, prior to analysis.

Polluted Atmosphere

Not surprisingly, a bewildering range of constituents can be detected in polluted rain. As a result, a general description will not be attempted here; instead the focus will be on one topical aspect: that of rainwater acidity.

The pH of 5.6 often quoted as being characteristic of pure water in equilibrium with atmospheric CO_2 is not to be seen as having any overwhelming chemical significance with regard to clouds or rain: it is simply a commonly chosen reference point. It is clear from the short discussion on cloud formation and scavenging given earlier that all cloud water and rainwater will contain acid-base constituents other than CO_2, the added material coming from trace gases and the atmospheric aerosol. Thus pH values other than 5.6 should not be unexpected (for more details see Morgan, this volume, and (6)).

Leaving aside the importance or otherwise of pH values near 5.6, it has become widely accepted in recent years that sites located in, or downwind of, urban/industrial areas often experience rainwater pH values one or more pH units lower than sites located in areas remote from, or upwind of, populated areas.* The populated areas showing highest precipitation acidity (lowest pH) are regions of Europe, and the northeastern corner of the US, where acid levels appear to have increased over the last three decades or so in parallel with increasing emissions of SO_2 and NO_x (11). Typical of current levels of H_2SO_4 and HNO_3 found in areas of concern are the values listed for Ithaca, NY, in Table 7.

The ions H^+, NH_4^+, SO_4^{2-}, and NO_3^- dominate over the sea-salt and continental components, a typical occurrence for inland areas subject to acidic rain, as is the balance between the sums $H^+ + NH_4^+$ (117 μmol dm^{-3}) and $2SO_4^{2-} + NO_3^-$ (110 μmol dm^{-3}).

Levels of acidity in rain commensurate with Australia's small population and relatively low SO_2 and NO_x emissions have recently been observed in the Sydney basin (Table 7). On occasions of high acidity ($H^+ + NH_4^+ \approx 100$ μmol dm^{-3}), NO_3^- levels were also very high (also ~100 μmol dm^{-3}), a fact consistent with the dominance of NO_x over SO_2 in the Sydney atmosphere (Australian fuels are very low in S: 0.3-0.5% S in coals and 0.1-0.5% S in oils). More complete data from Sydney therefore may well support the type of relationship between SO_2/NO_x levels and SO_4^{2-}/NO_3^- levels apparent in the US (11). A further 10 months of pH data from Sydney also exhibit the summer maximum and winter minimum in H^+ reported overseas.

Preliminary analysis of the Sydney data suggests that ammonia is of considerable importance to the overall acid-base balance.

---

*One notable exception occurs in some recent data from Amsterdam Island, an oceanic site located at 38°S for which pH values as low as 3.8 are reported (Galloway, Keene, and Likens, personal communication). No explanation is available as yet. This result is in stunning contrast to the Australian maritime data summarized in Table 1 and bears careful investigation.

TABLE 7 - Compositions of precipitation at Ithaca and Sydney (μmol dm^{-3}).

Place	pH	H^+	Na^+	K^+	Mg^{2+}	Ca^{2+}	NH_4^+	Cl^-	NO_3^-	SO_4^{2-}
Ithaca[a]	4.03	93.3	5.2	0.64	1.0	5.5	22.7	8.7	33.1	38.4
Sydney[b]	4.44	36.3					18.2	116	13.7	n.d.[c]

[a] 45 samples (19).
[b] 12 sites, 294 samples, November 1980-March 1981.
[c] Not determined.

For light to moderate rain events (<20mm) the data yielded the following results when subjected to regression analysis:

$$p(H^+ + NH_4^+) = 4.23 + 0.012\ h, \tag{6}$$

$$pNH_4^+ = 4.45 + 0.041\ h, \text{ and} \tag{7}$$

$$pH = 4.99 - 0.023\ h, \tag{8}$$

where $pX = -\log_{10}X$ and h is rainfall per event (in mm). In each case the relationship was significant at the 0.01 level. Each slope was also different from zero at the 0.01 level.

The decrease in ionic concentration with increased rainfall amount evident from Eqs. 6 and 7 is a phenomenon that has been widely observed and is accepted as being a consequence of the usual scavenging processes. However, H^+ (Eq. 8) exhibited the unusual behavior of increasing with precipitation amount. The explanation is evidently that the rate of increase of pNH_4^+ was three times that of $p(H^+ + NH_4^+)$. In consequence, a net decrease in pH was observed. If the balance $H^+ + NH_4^+ = 2SO_4^{2-} + NO_3^-$ occurs in Sydney, as it does at Ithaca and other overseas sites, then very different source/sink balances are indicated for atmospheric ammonia and the acids H_2SO_4, HNO_3, or their precursors. However, no quantitative description is available at present. It should be noted that a similar case can be found in the literature, although in that case the alkaline component exhibiting rapid decrease was $CaCO_3$ aerosol (7).

CONCLUDING REMARKS

Given the nature of a background paper and the scope of the

present topic, the treatment given here should be seen as illustrative, rather than exhaustive. Emphasis has been put on the primary links between the composition of atmospheric trace gases and aerosols and the composition of precipitation. These links are stressed in each of the cases discussed: remote maritime, remote continental, and polluted precipitation.

However, a quantitative description of the processes leading to a particular observed composition must also recognize the influence of other factors, including meteorological and seasonal factors; the microphysics and dynamics of the cloud system; the intensity, duration, and size spectrum of precipitation; the oxidation capacity of the gaseous and aqueous phases; and for some gases, equilibrium considerations (Henry's Law). There are also the ever present problems of site representativeness and contamination, not to mention analytical imprecision. The brief mention made here of some of these factors is no reflection on their relative importance, but arises from a lack of space and the knowledge that they are discussed more properly elsewhere (see Morgan and Slinn, both this volume).

The present work would also be incomplete without further mention of two topics cited at the outset: trace metals and organics in precipitation. These topics involved some discussion at the workshop but again have not been pursued here because of lack of space. An excellent basis for discussion can be found in a workshop report entitled Toxic Substances in Atmospheric Deposition: A Review and Assessment (10).

REFERENCES

(1) Andreae, M.O. 1982. Marine aerosol chemistry at Cape Grim, Tasmania and Townsville, Queensland. J. Geophys. Res., in press.

(2) Ayers, G.P., and Gras, J.L. 1982. The concentration of ammonia gas in Southern Ocean air. Paper presented at the Second Symposium on the Composition of the Nonurban Troposphere, Williamsburg, Virginia, May 1982.

(3) Baseline Air Monitoring Report 1978. Canberra: Australian Government Publishing Service.

(4) Bettany, E.; Blackmore, A.V.; and Hingston, F.J. 1964. Aspects of the hydrologic cycle and related salinity in the Belka Valley, Western Australia. Aust. J. Soil Res. 2: 187-210.

(5) Briner, G.P., and Peverill, K.I. 1976. Rain as a source of fertilizer in Australia. Research Project Series No. 16. Victoria: Department of Agriculture.

(6) Charlson, R.J., and Rodhe, H. 1982. Factors controlling the acidity of natural rainwater. Nature 295: 683-685.

(7) Dawson, G.A. 1978. Ionic composition of rain during sixteen convective showers. Atmos. Env. 12: 1991-1999.

(8) DeLuisi, J.J., ed. 1981. Geophysical Monitoring for Climatic Change No. 9, Summary Report 1980. U.S. Department of Commerce.

(9) Douglas, I. 1968. The effects of precipitation chemistry and catchment area lithology on the quality of river water in selected catchments in eastern Australia. Earth Sci. J. 2: 126-144.

(10) Galloway, J.N.; Eistenreich, S.J.; and Scott, B.C., eds. 1980. Toxic Substances in Atmospheric Deposition: A Review and Assessment. Washington, DC: U.S. EPA Office of Pesticides and Toxic Substances.

(11) Galloway, J.N., and Likens, G.E. 1981. Acid precipitation: The importance of nitric acid. Atmos. Env. 15: 1081-1085.

(12) Garland, J.A. 1981. Enrichment of sulphate in maritime aerosols. Atmos. Env. 15: 787-791.

(13) Hales, J.M., and Drewes, D.R. 1979. Solubility of ammonia at low concentrations. Atmos. Env. 13: 1133-1147.

(14) Huebert, B.J., and Lazrus, A.L. 1980. Tropospheric gas-phase and particulate nitrate measurements. J. Geophys. Res. 85: 7322-7328.

(15) Hingston, F.J., and Gailitis, V. 1976. The geographic variation of salt precipitated over Western Australia. Aust. J. Soil Res. 14: 319-335.

(16) Hutton, J.T. 1968. The redistribution of the more soluble chemical elements associated with soils as indicated by analysis of rainwater, soils and plants. Proceedings of the 9th International Conference on Soil Science, Adelaide.

(17) Hutton, J.T. 1976. Chloride in rainwater in relation to distance from ocean. Search 7: 207-208.

(18) Hutton, J.T., and Leslie, T.I. 1958. Accession of non-nitrogenous ions dissolved in rainwater to soils in Victoria. Aust. J. Agric. Res. 9: 492-507.

(19) Miller, J.M.; Galloway, J.N.; and Likens, G.E. 1978. Origin of air masses producing acid rain at Ithaca, New York. Geophys. Res. Lett. 5: 757-760.

(20) Probert, M.E. 1976. The composition of rainwater at two sites near Townsville, Qld. Aust. J. Soil Res. 14: 397-402.

(21) Pruppacher, H.R., and Klett, J.D. 1978. Microphysics of Clouds and Precipitation. Dordrecht, Holland: D. Reidel Publishing Company.

(22) Twomey, S. 1977. Atmospheric Aerosols. Amsterdam: Elsevier Publishing Company.

(23) Wetselaar, R., and Hutton, J.T. 1963. The ionic composition of rainwater at Katherine, N.T., and its part in the cycling of plant nutrients. Aust. J. Agric. Res. 14: 319-329.

(24) Williams, W.D., and Seibert, B.D. 1963. The chemical composition of some surface waters in central Australia. Aust. J. Freshwater Res. 14: 166-175.

Atmospheric Chemistry, ed. E.D. Goldberg, pp. 57-90. Dahlem Konferenzen 1982.
Berlin, Heidelberg, New York: Springer-Verlag.

Some Influences of the Atmospheric Water Cycle on the Removal of Atmospheric Trace Constituents

W. G. N. Slinn
2215 Benton Avenue, Richland, WA 99352, USA

Abstract. Research needed to solve practical precipitation-scavenging problems is outlined. Topics are organized, in the main, into three space scales (local, regional, and global), two time scales (acute and chronic), and separately for particles and gases.

INTRODUCTION

Great: after finishing writing the text (it's best for me to leave the introduction to last), now I see the typing instructions. I knew that there was a page limit, but the way they want this typing done, my old 2-for-1 rule (handwritten for typed pages) won't apply. But I'll be damned if I'm going to rewrite -- and probably damned if I don't. But I don't have the time! Well, phooey on the page limit: if readers want less, let them read less; if the organizers want fewer pages, let them print only the first 15, or every second page, or whatever... But maybe I can do it yet! More than 400 years ago, Tusser reviewed the whole of atmospheric chemistry in two lines:

> Except wind stands as never it stood,
> It is an ill wind turns none to good.

[I see the summary this way: perhaps even global air pollution "turns (some) to good"; perhaps with it, we will better appreciate that we all live in a single world... And how better to emphasize the contrast between the unity of the rest of Nature and the disunity caused by Man than to hold a workshop on atmospheric chemistry in the walled city of Berlin?]. And as for a review of the atmospheric water cycle, I doubt that anyone will improve on Longfellow's five lines, written more than a century ago:

> Nothing that is can pause or stay;
> The moon will wax, the moon will wane,
> The mist and cloud will turn to rain,
> The rain to mist and cloud again,
> Tomorrow be today.

So that leaves only precipitation scavenging to review, and I offer this: while the winds blow around the freeways in the air, mocking our stupid walls and border stations, precipitation keeps the streets clean! That completes this paper, well within the page limit. For interested readers, additional details can be found in the appendix -- and no one sets page restrictions on an appendix!

APPENDIX

Without natural atmospheric cleansing processes, the air would soon become too polluted for the survival of most - if not all - life forms. With these cleansing processes, many - but not all - air pollutants are cycled through the air in about a week. Particles and reactive gases such as H_2SO_4, I_2, etc., are rapidly scavenged by precipitation; other pollutants, such as halocarbons, ^{85}Kr, CO, etc., pass through this water filter, and (though vacuumed by storms and, thereby, mixed more uniformly in the atmosphere) they then must await the action of other, typically slower cleansing processes such as chemical transformations, radioactive decay, and/or dry removal at the earth's surface. If a pollutant is removed by more than one process, if separate removal rates can be identified for each (first-order) process, and if a number of other conditions are satisfied (11), then the overall tropospheric residence (or turnover) time for the pollutant can be found from

$$\tau^{-1} = \tau_w^{-1} + \tau_d^{-1} + \tau_{ph}^{-1} + \tau_{ch}^{-1} + ,$$

where the subscripts w, d, ph, and ch are abbreviations for wet, dry, physical, and chemical removal. However, here I will ignore these other removal processes; here, the focus will be on precipitation scavenging.

I will try to keep the focus fairly sharp and therefore omit much of the standard description of precipitation scavenging: elsewhere, just a sketch of this standard material required an order-of-magnitude more text (17). My plan is to emphasize research needed to help solve real and perceived, practical problems. And in an attempt to organize these suggestions, I identify problems at three space scales (local, regional, and global), each with two time scales (acute and chronic), and treat particles and gases separately. This gives a potential for 3x2x2 = 12 subdivisions. The subdivisions, however, are not even, and indeed, will be abandoned when found to be too constraining. I hope this unevenness and abandonment of a classification scheme will not distract; I am convinced that there is value in versatility at abandoning artificial divisions of a whole. Also, as a final introductory caveat: the workshop organizers have said that these background papers "should include open questions and provocative statements" - and I don't know how anyone can explore the path of open environmental questions very far without stumbling onto provocative political and philosophical problems.

I. LOCAL SCALE 1. Acute Doses (i) Particles

I use the word "acute" as an abbreviation to describe a short-term release of pollution (e.g., the release of radioactivity from a nuclear-reactor or transportation accident). For particles, the appropriate theory was developed by Chamberlain, 30 years ago. In this theory, the number density of particles, $n(a;\vec{r},t)$, of a given size class a, decreases by scavenging via $\partial n/\partial t = -\psi n$, where the scavenging rate, ψ, is found by integrating over the collection by hydrometeors of each size class, here identified with length scale ℓ:

$$\psi(a;\vec{r},t) = \int_0^\infty d\ell\ N(\ell;\vec{r},t)\ [v_t(\ell) - v_g(a)]\ A_x\ E(a,\ell), \qquad (1)$$

in which the v's are settling speeds, A_x is essentially the cross-sectional area of the hydrometeor (perpendicular to $\vec{v}_t$ and including the size of the particle), and E is the collection efficiency. If E = 1, then all particles in the path of a hydrometeor will be captured by it. E can be larger than unity (e.g., if electrical forces are strong), but usually $E<1$, and a major problem is to specify E accurately.

How well is E known? Figure 1 is illustrative for the case of rain scavenging. The lowest curve on this figure, labeled t = 0 and R = 0.5 mm, was guided by many sets of laboratory data (few of which are in agreement, and contains theoretical descriptions of Brownian diffusion (which is expected to dominate for particles with $a \lesssim 0.1$ μm), interception, and inertial impaction (which dominates for particles with $a \gtrsim 1$ μm). In

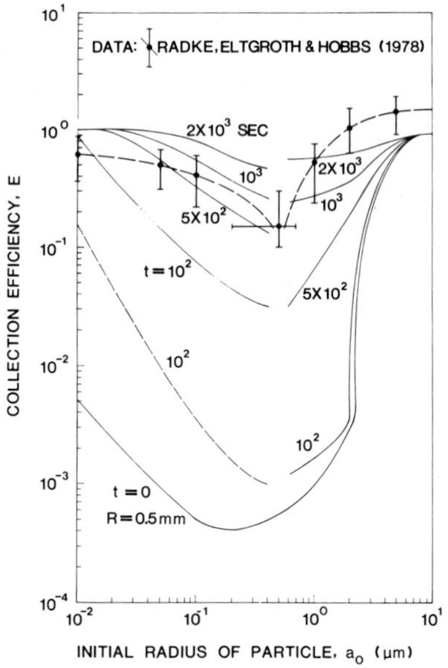

FIG. 1 - Measured and modeled, particle/raindrop collection efficiency (reprinted from (17)).

contrast to this laboratory-data/theoretical curve, however, and in places three orders-of-magnitude higher (!), is a field-data curve through the average of seven sets of measurements by Radke et al. (10). The difference between these two curves is a dramatic statement of how well (or better, how poorly) E is known. Of course, many have suggested ways to remove this discrepancy. For example, as indicated by the series of curves labeled 10^2, $5x10^2$, 10^3, and $2x10^3$ (seconds) on the RHS of Fig. 1, perhaps the larger-than-expected, field-measured E reflects particle growth by water-vapor condensation; thereby, the particles would have been larger when scavenged (wet) than when sampled (dry). That may be, but it is speculation. Similarly, for particles smaller than about 0.1 μm, perhaps they attached, for example, to plume droplets, which were subsequently scavenged.

However, as is indicated on the LHS of Fig. 1, the attachment rate would need to be about 10^2 times faster than by Brownian diffusion to lead to the measured E. Therefore, other authors suggest that electric or thermophoretic effects were important, and they choose electrical or temperature gradient parameters for their models to fit the data. (And most of these numerical codes have enough free parameters to fit an elephant!) Yet the data may be misleading, for any of many reasons: unsatisfactory performance of the particle-size analyzers, scavenging of precursor gases (so that the particles were not there to be scavenged during the rain), a change in air mass (by the downdrafts associated with the showers), etc. These speculations aside, some points are clear: more field data are needed, more measurements should be taken of potentially significant variables, and more effort should be devoted to data analyses and associated theoretical developments. True, it would be hoped that such obvious statements would be unnecessary, but unfortunately in these days of reduced support for science, such obvious statements seem to have become essential.

Figure 2 suggests similar uncertainties about E for snow scavenging of particles. If I had had more time for preparation of this paper, I would have added to Fig. 2 the recent field data obtained by Murakami et al. (9), which demonstrate an order-of-magnitude larger E for rimed vs. unrimed, dendritic snowflakes. As I plan to demonstrate in a later publication, I believe that the new data (and the data in Fig. 2) suggest many inadequacies in previous theoretical interpretations of snow scavenging. It will be necessary, I think, to develop a filter model for an individual dendritic snowflake: with this filter model varying between those developed for fiber filters to those for millipore filters - depending on the degree of riming. In addition, when inertial impaction (rather than interception) dominates, then the characteristic length scale (for the impaction parameter) must be defined carefully. But before we go too far with the theory, I for one would be very appreciative if the amount of good field data (viz., the data by Murakami et al.) were essentially doubled; i.e., if someone would count the particles on another 16 snowflakes! True, there may be about 10^{16} snowflakes falling per year, but that doesn't mean...!

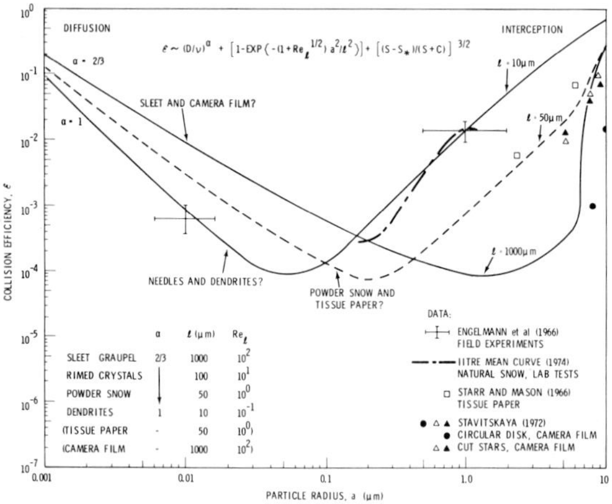

FIG. 2 - Measured and modeled, particle/snowflake collection efficiency (reprinted from (15), where references to data sources are given).

I conclude this subsection advocating the following additional research: (a) more careful laboratory studies in which, if field conditions are not simulated, then at least the conditions (drop circulation, humidity, electrical charges, particle growth, etc.) should be well defined; (b) more field studies, preferably by directly counting individual particles captured (rather than determining E indirectly via measuring differences in air concentrations of particles before and after precipitation), but the research by Rosinski and co-workers should be appreciated (13): when large particles are captured by drops, a host of small particles can be released from the surface of each large particle; (c) theoretical studies with emphasis on analytical as opposed to numerical studies, so that the results will be of more value to others; and (d) explorations of specific questions such as these: What is the influence of humidity on electrical effects? If a drop evaporates in the neighborhood of a solution droplet, how will this influence scavenging? What is the humidity in a plume, and how does this influence diffusiophoresis/Stefan flow? What is the relative importance of diffusiophoresis/Stefan flow and thermophoresis in the neighborhood of an ice crystal growing only at select sites on the crystal's surface? How does the pollution level, especially of alkali halides, influence charge separation in ice crystals and, thereby, the ice crystal collection efficiency? Many other questions could be asked, but perhaps these are enough to suggest that more research is needed if our understanding is to be enhanced.

I. LOCAL SCALE 1. Acute (ii) Gases

For the local space-scale and acute time-scale, general aspects of the rain scavenging of gases are fairly well understood (3). Compared to the case of particle scavenging, the main new feature of the theory is to account for gas saturation of raindrops, and their possible desorption of the gas as the drops emerge from beneath a plume. Thus, if $\chi(z)$ is the gas concentration in the air at some height z, and if $\chi_{eq}(z)$ is the air concentration that would be in equilibrium with the concentration, $\kappa(z)$, of the gas in a specific drop, then the drop's concentration changes according to

$$V \frac{d\kappa}{dt} = Ak_o \, [\chi - \chi_{eq}], \qquad (2)$$

where V is the drop's volume, A is its surface area, and k_o is an overall transfer velocity (which accounts for the transfer both through the air and within the drop). Use of Eq. 2 for each drop size class and integration over all drop sizes and all heights above z lead to a complicated integro-differential equation for the change in the air concentration, χ. Thus, in general, it is not possible to define a scavenging rate without reference to the plume's geometry.

Many cases, however, permit simplifications to the general theory. If the average, wind-driven gas flux in the plume ($\simeq\bar{u}\bar{\chi}$) is large compared to the average rain-borne flux ($\simeq p\bar{\kappa}\simeq p\alpha_*\bar{\chi}$, where p is the rain rate and α_* is an appropriate solubility coefficient) -- that is, if $\alpha_* \ll \bar{u}/p \lesssim 1\ \mathrm{m\ s^{-1}}/10\ \mathrm{mm\ hr^{-1}} \simeq 10^5$, which is true for most gases (but not for HTO or for SO_2 if the pH$\gtrsim$5), then the air concentration in the plume, χ, does not change substantially because of scavenging. Then, use of a given, fixed $\chi(z)$ in Eq. 2 leads to κ for each drop, to the flux of gas from the plume (a result that can be used to check the assumption that $\chi(z)$ is not changed substantially by the scavenging), and finally, if desired, to an overall scavenging rate. An additional, major simplification can be made if the plume is "sufficiently" diffuse. The sufficiency condition, different for different gases, can be seen from Eq. 2. For gases that form simple solutions in rainwater, the equilibrium concentration in Eq. 2 is given by Henry's Law; $\chi_{eq} = \kappa/\alpha$, where α is the Ostwald solubility coefficient (one of the many forms of the Henry's Law constant). Consequently, Eq. 2 predicts that drops will seek an equilibrium concentration in an e-fold equilibration time $\tau_{eq} = V\alpha/(Ak_o) = \alpha R/(3k_o)$, where R is drop radius. Use of estimates for the transfer velocity, k_o (which can be either gas- or liquid-phase controlled, depending on drop size and gas solubility), leads to estimates for the corresponding equilibration fall distances, $\lambda_{eq} = \tau_{eq}V_t$, shown in Table 1. If the plume's spatial gradients are small for the length scales indicated, then an adequate

TABLE 1 - Approximate fall distances in meters for drops of indicated size to attain their equilibrium concentration of the indicated gases. For the case of SO_2 (cf. the three papers by Barrie, Garland, and Hales in the Proceedings of the Dubrovnik Conference (see, e.g., (3)), it is assumed that the drops are initially pure water.

GAS	DROP DIAMETER (mm) 0.2	1.0
CO_2, CH_4	10^{-1}	10^{0}
DDT (if gaseous)	10^{-1}	10^{1}
SO_2 with χ = 100 μg m^{-3}	2	75
= 10 μg m^{-3}	5	230
= 1 μg m^{-3}	15	730

approximation is to assume, for any height in the plume, that drops of the indicated size are saturated with the gas, and therefore that the gas flux carried by the drops is simply $p\kappa_{eq} = p\alpha_* \bar{\chi}$, where $\bar{\chi}$ is the spatial-average, local χ, and the enhanced solubility coefficient, α_*, accounts for any possible rapid ionization of the gas within the drop.

Yet there remain some significant, unsolved, practical gas-scavenging problems. One, which seems rather amazing to me, is that chemists appear to be so uncertain about the solubility of gases, even in quite-pure water. Recent studies of the solubilities of NH_3 and NO_2 (e.g., (8)) demonstrate some of these uncertainties, and I am led to believe that these cases scarcely scratch the surfaces of the uncertainties, e.g., for organics and metal(loid)-bearing molecules. Of course there are also major uncertainties about chemical transformations of the gases in polluted rainwater, but I am not sufficiently knowledgeable to indicate potentially significant cases for the local-space and acute-time scales. However, I see that the reaction that has recently received much attention ($HSO_3^- \rightarrow SO_4^{2-}$) is typically too slow to be of concern, except for regional scale problems.

Other unsolved problems at the local scale, and ones with which non-chemists may feel more inclined to wrestle, include these:

One (highly relevant, for example, for radioactive I_2 released during a nuclear accident) is to account for the attachment of the released gas to ambient and released aerosol particles, which would normally be scavenged at rates different from the rate for the gas. A similar problem is to account for a release's distribution (especially for high-molecular-weight hydrocarbons) between its vapor and condensed phases, as a function of ambient particle concentrations and temperature. Then at low temperatures, of course there is the almost totally unexplored problem of gas scavenging by snow. It appears to be true that the molecules of most gases are rejected by a growing ice crystal, but for some gases (HTO, HF, NH_3) the "segregation coefficient" is not much smaller than unity, especially if the crystal's growth rate is large. In addition, as a snowflake falls through a gas plume, the gas molecules almost certainly attach to the surfaces of snowflakes (and can be dissolved if the snow is partially melted). This attachment may lead to significant gas scavenging because, as seen in recent laboratory studies with soils and building materials (6), about 0.1 g of SO_2 or NO_2 will adsorb on each projected m^2 of dry surface and ~1 g m^{-2} on moist surfaces -- and of course the surface area of falling snow is large.

Finally, there is a group of practical problems that, though they are at the fringes of what can be called gas scavenging research, I would still like to mention. The general area is to explore possible countermeasures for dangerous, accidental gas releases. Examples include the simultaneous release of alkaline material (e.g., NaOH) with the accidental release of acidic vapors (e.g., H_2SO_4 and HCl) and, similarly, of NH_3 or --- with SO_2 or ---. In other words, I become despondent that we talk a lot about air pollution but (as with the weather) do little about it. As a culminating example, should anyone be foolish enough to initiate chemical or biological warfare, it would indeed be satisfying if countermeasures were taken with another gas that could turn both to fertilizer (e.g., phosgene + ammonia → urea). [Note added later. The headline of today's local newspaper, (26 February 1982) states: "Enzyme neutralizes nerve gas," so it appears that this welcome research is underway.]

I. <u>LOCAL</u> 2. <u>Chronic</u>

I have little to say about the long-time-scale scavenging of either particles or gases local to their source. In modeling, the need continues for more informed decisions about the fraction of pollution deposited locally (typically small!) and the fraction polluting others (12). And in monitoring, the need continues for thorough studies, designed to answer specific questions. Some questions I would like to see answered include these: (a) What fraction of chronic particle deposition to forests is via wet vs. dry deposition, and in particular, how much of the sulfate deposition is actually deposited as SO_2 and then oxidized in (or on) the vegetation? (b) What are the reaction rates and solubilities for gases in (and on) vegetation? (c) Can chronic effects from releases of SO_2, NO_2, etc., be counteracted by simultaneous release of NH_3 or other buffering compounds (such as alkaline fly ash!)? (d) Similarly: is precipitation less acidic in agricultural areas where there are substantial NH_3 emissions, e.g., in Denmark (Rodhe, personal communication)? (e) Finally, and similar to questions asked for the short-time-scale case, there are questions dealing with particle growth and scavenging by rain and snow, gas solubilities and transformations in polluted rain, etc.

II. <u>REGIONAL SCALE</u> 1. <u>Acute</u>

The prototypic problem at these (rather ill-defined) time and space scales is to understand and describe the "intermediate and long-range transport" of, e.g., a release from a nuclear reactor. A major aspect of this problem is to describe realistically the release's transport and diffusion (not using just isobaric or mean, mixed-layer trajectories), but that aspect will not be addressed here. Instead, I will focus on precipitation scavenging.

For both particles and gases, a major problem area can be identified under a title such as "storm dynamics." What is needed for a particular case is to be able to describe the flow around, into, and through storms. However, for predictions, statistics are needed: the probability of storm occurrence, statistics of the aerial extent and precipitation for storms, and (to be described in more detail later) statistics on precipitation efficiency.

Some preliminary work has been done in this area of research (e.g., see Fig. 3), but an enormous amount is still needed. For example, the case of frontal storms has hardly been examined at all. I know of only one set of detailed studies of winds, precipitation particles, and precipitation efficiency for frontal storms (5), and yet, in latitudes where most pollution is released, frontal storms are typically the source of most precipitation. The reason for this lack of information is commonly said to be the expense of the field studies, but that excuse is shallow: a comprehensive study of frontal storms could be undertaken for under (U.S.) $\$10^6$ per year; in contrast, in the U.S., more than $\$10^{10}$ is spent per year playing coin-operated, computerized video games! The root cause of our lack of knowledge is

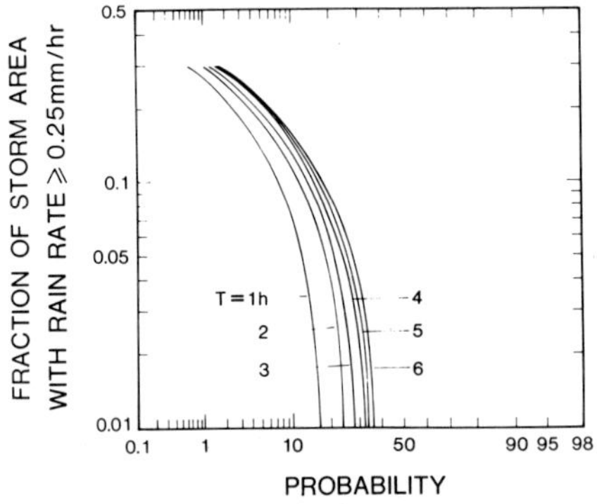

FIG. 3 - Estimated probability of rain in an area greater than the indicated area fraction of the total storm area, during the observation time T (adapted from (20) and reprinted from (17)).

therefore not a matter of money; it's a matter of priorities. From the scavenging viewpoint, I think that one of the most important aspects of this frontal-storm research is to investigate the downdrafts associated with the precipitation and the updrafts that must exist to satisfy continuity.

As well as dominating the trajectories of pollutants, these updrafts and downdrafts, associated with rainbands, further mock the dichotomy of "below-cloud" vs. "in-cloud" scavenging -- concepts that chemists, especially, seem to promote. Moreover, not only do "below-cloud" and, separately, "in-cloud" scavenging rarely exist, but even when they do, the theory for both cases is the same: for example (as was suggested earlier in this review), water-vapor condensation on particles, located anywhere in the atmosphere during precipitation, cannot be ignored. To show this unity more clearly, and to comment about other details, it may be useful to describe the cases of particles and gases, separately.

II. <u>REGIONAL</u> 1. <u>Acute</u> (i) <u>Particles</u>

First, consider some approximations for the scavenging rate, given by Eq. 1. Because of the previously-mentioned uncertainties in E, and because of similar uncertainties about hydrometeor size distributions (especially for predictions!), it therefore seems consistent to accept some rather crude approximations to the scavenging rate. Toward this end, notice the similarity between the following approximation for the rain-scavenging rate (cf. Eq. 1)

$$\psi_r \simeq \int_0^\infty dR\ N(R;\vec{r},t)\ v_t\ \pi R^2\ E(a,R), \tag{3}$$

and the expression for the rain rate (volume flux)

$$p = \int_0^\infty dR\ N(R,\vec{r},t)\ v_t\ \tfrac{4}{3}\pi R^3. \tag{4}$$

If the collection efficiency is not too strongly dependent on drop size (not a very good approximation for particles with a $\lesssim 0.1\ \mu m$, but for inertial impaction -- usually dominant for total mass scavenged -- then the governing parameter, the Stokes number, is almost independent of drop size), then it's easy to see from Eqs. 3 and 4 why I have suggested the approximation

$$\psi_r(a;\vec{r},t) \simeq c\ p(\vec{r},t)\ \bar{E}(a,R_m)/R_m, \tag{5}$$

where c is a numerical constant near unity, values for which I have suggested elsewhere, and R_m is a mean (e.g., mass-mean) raindrop radius, which typically depends weakly on p. Figure 4 illustrates that Eq. 5 fairly well describes the rain scavenging rate's dependence on rainfall rate. With similar liberties, a similar approximation for the snow scavenging rate is

$$\psi_s = \gamma p E(a,\lambda)/D_m, \tag{6}$$

where γ is near unity, p is the precipitation rate (rainwater equivalent), and D_m is a characteristic length scale, different for different crystal types (e.g., D_m = 27 μm for rimed plates, 10 μm for powder snow and spatial dendrites, and 3.8 μm for plane dendrites). I would like to repeat, here, that these approximations for the scavenging rates are suggested for both in- and below-cloud scavenging, provided account is taken of the height dependence of the precipitation rate, particle growth by water-vapor condensation, attachment to cloud particles, etc.

A scavenging rate -- e.g., the rain scavenging rate as given by Eqs. 1, 3, or 5 -- is basically a weighted-average collection

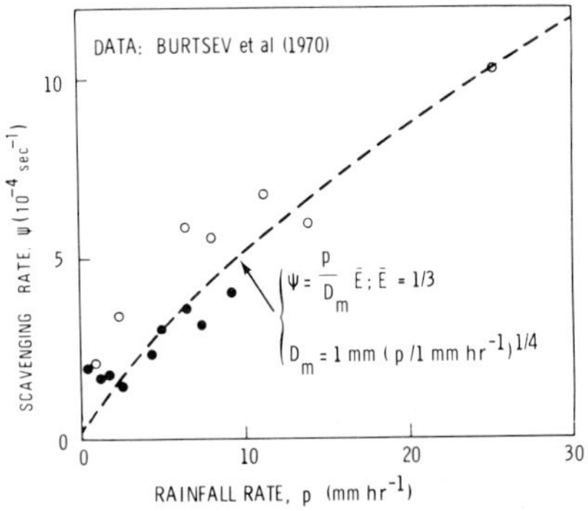

FIG. 4 - Comparisons of field data with ψ as given by Eq. 5 (reproduced from (15), where reference to the data source can be found).

efficiency, where the average is over all drop sizes. For application to predictions of total radioactivity or total particle-mass scavenged, however, it is necessary to use another weighted average: an average of the scavenging rate over all particle sizes. Thus, if f(a) is the probability density function describing the distribution of particle sizes (e.g., a log-normal distribution), then the nth moment of the scavenging rate (e.g., n = 3 gives the mass-average scavenging rate for spherical particles) is

$$\langle {}^{n}\psi(\vec{r},t)\rangle = \int_0^\infty \psi(a;\vec{r},t)\ a^n f(a)\,da \Big/ \int_0^\infty a^n f(a)\,da. \tag{7}$$

For applications, two important features of these frequently-used, particle-average scavenging rates should be noticed. First, as illustrated in Fig. 5a for a log-normal particle distribution (as well as a log-normal raindrop distribution and the E(a) curve shown earlier as the bottom curve in Fig. 1), the broader is the particle size distribution (i.e., larger geometric standard deviation, σ_g), then for particles with an

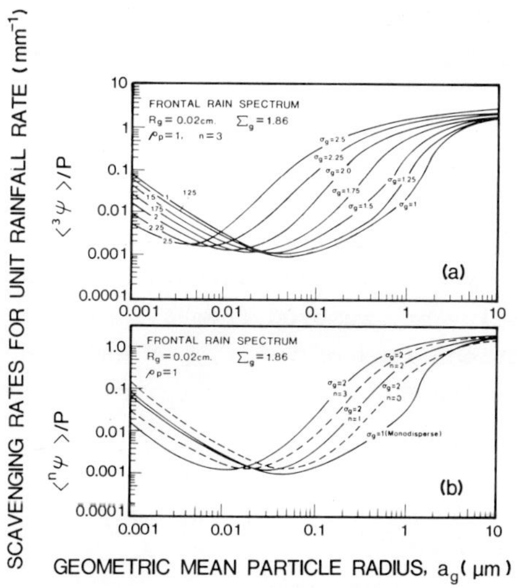

FIG. 5 - Particle-size average scavenging rate (reproduced from (17), where reference to the original source (Dana and Hales) can be found).

$a_g \gtrsim 0.1$ μm, the substantially greater is the mass-average scavenging rate. Physically, the result simply shows that a few large particles contribute very substantially to the total mass scavenged. Second, as illustrated in Fig. 5b for a fixed $\sigma_g = 2$, the particle-average scavenging rate is also substantially larger for all moments (n = 1,2,3, etc.) than for the case of a monodisperse aerosol. The case n = 2 would be appropriate to estimate the scavenging of total radioactivity distributed on particles in proportion to their surface area. Thus, even if particle growth by water-vapor condensation were not important, the particle-average scavenging rate (for a polydisperse aerosol with geometric-mean particle radius larger than about 0.1 μm) is much larger (in some cases, see Fig. 5, three orders-of-magnitude larger!) than for a monodisperse aerosol.

In some estimates for the regional-scale scavenging of accidental releases (e.g., of radioactivity), authors have used still-another average, beyond the averages over drop and particle size distributions. If an average is also taken over the height of a storm, then a "scavenging ratio" can be defined. This scavenging ratio, s, is the concentration of the pollutant in surface-level precipitation, κ_o, divided by its concentration in surface-level air, χ_o; i.e., $s = \kappa_o/\chi_o$. One way to derive an expression for s (say for the case of a monodisperse particle distribution, but then the result can be averaged over any other distribution) is to find κ_o by dividing the flux of pollutant, removed from all heights above the sampler, by the surface flux of water, p_o; and then divide by χ_o to get s. Thus, using Eq. 5 for rain scavenging (and similarly, Eq. 6 for snow scavenging),

$$s = \frac{\kappa_o}{\chi_o} = \frac{\int_0^\infty dz\ \chi(z)\psi}{\chi_o p_o} \simeq c\ \bar{E}(a) \int_0^\infty \frac{dz}{R_m} \frac{p(z)}{p_o} \frac{\chi(z)}{\chi_o}. \qquad (8)$$

Note from Eq. 8 that if $p(z) \simeq p_o$ (over a height of about 1 km), and if $\chi(z) \simeq \chi_o$, then for $R_m \simeq 1$ mm, $s \simeq 10^6\ \bar{E}(a)$, dimensionless. The uniformity of field results such as those shown in Fig. 6, in which the only remaining variation is from $\bar{E}(\bar{a})$, encouraged the use of these scavenging (or washout) ratios in a number of "environmental-impact" assessments.

My main point in introducing these scavenging ratios for particles, however, is to urge caution. As they appear in Eq. 8 and Fig. 6, the scavenging ratios represent not only averages over all hydrometeors, particles, and heights, but also (for the data in Fig. 6) the average is over a substantial number of storms, with the average found by weighting each s with the amount of precipitation from the storm. With so many averages taken, it is somewhat surprising that any variations remain! In contrast, for specific applications, substantial variations should be expected; put differently, for predictions, substantial uncertainties in the scavenging ratios should be incorporated into the analysis. The amount of this variability is suggested by the "error bars" in Fig. 7. The horizontal variation of each bar reflects not the standard deviation of the particle size distribution, but different mass-median diameters for the material measured at different sites; the vertical bar occurs at the mass-median radius appropriate for St. Louis, where the

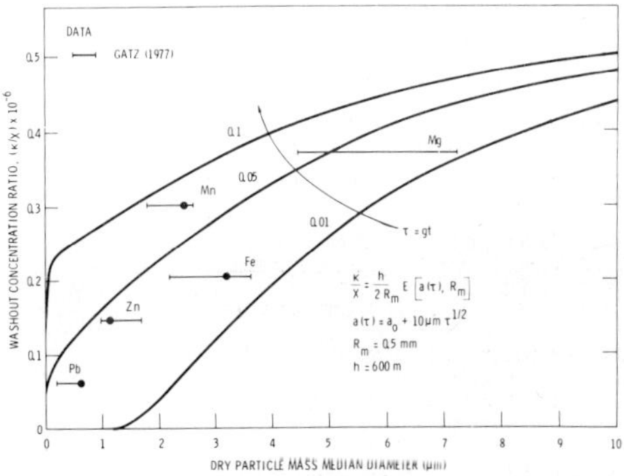

FIG. 6 - Comparisons of measured, mass-mean scavenging ratios as found by Gatz with theoretical expression given by Eq. 8, allowing for particle growth by vapor condensation and where h is the integral indicated in Eq. 8 (reproduced from (15), where reference to the data source can be found).

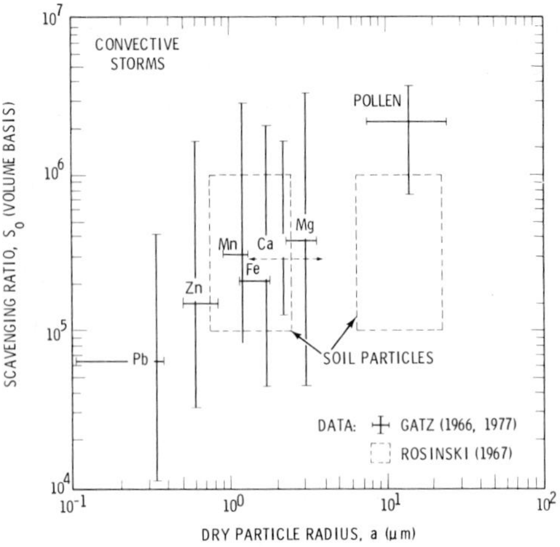

FIG. 7 - Indications of variations in measured scavenging ratios (reproduced from (16), where references to the data sources can be found).

scavenging ratios were measured; and the vertical variation spans approximately from 5 to 95% on a cumulative frequency distribution of the measured scavenging ratios. Moreover, these variations were found for a particular type of storm (viz., convective storms, which generally yield smaller s values than frontal storms - see (17)), and for the general source/sink spatial configuration in St. Louis (which should lead to a generally-characteristic ratio for χ in clouds to χ_o at the surface). Consequently, for the regional space-scale and short time-scale (as opposed to the long-term-average case, to be mentioned later), I recommend restricted and cautious use of scavenging ratios. A useful analogy may be this. Scavenging ratios are like essentially all bureaucracies: designed for the general case, but crude, cumbersome, ineffectual, and sometimes very costly when applied to special cases -- and special cases are all that ever occur!

II. <u>REGIONAL</u> 1. <u>Acute</u> (ii) <u>Gases</u>

I don't have much to say about scavenging of gases for the regional space-scale and event time-scale. Not that there aren't major problems, but because many are the same as for the local-scale for gases and the regional-scale for particles. For example, for the release of HCl during a space-shuttle launch, or of I_2 from a reactor accident, then in addition to the problems as for particles (transport, diffusion, storm dynamics, precipitation statistics, possible precipitation modification, etc.), there are questions such as: How much of the gas is transformed by chemical reactions, both in the air and in cloud water? How much of the gas attaches to particles? How much will "off-gas" from hydrometeors? Etc. But a specific point that may be useful to emphasize is this. For gases well-mixed in the atmosphere and that form simple solutions in rainwater, then (as described earlier) the scavenging ratio is simply the Ostwald solubility coefficient: $s = \kappa_o/\chi_o = \alpha$. For gases that ionize substantially in water (e.g., $SO_2 \cdot H_2O \rightleftarrows H^+ + HSO_3^-$), an enhanced solubility coefficient, α_*, can be defined and $s = \alpha_*$ (which depends on the drop's pH and/or the air concentration, χ_o). But caution! The χ_o used in s is the surface-level concentration of the gas in the <u>downdraft</u> air, associated with the rain shaft, and this χ_o can be substantially different from the (usually-measured!) concentration in the updraft: the updraft contains moist and (typically) polluted air; the downdraft (e.g., for a cumulonimbus cloud) can be from the mid-troposphere and is usually relatively clean. Consequently, if the usually-measured χ_o is used, the actual scavenging ratio, $s = \kappa_o/\chi_o$, can be an order-of-magnitude or more different from the solubility coefficient.

II. <u>REGIONAL</u> 2. <u>Chronic</u>

The prototypic problem here, of course, is the "acid-rain question." This question has received an enormous amount of attention during the past decade or so (but few answers!), and I will not review what has been reviewed in so many other reviews. Thus, I will not address the important, unsolved problems of the transformation of SO_x and NO_y in cloud water - in part because I am not a chemist. Also, I know essentially no biology, and therefore know essentially nothing about biological processes in clouds and precipitation. Instead, I will address some topics that are relevant for both particles and gases, and that have been given inadequate attention. I will place these topics into three categories: (i) modeling, (ii) statistics, and (iii) field studies. Thereby, I am abandoning a part of the previously-advertised structure for this paper, with its separate subsections for particles and gases. But the divisions were artificial anyway.

II. REGIONAL 2. Chronic (i) Modeling

My major point about modeling of acidic rain is that, in essentially all "long-range-transport" models, nonlinearity -- or the concept of a "saturated air shed" -- is ignored. In contrast, the actual chemistry is almost certainly nonlinear: as the pH of cloud water decreases, less SO_2 is absorbed; as the oxidant is consumed (e.g., by NO_2), less SO_2 is oxidized; as the SO_2 is oxidized to H_2SO_4, perhaps some nitrogen compounds are reduced and re-released from particles and cloud water to the air; as more cloud condensation nuclei are formed, the clouds and precipitation can be modified; as the acids dry deposit, dry deposition is inhibited; and so on. As I said, these nonlinearities are ignored in most models, and yet they may have substantial practical and financial significance. The picture I have in mind is of the U.S., for example, investing substantial funds to control SO_2 emissions, and receiving a postcard of thanks from seagulls in Greenland! -- while the Canadians accrue very little benefit, until U.S. and Canadian emissions are drastically reduced. Granted that the right-brain's pictures are not constrained by realism(!), but even my analytical left-brain supports the following: for the acidic-rain question, nonlinear chemistry is needed, yet poorly modeled, and we do not know how far some airsheds may already have been polluted past linearity.

II. REGIONAL 2. Chronic (ii) Statistics

The statistics I have in mind, needed for modeling, are climatological statistics of storms. However, in general, it is not a matter of just perusing reams of old data to obtain these statistics, because, in general, the data are not available. That is, a very substantial effort will be needed (10 to 100 scientists working perhaps 10 years) to generate this information. This is a substantial scientific effort, but of course a miniscule effort compared to what our societies are devoting to less useful (even to useless, or worse-than-useless) tasks. The needed information concerns the precipitation efficiency of storms, the (Lagrangian) frequency of storms, and the statistics of nonprecipitating clouds.

This last case -- the statistics of nonprecipitating clouds -- is one of the most obvious, and one of the most embarrassing. How many times, I wonder, has the following (or similar) been requoted: "Junge has estimated that, in midlatitudes, particles will be subjected to about 10 condensation-evaporation cycles (i.e., clouds) before being removed by precipitation." I have quoted it a number of times. The estimate is from a 20-year-old final report on a U.S. Army contract, and I don't think the analysis has ever been published in the open literature -- nor, to my knowledge, has it been extended. This type of information (giving statistics for the number of nonprecipitating clouds by which a given pollutant is modified) seems to be absolutely essential if we wish to know the transformations of gases, evolutions of particle-size distributions, changes in nucleating abilities of cloud-active particles, etc. I am certain that the number of such clouds (which influence the pollution and therefore its removal) is strongly dependent on location and season, and that different nonprecipitating clouds modify pollutants differently. In contrast to what it should be, our knowledge in this area is nothing less than embarrassing.

Then there is the question about the frequency of *precipitating* storms as experienced by the pollution (i.e., in a Lagrangian sense). That the Lagrangian time between storms is different from the time between storms as recorded at, say, a weather station (Eulerian sense) is obvious if one considers extremes: at one extreme, say for *continuous* orographic precipitation at a mountain weather station, the Eulerian time between storms would be zero (even though the air passes through the storms and may not experience another precipitating storm for a week or so); yet at the other extreme, say for a weather station located on the other side of the mountain barrier just mentioned, it may essentially never rain (i.e., a very long Eulerian time between storms), but the air will still experience another precipitating storm in a week or so (Lagrangian time). And though there surely exist thousands of studies of the Eulerian statistics of storms, yet I know of only two (elementary, crude) studies of the needed Lagrangian statistics: both use inadequate, low-level (e.g., 850 mb) trajectories. In reality, in contrast, if some pollution becomes involved in the precipitation-formation process, then its trajectory can continue in the stratosphere! Yet now I see the results of these two studies (which cost less than a few tens of thousands of dollars to perform) being used in numerical models that could influence very large expenditures (~U.S. $\$10^{10}$) on pollution control. This illustration, as with so many others, leads me to question the collective wisdom of modern humans.

But perhaps of even more importance than the statistics of nonprecipitating storms, and the Lagrangian frequency between precipitating storms, is statistical information about individual, precipitating storms: their precipitation efficiencies, the durations that pollutants are exposed to precipitation and precipitation-formation regions within the storms, the behavior of pollutants in such storms, and so on. There is inadequate space here to comment on these topics adequately, but let me at least devote a paragraph or two to the precipitation efficiency and its significance in scavenging.

One of the complicating features of precipitation efficiency studies is that there is an unfortunate plethora of "efficiency" definitions. There is, for example, the cloud-water-removal efficiency:

$$\varepsilon_{cw} = \frac{\text{water out}}{\text{water available}} = \frac{\text{precip. out}}{\text{max. condensed}} = \frac{\text{vapor in-vapor out}}{\text{vapor in-vapor at "top"}}, \qquad (9)$$

where by "top" I mean the level in the storm where most water is condensed. Data are also available for a reduced, cloud-water-removal efficiency, ε'_{cw}, used as an estimate of the increase in ε_{cw} for a frontal storm, caused by an orographic barrier. Data are now also available for ε_{cw} for individual rainbands in a frontal storm, and I will call these ε_{cw}^{RB} (RB for rainband) to distinguish them from ε_{cw}^{OR} for isolated orographic clouds, and from $(\varepsilon_{cw}^{OR})'$ for the enhancement of ε_{cw} in a frontal storm caused by an orographic barrier. In looking through the past thirty years of appropriate journals, a colleague (M.A. Wolf) and I have found only four studies of these cloud-water-removal efficiencies, and the early studies, of $(\varepsilon_{cw}^{OR})'$, were quite crude. Table 2 summarizes the data; if anyone knows of more, I would be grateful to receive copies of the reports.

For cumulonimbus storms (Cbs), additional data are available, but these are usually reported in terms of still-another efficiency, unfortunately called the precipitation efficiency: ε_p^{Cb} = (precipitation out) ÷ (vapor flow into the cloud base). A more informative efficiency uses, in the denominator, the net vapor flux at the cloud base. Figure 8 summarizes available data. The wind shear is wind speed at cloud top (say 25 m s^{-1}), less wind speed at cloud base (say 5 m s^{-1}), divided by cloud height (say 10 km, and then for this case, the wind-speed shear would be 2×10^{-3} s^{-1}). It could be expected that ε_p^{Cb} would return to a value near zero for no wind shear, since without shear, a Cb's downdraft will suppress its updraft. On the other hand, for large wind shear, Marwitz

TABLE 2 - Cloud-water-removal efficiencies. For references to data, see (19) (original reports are by Elliott and Hovind (1964), Dirks (1972), Marwitz (1974), and Hobbs et al. (1980)).

Notes		Data		
San Gabriel Mts. (~8000'),	31 cases, stable storms	$(\varepsilon_{cw}^{OR})'$	=	26%
,	8 cases, unstable		=	27%
Santa Ynez Mts. (~4000'),	21 cases, stable	"	=	17%
,	22 cases, unstable	"	=	26%
Medicine Bow Mts. (~3500'),	Cld. Top			
	T = -23°C, $\bar{u}$ = 12 ms^{-1}	ε_{cw}^{OR}	=	65%
	T = -19°C, $\bar{u}$ = 22 ms^{-1}		=	55%
	T = -35°C, $\bar{u}$ = 20 ms^{-1}	"	=	25%
San Juan Mts. (~4500'),	T = -28°C, $\bar{u}$ = 30 ms^{-1}	"	=	62%
Warm Sector Rainband,	Precip = 100-123 kg s^{-1} m^{-1}	ε_{cw}^{RB}	=	40 - 50%
Wide cold-frontal band No. 1,	Precip = 40-50 kg s^{-1} m^{-1}		=	80 -100%
Wide cold-frontal band No. 2,	Precip = 6 kg s^{-1} m^{-1}	"	=	20%
Narrow cold band,	Precip = 11-18 kg s^{-1} m^{-1}	"	=	30 - 50%

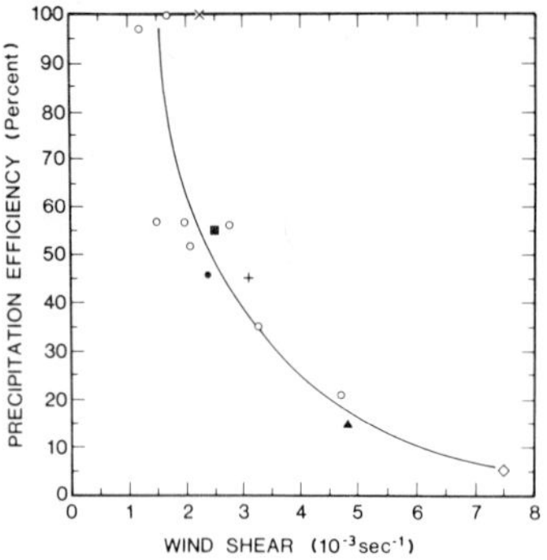

FIG. 8 - A summary of available data for the precipitation efficiency of cumulonimbus storms (reproduced from (17), and adapted from (2), where references to the data sources can be found).

suggests that the strong winds aloft will blow the ice off the Cb's anvil (leading to "orphaned anvils"!) and/or cause the precipitation to fall outside the saturated cloud environment. The possibility that, at any instant, ε_p^{Cb}>100% points to inadequacies in the definition, and to difficulties in the field studies.

The significance of these precipitation efficiencies is simply this: if a storm "dumps" only 50% (say) of its precipitable water, then it would almost certainly be unrealistic to assume (as is done in essentially all scavenging models!) that the storm will scavenge essentially 100% of its pollution. True, it can -- if the pollution is incredibly active as a cloud condensation nucleus -- but the modeling method used so frequently (viz., with a Eulerian time of a few hours of rain, and a scavenging rate of, say, 10^{-4} s^{-1}) is incredibly naive: I doubt that the Eulerian time gives even a crude estimate of the time the pollution experiences precipitation (a Cb can sit above a mountain for most of an afternoon, but the air will be processed by the storm in about 10^3 s). Stated mathematically, define ε as the efficiency of a storm to remove a specific pollutant. If the volume of air into the storm's updraft during unit time is $\dot{V}$, and if the concentration of the pollution, there, is χ_u, and of water vapor is ρ_{wv}, then the storm's scavenging efficiency (using $\kappa_o\dot{P} = (\kappa_o/\chi_o)\chi_o\dot{P} = s\chi_o\dot{P}$ for the outflow, where $\dot{P}$ is the precipitation outflow) will be

$$\varepsilon = \frac{\text{outflow}}{\text{inflow}} = \frac{\kappa_o\dot{P}}{\chi_u\dot{V}} = \frac{s\chi_o}{\chi_u}\frac{\rho_{wv}}{\rho_w}\frac{\rho_w\dot{P}}{\rho_{wv}\dot{V}} = \frac{s}{s_{wv}}\frac{\chi_o}{\chi_u}\varepsilon_p, \quad (10)$$

where ε_p is the precipitation efficiency and s_{wv} is the scavenging ratio for water vapor. (It is relatively easy to see (cf. (18)) that s/s_{wv} is essentially the ratio of hydrometeor collection of polluted, to other, cloud particles.) Thus, from Eq. 10, for $s_{wv} \simeq s$ and $\chi_o \simeq \chi_u$, then $\varepsilon \simeq \varepsilon_p$, and the efficiency of the storm to remove pollution will be about the same as its efficiency to remove its ingested water. The amount of additional research needed before we obtain a significantly better understanding of these topics is substantial. A derogatory summary of the past 30 years' research is that we now know that storms remove particulate mass with an efficiency of about 1/2±1/2!

II. <u>REGIONAL</u> 2. <u>Chronic</u> (iii) <u>Field Studies</u>

I will try to be brief. Not only have I grossly violated the page-number recommendation for this report, but I am quite disgusted with having written the same "research needs" so many times, for so many "workshops" (writers' workshops!) during the past decade. In fact, these workshops and the related methods by which environmental research is funded (at least in the U.S.) point to a truly major problem needing solution. I mention this problem here because, like pollution, it seems to be spreading to other nations. The problem was caused, I think, by avarice and cowardice in the previous generation of U.S. scientists, who accepted the system of proposal writing and reviewing when it was first introduced. Now that many of the scientists are unfunded (since the system now funds the most political proposals), perhaps they see their error, and we can all help solve the problem -- by refusing to participate. If anyone is to write proposals, it should be the administrators, writing to the scientists! Be that as it may, specific field

studies needed are these: (a) Pollution evolution in nonprecipitating clouds. I know of only three studies of this type (all performed as part of larger studies, and almost while the sponsor was not looking!). If the studies need more description, perhaps this will suffice: perform atmospheric chemistry, aerosol, and cloud particle studies not in comfortable laboratories, but in the laboratories that Nature provides; i.e., in clouds.
(b) Pollution budgets (or inventories) for precipitating storms. For a dozen years I have been describing, advocating, and proposing such studies, to be done starting with simple, orographic storms. Recently, finally, the U.S. Environmental Protection Agency began funding two such studies (known by the acronyms APEX and OSCAR), but the inventories are incomplete, and the studies have jumped directly to the most complicated storms (in part, I presume, for political reasons). Given this history, I am beginning to doubt that these studies will be done well during my lifetime. (c) Inadvertent weather modification studies. Some field studies of this type have been performed (e.g., in conjunction with METROMEX), but there is an enormous amount left to do, at all scales: at the local scale, there is the possibility that large moisture and heat releases (e.g., during a nuclear accident) could trigger precipitation during sensitive meteorological conditions; at the regional scale, releases of condensation and freezing nuclei can influence precipitation growth and scavenging; and at larger scales, I recall the estimate by Robinson that the decrease in ground-level solar energy, caused by particles released from fossil-fuel combustion in the Northeastern U.S., is very nearly equal to the energy generated by the fuels! (d) Countermeasures and chemical modification studies. Examples of this type of field study include seeding clouds to dump radioactivity or other noxious material where it would do least harm (or overseeding, to delay precipitation, if that would be desirable, and "seeding" acidic clouds with alkaline material to suggest other ways of avoiding acidic rain. There are a number of reasons why I advocate this second example: (i) to enhance understanding of cloud chemistry, (ii) to promote consideration of other ways to

fight acidic rain, and (iii) because fighting acidic rain by first neutralizing emissions may be "best." In this regard, note that the total-pollution loading need not be very high for the precipitation's pH to be low (and therefore, doubling the pollution loading, with alkaline material, may be tolerable in some regions), and note that there are suggestions that a major cause of decreasing pH in eastern North America during the last few decades (if there has been a decrease!) may be just from changing agricultural practices, leading to less re-suspension of Ca^{2+} from soils, and from filtering (alkaline) fly ash from the stacks. Of course I am aware that some will oppose fighting pollution by polluting more. However, when progress is blocked, there comes a time when there is marginal value in continuing to bang one's head against a brick wall. Many times, in this nonlinear world, the shortest distance between two points is not a straight line.

III. GLOBAL SCALE

For the global space scale, I will abandon the subdivisions even into acute vs. chronic time scales. Happily, with the end of most atmospheric, nuclear bomb tests, we no longer have much interest in global consequences of short-term releases. But interest in continuous releases has grown, and lessons learned during the radioactivity fallout era should not be forgotten. For example, let us not forget the great variability even of the tropospheric residence time of particles: without my going into details, let me remind you that the value for the mean, tropospheric-residence time given in Table 41 of Junge's well-known book is 18.9 days; and the standard deviation about this mean is 13.4 days! We should remember this variability when considering continuous releases, because it helps to bring the oft-quoted estimate of a mean tropospheric-residence time of about a week (for pollution that becomes involved in the atmospheric water cycle) into harmony with the detection of such pollution, in the Arctic and Antarctic, many weeks -- or even months -- downwind of the pollution sources.

I would like to spend a little more time addressing this question of mean residence time and variations. Suppose that, initially, the amount, q_o, of some pollution was released to the atmosphere, and that this pollution was scavenged only by precipitation. To estimate the amount, $\tilde{q}(t)$, still airborne at time t, suppose that the random fraction $\tilde{\varepsilon}_i$ $(o \le \tilde{\varepsilon}_i \le 1)$ of $\tilde{q}(t)$ is removed by storm i. Then after encountering $\tilde{n}$ storms, the amount remaining airborne would be the random variable (18):

$$\tilde{q}_i(t)/q_o = \prod_{i=1}^{\tilde{n}} (1-\tilde{\varepsilon}_i). \tag{11}$$

From the central limit theorem applied to $\ell n(\tilde{q}/q_o)$, then for large $\tilde{n}$ (e.g., n = 2 or 3!), the probability density function (pdf) for $\tilde{q}$ will tend (regardless of the pdf for $\tilde{\varepsilon}$) to the log-normal distribution

$$f(q/q_o) = [(q/q_o)\sigma_T\sqrt{2\pi}]^{-1} \exp\left\{-[\ell n(q_o/q)+\mu_T]^2/2\,\sigma_T{}^2\right\}, \tag{12}$$

where $\mu_T = \bar{\upsilon}t\mu_\varepsilon$ and $\sigma_T^2 = \bar{\upsilon}t\sigma_\varepsilon^2$ (with $\bar{n} = \bar{\upsilon}t$, $\bar{\upsilon}$ = the average, Lagrangian frequency of storm encounters, and with μ_ε the mean of $\ell n(1-\varepsilon_i)$ and σ_ε^2 its variance). There are, I think, a number of important features of this result. One is that, from Eq. 12, an estimate can be made for the coefficient of deviation of fluctuations in concentrations of particles and trace gases subjected to scavenging (7,18). Another result is the reappearance of two variables described in the previous section: the overall scavenging efficiency of storms, ε_i, and the Lagrangian time between storms, υ^{-1}. Also, if this Lagrangian frequency is to be estimated realistically, it is clearly necessary to describe, realistically, air trajectories through storms. As mentioned earlier, some of these trajectories (e.g., through deep Cbs and frontal storms) will transport pollution into the stratosphere (leading to a very long time before the next storm encounter), and provide the possibility for midlatitude pollution reaching much higher latitudes (via the stratosphere, and with subsequent subsidence near the poles). Moreover, the result emphasizes the substantial variability in the amount of pollutant remaining airborne: it is a "long-tailed" log-normal distribution.

Of course there are other global-scale problems needing solutions, but I see the main problems being much broader in scope, and therefore, so must be the scope of the solutions. Many of the problems fall into the two broad fields of biology and climate. Thus, illustrative climatic problems are: (a) An increase in the "background concentration" of cloud condensation nuclei, which appears to be occurring, may cause average clouds to have more but smaller droplets; this could increase average cloud albedo, and thereby decrease the earth's average temperature. (b) Similarly, an increase in "background concentrations" of some gases (certainly H_2O and CO_2, but also N_2O and CH_4) and particles (especially soot particles) could also be influencing the earth's radiation budget. And from the field of biology, of course there are concerns about the global-scale acidification of precipitation, the influence of various gases on the ozone shield, and the generally-unknown biological consequences of heavy-metal and anthropogenic-hydrocarbon releases, especially those hydrocarbons with which known life forms have no experience. As stated before, the scope of these problems forces broad-mindedness if we are to find solutions, and to conclude this paper, I want to record a few thoughts (grouped under the headings: emission controls, monitoring, and modeling) that may stimulate some additional general thoughts in the minds of readers. And I have already warned readers that the invitation for "open questions and provocative statements", in atmospheric chemistry, inevitably leads to politics and philosophy.

Under the subject of emission controls, I can see nothing but pollution until a different philosophy has widespread acceptance. For the international "acid-rain" problem, for example, I can see little relief for nations downwind of major pollution sources, until first the downwind nations emit less SO_2 and NO_x, on a per-capita basis, than those upwind. In contrast, as examples, Canada releases about twice as much SO_2 per person as the U.S., and in 1973, the Scandinavian countries (except Norway) released about twice as much, per person, as

France and West Germany (14). If a "per-capita condition" is not met, then the morality of proposed solutions has no stronger foundation than the direction of the prevailing wind! Indeed, since the wind direction shifts, so could the arguments: the more populated countries could argue a case of greater total damages -- resting their case on the morality of abundant population! Eventually, though, when the per-capita condition is met (e.g., now, by Sweden), then the root problem can be seen to be too many people devouring too many resources with too little respect for others or for Nature -- and essentially no perspective.

To solve such problems will indeed require a different perspective on life. Elsewhere, I have encouraged recognition that the prime, known goal of all life is simply: to continue. (But this simple "fact" has been twisted almost beyond recognition by the stress of thousands of years of what Spinoza called "confused thought.") Consequently, there is a great need for widespread recognition and acceptance of what I call rational survivalism, which includes mutual altruism. Thus, for example, clearly it is immoral to bring another child into this world -- with morality based on any of the many versions of the same "survival code" (as given, e.g., by Confucius, Buddha, Christ, Mohammed, etc.) -- and it will continue to be immoral so long as the addition of new life hinders the survival of the species. Similarly, it is immoral to burn any fossil fuels (with morality, as always, based on the survival of the species) unless, thereby, we leave our progeny something better. Pure altruism, however, has almost certainly been bred out of the species by evolution (4), but kin and mutual altruism have not. Therefore, I think the answers to our pollution problems (and much more) can only be found in widespread recognition and acceptance that we are but one link in a long but brittle chain of life, that there is survival value to each if all help each other, and that all humans are bound to one another no more distantly than as fiftieth cousins. As cooperating cousins with perspective, perhaps we can someday establish a world court of environmental scientists who would pass judgment on the morality of releasing any specific chemical to the environment, with morality based (as always) on the survival of the species.

For monitoring, clearly the need is again for morality, cooperation, and perspective. We must overcome the all-too-prevalent inability of our "leaders" to see beyond the next election -- or the next revolution. Unless we develop adequate monitoring stations and data storing and reporting procedures (e.g., for hydrocarbons and metals in rain over the mid ocean; for trace gases measured, e.g., from satellites; for biological integrators on land and at sea; etc.), the governments of the world may be in for the biggest revolution since evolution began. And to those who complain that people in governments could commit horrible atrocities if guided by a moral code that sets as its highest goal "only" the survival of the species, I would first respond, facetiously: then at least the code is as general as others that have been applied. But more seriously: it seems obvious that survival has always been the goal, albeit buried in symbols and rhetoric. And to those who complain that additional descriptions are needed, I respond with Sartre: "There is no way to escape responsibility for one's actions. Even if an angel whispers in your ear, it is still you, yourself, who must decide whether it is, or is not, an angel." Besides: that we must grope to find our way makes life all the more intriguing -- and dangerous.

Finally, on the topic of modeling, I become really quite upset. By modeling, I don't necessarily mean massive computer programs: I mean understanding, and many times this understanding can best be displayed on the familiar "back-of-the-envelope." My concern is that our commitments to modeling (via people, programs, and funding) are pitifully inadequate, by at least an order-of-magnitude. Moreover, I think much of our modeling misleads us. The one phrase in this whole report that I want to underline is from the philosopher Alan Watts: It's not a linear world! Especially in western societies, our hierarchical, left-brain-dominated thinking promotes acceptance of linearity, and our linear thinking leads to complacency. I already commented on inadequacies of linear thoughts about acidic rain; another example is for CO_2. Linear extrapolations, even from nonlinear analyses, suggest that if the CO_2 concentration increases by x%, then the earth's temperature will increase by y%, and the social consequence will be z. But it is not a linear world: push the climate too far, and we should expect a jump instability to another state (either ice-free or ice-covered) -- and expect that nothing can be done to stop it (see Budyko (1) for a simple model of this nonlinearity); feed some little bug somewhere with a new hydrocarbon, or release a new species from a laboratory, and expect the species to sweep over the earth, dominating the biosphere (Lovelock's homeostasis hypothesis was never meant to placate humans); and disturb the social balance too much, and expect a revolution. In Nature, including "human nature," linearity is the exception; nonlinearity is the rule.

Am I an alarmist? I hope so. I think we should be scared stiff by our stupidity -- instead, we seem to be too stupid even to be scared. It took Nature a billion or more years to develop the human species, which we claim to be the crowning achievement of evolution. And we will likely ruin it all, proving uncontestably that we're not. We spend a trillion dollars per year on defense against each other, and negligible on defense against ourselves. Given this fact, I think we must seriously question if we humans are higher on the evolutionary ladder than is the wind. But maybe there is no ladder. Maybe it is as Matthew Arnold said more than a century ago:

> Nature, with equal mind,
> Sees all her sons at play;
> Sees man control the wind,
> The wind blow man away.

And so it may be in Berlin: in time, the wind will even blow the wall away.

Disclaimer. On the first page of this paper, I have purposefully not identified my employer - to emphasize that the opinions expressed herein are personal. Similarly, no agency sponsored these personal opinions. However, I would like to express my gratitude for the kindness bestowed on me by many individuals (whose names it may be best not to list) and for the critical comments of a number of attendees at the Dahlem Workshop (Ayers, Brinckman, Crutzen, Walker, and especially Rodhe - and others whose last names I didn't catch or whose signatures I can't decipher!). And to (silent) critics, I want to express agreement: it does seem safer to be silent. Permanent silence, which is near, should therefore be quite safe -- for the individual.

REFERENCES

(1) Budyko, M.I. 1971. Climate and Life. Leningrad: Hydrological Publishing House.

(2) Foote, G.B., and Fankhauser, J.C. 1973. Airflow and moisture budgets beneath a NE Colorado hailstorm. J. Applied Meteorol. 12: 1330-1353.

(3) Hales, J.M. 1978. Wet removal of sulfur compounds from the atmosphere. Atmos. Env. 12: 681-690.

(4) Hardin, G. 1977. The Limits of Altruism: An Ecologist's View of Survival. Bloomington and London: Indiana University Press.

(5) Hobbs, P.V., et al. 1980. The mesoscale and microscale structure and organization of clouds and precipitation in mid-latitude cyclones. J. Atmos. Sci. 37: 568-596.

(6) Judeikis, H.S., and Wren, A.G. 1978. Laboratory measurements of NO and NO_2 deposition onto soil and cement surfaces. Atmos. Env. 12: 2315-2319.

(7) Junge, C.E. 1974. Residence time and variability of tropospheric trace gases. Tellus 26: 477-488.

(8) Lee, Y.-N., and Schwartz, S.E. 1981. Evaluation of the rate of uptake of nitrogen dioxide by atmospheric and surface liquid water. J. Geophys. Res. 86: 11,971-11,983.

(9) Murakami, M.; Hiramatsu, C.; and Magono, C. 1981. Observation of aerosol scavenging by falling snow crystals at two sites of different heights. J. Meteorol. Soc. Japan 59: 763-771.

(10) Radke, L.F.; Hobbs, P.V.; and Eltgroth, M.W. 1980. Scavenging of aerosol particles by precipitation. J. Appl. Meteorol. 19: 715-722.

(11) Rodhe, H. 1978. Budgets and turn-over times of atmospheric sulfur compounds. Atmos. Env. 12: 671-680.

(12) Rodhe, H. 1980. Estimates of wet deposition of pollutants around a point source. Atmos Env. 14: 1197-1199.

(13) Rosinski, J., and Langer, G. 1974. Extraneous particles shed from large soil particles. J. Aerosol Sci. 5: 373-378.

(14) Semb, A. 1978. Sulphur emission in Europe. Atmos. Env. 12: 455-460.

(15) Slinn, W.G.N. 1977. Some approximations for the wet and dry removal of particles and gases from the atmosphere. J. Water, Air, Soil Pol. 7: 513-543.

(16) Slinn, W.G.N. 1978. Some comments on parameterizations for resuspension and for wet and dry deposition of particles and gases for use in radiation dose calculations. Nuclear Safety 19: 205-219.

(17) Slinn, W.G.N. 1982. Precipitation scavenging. In Atmospheric Sciences and Power Production, ed. D. Randerson, Ch. 11. Oak Ridge, TN: U.S. DOE Technical Information Center (and to be available from Springfield, VA: National Technical Information Service).

(18) Slinn, W.G.N. 1982. Estimates for the long-range transport of air pollution. J. Water, Air, Soil Pol. 14: 133-157.

(19) Slinn, W.G.N., and Hales, J.M. 1982. Wet removal of atmospheric particles. In Fine Particles in the Atmosphere, ed. A.P. Altshuller. Ann Arbor: Ann Arbor Science Publishers.

(20) Watson, C.W.; Barr, S.; and Allenson, R.E. 1977. Rainout Assessment. Report LA-6763. Los Alamos, NM: Los Alamos Scientific Laboratory.

Group on Aqueous Chemistry in the Atmosphere

Standing, left to right:
George Slinn, Bernd Schneider, Jim Morgan, Greg Ayers, and Bob Duce.

Seated, left to right:
Ollie Zafiriou, Hans Georgii, Tom Graedel, Dieter Klockow, and Henning Rodhe.

Atmospheric Chemistry, ed. E.D. Goldberg, pp. 93-118. Dahlem Konferenzen 1982.
Berlin, Heidelberg, New York: Springer-Verlag.

Aqueous Chemistry in the Atmosphere
Group Report

T. E. Graedel, Rapporteur
G. P. Ayers, R. A. Duce, H. W. Georgii, D. G. A. Klockow, J. J. Morgan,
H. Rodhe, B. Schneider, W. G. N. Slinn, O. C. Zafiriou

Abstract. This report of the working group on aqueous atmospheric chemistry assesses the state of understanding of the chemical processes that occur within cloud droplets, raindrops, and water-coated aerosol particles. Among the areas judged likely to be of importance but where understanding is presently meager are: knowledge of the details of the atmospheric water cycle, the convective transport of trace components through clouds, the incorporation of trace gases into atmospheric water, the role of catalysis in atmospheric water chemistry, the role of photochemical processes, and the efficiency of gas and aerosol particle scavenging by the Earth.

INTRODUCTION

Research on the chemistry of atmospheric water might be said to have begun with the work of Rossby and his colleagues in the early 1950s, followed by that of Junge and co-workers some years later. Their work demonstrated that the atmospheric water cycle plays an important role in the cycles of several of the Earth's elements: atmospheric particles form nucleating centers for water droplets, the droplets provide a medium for complex chemical reactions, and the deposition of rain and snow upon the earth is a deposition of trace compounds as well.

Despite the identification by Rossby and Junge of the major inorganic ions in rainwater, appreciation of chemical diversity and chemical reactions in atmospheric water has come much more slowly. This may have been due in part to the limitation for many years to ground level sampling. (Note the rapid progress that was achieved by dynamic meteorologists when data acquired within the atmosphere rather than at its lower boundary became available.) Another major limitation has been the difficulty of performing detailed analytical chemical studies on small, easily contaminated samples. These difficulties are now rapidly being overcome, and many scientific processes of interest are being investigated. The details of many of the processes remain obscure, but some major influences can be identified and some critical problems in understanding the role of water in transport, conversion, and removal of tropospheric trace components can be listed and discussed. This we attempt to do in the present report.

CRITICAL PROBLEM AREAS

In the discussion following this section of the report, we evaluate the state of knowledge of various aspects of the chemistry and cycling of atmospheric water. For the convenience of the reader, the critical problem areas that are identified by those discussions are presented here (not in order of priority). Details are contained in the appropriate sections of the report.

Knowledge of the Details of the Atmospheric Water Cycle Itself

Without a good understanding of the temporal and spatial changes of the medium within which the chemistry occurs, we are deprived of an effective starting point for our considerations. Among the needed information are the precipitation efficiencies of clouds and the (Lagrangian) frequency of storm occurrences.

Cycling of Trace Components Through Clouds

There is no question but that trace components participate in condensation/evaporation processes in the atmosphere and there is every reason to anticipate that such processes will have marked effects on the chemistry, but little information on these topics is available.

Incorporation of Trace Gases into Atmospheric Water

The gas fluxes into these systems are quite uncertain. We have no good idea of the efficiency of gas scavenging by ice crystals, many gas solubilities are uncertain or unmeasured, and the potential of organic surface films (if present) to inhibit gas transport remains to be evaluated.

Catalysis in Atmospheric Water Chemistry

Catalysis appears likely to play a major role in the chemistry, but the principal catalysts are uncertain and catalytic mechanisms in these systems are virtually unstudied.

Photochemistry in Atmospheric Water

Photons are present within the atmospheric water medium over a wide energy range, so photochemical processes seem very likely. None of the systems of possible interest has yet been studied, however.

Gas and Aerosol Particle Deposition

Numerical estimates for wet and dry deposition are too uncertain to permit reliable trace component budgets to be constructed, even if source identification were more definitive than is the case.

Measurement Uncertainties

Many difficulties exist with techniques for measurements of extremely small trace component concentrations in samples that are themselves minute. Particularly tenuous are current techniques for the measurement of hydrogen peroxide and of a variety of organic compounds.

Nonlinearities

This is a problem of perception, rather than one of information or technique. Until aqueous phase atmospheric chemistry indicates otherwise, we should anticipate that significant nonlinear behavior between gas-phase precursors and liquid-phase products is present.

INCORPORATION OF CHEMICAL COMPONENTS INTO ATMOSPHERIC WATER

Nucleation Scavenging

A quantitative understanding of the processes that transfer chemical species into cloud systems requires that current theories concerning nucleation are verified in the field. The physics of the nucleation process appear to have been verified experimentally in the laboratory and in the field. However, a paucity of experimental data on cloud water composition and concurrent sub-cloud aerosol/gas composition has precluded a similar verification regarding the chemistry involved. This verification should be attempted, in order to confirm the understanding of this step in the atmospheric water cycle.

Cloud Droplet and Ice Crystal Scavenging

The scavenging of gases in clouds has been investigated only to a small degree. Chameides and Davis (1) have reviewed the sparse data for gas removal by cloud droplets and performed calculations based partly on those data. For soluble species, their analysis suggests that the scavenging process can represent a significant loss mechanism and lead to vigorous aqueous phase chemistry. Data for scavenging of gases by ice crystals are essentially nonexistent. Particle capture by cloud droplets and ice crystals has been investigated theoretically; data are sparse. Some laboratory studies of thermophoresis and diffusiophoresis have been performed, but more (using wind tunnels) would be valuable. Solving the experimental difficulties of distinguishing captured from uncaptured particles within clouds will be necessary before particle capture within clouds can be considered understood.

Scavenging by Aerosol Particles

The incorporation of gases into aerosol particles is treated by the physics of diffusion in a formulation that includes a parameter, α, designated the "sticking coefficient." This parameter is an empirical representation of the probability of transport across or reaction at the interface between the gas phase and the condensed phase. It is difficult to determine experimentally, but suggested values cover the approximate range 10^{-5} to > 10^{-2}. Innovative experimental techniques for measuring α in different systems are needed.

A further complication in assessing scavenging of gases is the potential inhibition of gas transport by organic films on the aerosol particle surface. An assessment of this possibility suggests that the film will be rate-limiting in certain situations if it forms a continuous surface layer and has the necessary molecular geometry for close packing. Evidence for film structure and continuity is indirect and inconclusive; innovative laboratory studies are clearly needed.

TRANSFORMATION OF COMPONENTS IN ATMOSPHERIC WATER

Occurrence of Trace Components: Inorganic

The inorganic components most frequently determined in aqueous atmospheric samples include H_3O^+, Ca^{2+}, Mg^{2+}, NH_4^+, Na^+, and K^+ among the cations and SO_4^{2-}, NO_3^-, Cl^-, F^-, and Br^- among the anions (see Ayers and Morgan, both this volume). Many of these components are directly injected into the atmosphere as coarse dirt or sea-salt constituents; others appear to be formed in the gas phase. These cations and anions typically account for more than 99% of the inorganic ions in rain. However, certain inorganics present in low concentrations may be of crucial importance for understanding mechanisms of reaction in atmospheric aqueous systems. Among the frequently determined minor constituents are iron, manganese, aluminum, and silica. Information is rarely available on dissolved vs. particulate fractions of components in cloud- and rainwater. For improved understanding of chemistry and chemical transformation in atmospheric water it is desirable to obtain information on the concentrations of such species as nitrite, bisulfite (S(IV)), hydrogen peroxide, and ozone.

Some perspective on the potential impact of trace gases on aqueous atmospheric chemistry is provided by the realization that there are many gases in the atmosphere, even in "clean" conditions, that are either so abundant (e.g., O_2), soluble (e.g., NH_3), or chemically reactive (e.g., $HO_2\cdot$) that they are candidates for significant interactions in the aqueous chemistry of aerosols, cloud droplets, and precipitation. Table 1 lists some gases known or suspected of involvement in atmospheric liquid phase chemistry.

Occurrence of Trace Components: Organic

As with inorganic components, a wide variety of organic compounds has been found in atmospheric water. In polluted atmospheres one expects to find significant concentrations of easily scavenged organics, such as organic acids, aldehydes, and alcohols. Some of the most likely to be of interest are indicated in Table 2. Low molecular weight organic acids, especially formic and acetic, have been determined at nanomolar to micromolar levels. Formaldehyde in fog in southern California has been found in the 10^{-4} to 10^{-3}M concentration range, while reported CH_2O levels in rain and snow are lower. Information on the identity and representative concentration levels of organic compounds in atmospheric water is of potentially great significance because of the role of organics as reductants and because of coordination reactions of ligands and metals.

Interactions and Reactions: Oxidation/Reduction Reactions

As seen from the presentation above, we have a moderate level

TABLE 1 - Inorganic substances that may be involved in chemical reactions in aqueous droplets.

O,H	N,H,O	S,H,O
O_2, O_3, H_2O_2	NH_3	H_2S, $(CH_3)_2S$
$OH\cdot$, $HO_2\cdot$	HNO_2, $NO_2\cdot$	SO_2
	HNO_3, $NO_3\cdot$	H_2SO_4
I,O,H		Miscellaneous
HI, $IO_2\cdot$(?), $IO\cdot$(?)		HCl, HBr

TABLE 2 - Organic substances that may be involved in chemical reactions in aqueous droplets.

Compounds	Radicals
CH_3OOH, H_2CO, $HCOOH$ (methyl oxygenates)	$CH_3O_2\cdot$
R_nCOOH ($R_n = C_1$-C_7) (organic acids)	complex species
$HOOCR_nCOOH$ ($R_n = C_1$-C_8) (dicarboxylic acids)	$HOCH_2OO\cdot$
$CH_3C(O)O_2NO_2$ (peroxyacetyl nitrate)	
$(CH_3)_2SO$, CH_3SO_3H, $(CH_3)_2S$ (sulfides, sulfoxides)	

of knowledge concerning components present in atmospheric water. The principal oxidizing agents are thought to be H_2O_2, O_3, and O_2 (probably in conjunction with catalysts), but measurements of the concentrations of these species are of doubtful validity and chemical mechanisms are uncertain. A few systems of possible significance are presented in Table 3.

Some of the considerations and complexities typical of those to be expected in atmospheric water chemistry are illustrated by the system we believe to be best understood (although quite imperfectly at present): the solution oxidation of SO_2. The driving forces for redox reactions potentially important for production of acidity from SO_2 oxidation in atmospheric liquid water are generally large, but the reactions are kinetically controlled, either because of limited aqueous solubility or large redox reactive activation energies, or both. Thus, O_3 has a rather small solubility and pH-dependent aqueous phase kinetics for SO_2 oxidation, the rate being lower at low pH. O_2 concentrations in cloud and rain are 5-6 orders of magnitude greater than for O_3, but SO_2 oxidation rates depend strongly on the presence of catalysts and upon sufficiently high pH. Hydrogen peroxide appears to combine favorable properties of high intrinsic solubility and an SO_2 oxidation rate which is catalyzed by H_3O^+ and other acids. If gas-phase H_2O_2 concentrations are high (0.1 to 1 ppbv), or if H_2O_2 is produced in atmospheric

TABLE 3 - Oxidation/reduction reactions of potential interest in atmospheric water.

Discrete Processes	
$O_3 + 2H^+ + 2e^-$	$= O_2 + H_2O$
$O_2 + 4H^+ + 4e^-$	$= 2H_2O$
$NO_2 + e-$	$= NO_2^-$
$NO_2^- + 2H^+ + e^-$	$= NO + H_2O$
$NO_3^- + 4H^+ + 3e^-$	$= NO + 2H_2O$
$O_2 + 2H^+ + 2e^-$	$= H_2O_2$
$NO_2 + H_2O$	$= NO_3^- + 2H^+ + e^-$
Coupled Process A	
$HSO_3^- + H_2O$	$= SO_4^{2-} + 3H^+ + 2e^-$
$H_2O_2 + 2H^+ + 2e^-$	$= 2H_2O$
$HSO_3^- + H_2O_2$	$= SO_4^{2-} + H^+$
Coupled Process B	
$NO_2 + 2H^+ + 2e^-$	$= NO + H_2O$
$HSO_3^- + H_2O$	$= SO_4^{2-} + 3H^+ + 2e^-$
$NO_2 + HSO_3^-$	$= SO_4^{2-} + NO + H^+$

water by redox or chain termination processes, then peroxide oxidation of SO_2 may prove to be a dominant path for production of sulfate in an aqueous phase. An improved picture of the production of acidity requires a greater understanding of the processes for incorporating H_2O_2 and $HO_2\cdot$ in water and the pathways for production of reactive oxidants for SO_2 oxidation (as well as NO_2 oxidation) in aqueous solution.

It is important to emphasize that except for proton-transfer reactions and ion-pair formation/dissociation, equilibrium conditions cannot be assumed with confidence.

In such a complex medium, it <u>may</u> prove possible to identify the few "important" steps and so understand the system(s) on the basis of a reduced reaction set. More likely, many aerosol/droplet processes will be difficult to describe by specific reactions. Empirical rate laws may then be highly useful (when valid) for environmentally interesting situations.

A further complication is the potential existence of photochemical processes within aqueous atmospheric media. A number of molecules and radicals expected to be present in these solutions, including O_3, H_2O_2, NO_2^-, NO_3^-, and a number of organic compounds, have absorption spectra within that of solar tropospheric radiation. Given the broad range of organic and inorganic compounds present in these systems, photochemical processes of interest would be expected to be present. Thus far, however, no experiments expressly designed to investigate this point have been performed.

To sum up, oxidation/reduction reactions in atmospheric water are potentially very complex. They involve interplay among atoms, molecules, ions, radicals, and solar photons. These systems, while not presently understood, may have significant possibilities for nonlinear physicochemical behavior. Further details are provided by Morgan (this volume).

Interactions and Reactions: Complexation and Chelation

Relatively little is known concerning the presence of metal complexes and chelates in aqueous solutions in the atmosphere, although they must occur. Among the most likely single metal complexes formed under acidic conditions in the atmosphere would be chloride, nitrate, and sulfate complexes and perhaps those of the lighter carboxylic acids. Chelation of metals to nitrogen, carboxyl, and organic alcohol and phenolic functional groups of larger organic molecules are also of possible importance. Formation and dissociation of such complexes and chelates will, of course, be dependent upon their formation constraints and the concentrations of the various species in the solution. The latter will depend upon local source strengths and, for aerosol particles, ambient relative humidity.

Perhaps the primary importance of metal complex and chelate formation in atmospheric aqueous solutions is related to the catalytic activity of the metals. The catalytic nature of metals depends upon the specific chemical (and physical) forms of the metal in the solution. It would be very difficult to estimate this catalytic ability for a particular metal in an environmentally realistic way unless careful modeling or laboratory studies taking into consideration formation of complexes and chelates are undertaken.

Interactions and Reactions: Catalysis

Catalytic processes are known to occur in aqueous atmospheric solutions. The examples that have been recognized involve the oxidation of SO_2 to sulfate: it has been demonstrated that this process is enhanced by the presence of manganese ions and of elemental carbon. Our knowledge of the chemical pathways involved is insufficient, however, to allow the catalytic reactions to be specified.

The potential for the transition metals to function as catalysts depends upon a number of factors: abundance, chemical form, stable oxidation states, bonding properties, and solubility. An assessment of these factors suggests that the following transition metals have the potential to serve as homogeneous catalysts: copper, manganese, and (perhaps) vanadium. For heterogeneous catalysis, solubility is not a consideration and the list of possible active metals is longer: iron, copper, titanium, manganese, vanadium, and (perhaps) chromium, as well as elemental carbon.

The transition metals may be present in a variety of chemical forms, including organic complexes of unknown nature. In some cases, the organic complexes are the most effective catalysts (as is universal in organometallic enzymes, to cite an analogy), while in other cases the organic forms may be inactive. Thus the organic interaction can dramatically enhance or inhibit oxidations. For example, cobalt-phthalocyanine complexes are extremely effective in catalyzing O_2 oxidation of reduced sulfur compounds. In contrast, complexation of reduced forms of metals (e.g., Fe(II), Mn(II)) can greatly retard their oxygenation kinetics. Specific and nonspecific effects of complexation in aqueous atmospheric media need critical examination.

Given the information discussed above, it is anticipated that catalytic reactions will be important (and perhaps dominant) for many aqueous atmospheric processes. However, very few experimental studies have been performed and laboratory investigations of potentially interesting chemical systems are badly needed to help assess the role of catalysis in aqueous atmospheric chemistry.

Interactions and Reactions: Atmospheric Biochemistry

Bacteria are ubiquitous on the ocean surface, are injected into the air by bubble bursting at wave crests (a process that enriches their abundance by a factor of 10^2 - 10^3 relative to that of seawater), and are found in rain samples. In the latter, there are preliminary studies to suggest bacterial production of CO on a time scale of ca. 30 minutes. Bacteria are also widely found in aerosol particles over land (the so-called "viable particles"). With diameters that may be less than 1 μm, bacteria can be present on aerosol particles whose lifetimes are as long as several days (a time scale approximating that of many bacterial processes).

The influence of bacterial biochemistry in atmospheric water will not easily be determined. Nonetheless, it appears possible that the chemical cycles of at least a few components in atmospheric water may be dominated by bacterial influences.

The Chemical Data Base: Pertinent Equilibrium Data

Equilibrium for gas solubility ("Henry's Law constants") and protolysis reactions may be established sufficiently rapidly to justify equilibrium models for aqueous-gas atmospheric systems. Useful two-phase equilibrium data for O, N, and S species, including free energy and enthalpy data (or equilibrium constants at several temperatures) for oxides and oxyacids of sulfur and nitrogen and their aqueous anions as well as oxidants such as O_2, H_2O_2, and O_3 and carbon compounds such as carbon dioxide and formaldehyde, have been reviewed recently. The simplest equilibrium models for rainfall composition that result from these studies demand gas solubility and acid-base dissociation constants for the system $NO - NO_2 - SO_2 - HNO_3 - H_2SO_4 - NH_3 - CO_2 - H_2O$. Kinetic analysis reveals that the $NO - NO_2 - H_2O$ portion of the mixed-phase system approaches equilibrium very slowly. As discussed elsewhere, there appears to be some question concerning the applicability of NH_3 solubility data to atmospheric liquid water.

Recent ideas on the mechanism of aqueous oxidation processes for reduced sulfur, S(IV), have placed emphasis on coordination reactions between metals and liquids, e.g., Fe and Mn complexed by SO_3^{2-}, and between organic and inorganic species, e.g., CH_2O and HSO_3^-. There will thus be an increasing need for reliable ΔG^o and ΔH^o data for metal ion hydrolysis, metal ion solubility, and complex formation in order to couple equilibrium and kinetic models for aqueous atmospheric chemistry.

<u>The Chemical Data Base: Kinetic Parameters, Product Studies</u>

The virtual certainty that equilibrium will not be quickly established for many cases of interest in aqueous atmospheric chemistry implies a need for information on the kinetic parameters and products of a large number of solution reactions. Extensive reviews have recently established that many of the problems of interest have not been investigated experimentally. The rate constants for a number of chain-initiating reactions have been determined, but rarely the temperature dependence of the rate constants and seldom the reaction products. Without product studies, chemical chains often cannot be placed in perspective with competing processes. Perhaps the greatest lack of information concerns the reactions of the transition metal ions. A substantial laboratory effort in all aspects of the aqueous solution chemistry of atmospheric components is required.

The Chemical Data Base: Gas Solubilities

A parameter often of importance to a description of gas- to aqueous-phase transfer processes in the atmosphere is the equilibrium solubility of the gaseous species. Experimental values are available for some gases of interest. Although a less desirable approach, solubilities may be estimated from thermodynamic data. The determination of solubilities for a number of atmospheric gases are needed; examples are CH_3OOH and PAN. Temperature-dependent studies will be particularly important.

The case of ammonia raises the possibility that in some instances either the equilibrium solubility or the rate of dissolution may be different for distilled water and atmospheric liquid water. In the ammonia case, it is not clear whether the problem stems from uncertainty as to the correct value for the equilibrium constant, or from a perturbation to the dissolution kinetics caused by some atmospheric process that does not occur in distilled water. A note of caution is therefore warranted, at least with regard to aqueous ammonia in the atmosphere, and perhaps with regard to all reactive atmospheric gases.

Chemical Nonlinearities

Chemical systems are inherently nonlinear, since the concentration of a given product is often dependent on the concentration of its precursor raised to a power different from unity, or because different chemical channels become effective under different physical conditions. We expect this type of nonlinear behavior in the chemistry of atmospheric water as with other chemical systems. An example of nonlinear behavior is provided by observations of precipitation chemistry in northern Europe. It has been observed that the concentration of sulfate in north European precipitation does not seem to have increased significantly during the sixties and early seventies despite an increase of the SO_2 emissions during that period. This lack of increase might be caused by a slowing down of the effective oxidation rate associated with changes in the chemical climate of the region. For non-winter time conditions the gas-phase competition for OH· radicals caused by increased emissions of NO_x may have played a role. In the winter, when polar radiation levels are low, liquid phase oxidation rates may have undergone modification. An increase in dry deposition within the region is possible, as is transport to longer distances.

As has been demonstrated in gas-phase atmospheric chemistry and in countless other chemical systems, nonlinear behavior can be understood in a complex system only when that system is well understood. Thus, if we focus on too narrow a goal (e.g., the conversion of SO_2 to sulfate), the result may be much slower progress than if broad goals are chosen, allowing the parameters of the system to be more thoroughly explored.

CONDENSATION/EVAPORATION WITHIN THE ATMOSPHERIC WATER CYCLE

Most clouds do not produce precipitation. The cloud droplets evaporate and give back the dissolved and undissolved material to the atmosphere. Within a cloud, drops may go through condensation/evaporation cycles, especially if ice is present. In order to compare gas-phase transformation processes with those occurring in the liquid phase, one has to know how often a molecule (or a particle) finds itself in water and, perhaps, for how large a percentage of the time this occurs. These factors are not easy to estimate from standard meteorological data. It is not sufficient to know the fraction of the air volume that on the average is covered by clouds. It is also a question of the rate of movement of air parcels in and out of clouds.

Observations of inhomogeneities in liquid water content of many clouds imply inhomogeneities in the chemical processes within cloud droplets: wide variations in ionic strength in droplets in "wet" and "dry" parts of the cloud may be envisaged. Furthermore, if "dry" patches result from the entrainment of clear air from above cloud, or from the sides, the overall chemical mass balance will be altered during mixing processes that subject cloud droplets to alternate evaporation/condensation cycles within cloud.

We know little of these processes, but are vitally interested in their consequences. What happens chemically and physically with the aerosol particles during such a cycle? How many such cycles would an average aerosol particle experience during its lifetime in the atmosphere? It has been suggested that water molecules go through ~10 cycles before being deposited. The same may be true for aerosol particles. We would like to know: a) What happens during one cycle?, b) How often does a particle go through such a cycle?, and c) How many cycles would an "average" particle go through during its lifetime (or, alternatively, taking a particle residing in the atmosphere, how many cycles would it have gone through?).

WET AND DRY REMOVAL PROCESSES

Wet Deposition

At distance scales from regional to global, one major application of wet removal (or precipitation scavenging) studies is to estimate the "residence" or "turnover" - times of atmospheric trace constituents. If a number of conditions are satisfied, the average residence time for a specific atmospheric constituent can be estimated from

$$\frac{1}{\langle\tau\rangle} = \frac{1}{\langle\tau_w\rangle} + \frac{1}{\langle\tau_d\rangle} + \frac{1}{\langle\tau_c\rangle} + \frac{1}{\langle\tau_p\rangle} + \dots \tag{1}$$

in which subscripts w,d,c,p,... identify residence times if only wet, dry, chemical, physical... processes were separately active.

The present state of knowledge of τ_w is described by Slinn (this volume). Field experiments designed to determine τ_w are complex, expensive, and often difficult to interpret. As an example of the precision of current determination of τ_w, let $\bar{E}$ be the average scavenging efficiency for a particular component to be scavenged, the average being taken over appropriate storms. (E is the fraction of the component entering a storm that is scavenged by it.) Let $\bar{\nu}$ be the average frequency that a given air parcel is present within precipitating storms. Then it is easy to see that

$$\langle\tau_w\rangle = (\bar{E}\bar{\nu})^{-1}. \tag{2}$$

As illustrations of Eq. 2, suppose $\bar{E} = \frac{1}{2} \pm \frac{1}{4}$ and, for the boundary layer in temperate latitudes, take $\bar{\nu}_\omega = (2 \pm 1 \text{ days})^{-1}$ for the winter season and $\bar{\nu}_s = (4 \pm 2 \text{ days})^{-1}$ for the summer season. Then from Eq. 2, and ignoring any seasonal dependence of $\bar{E}$, we obtain

$$\tau_w = \left\{ \begin{array}{l} \text{4 days (uncertainty range 1 - 12 days), winter} \\ \text{8 days (uncertainty range 3 - 24 days), summer} \end{array} \right\}. \qquad (3)$$

(For the middle and upper troposphere, τ_w values are expected to be significantly larger.) To improve on these crude estimates, substantial effort will be needed to obtain more realistic values of $\bar{E}$ and $\bar{\nu}$. This effort can be expected to require large-scale field experiments involving extensive measurements within clouds, perhaps by Lagrangian techniques.

Alternative methods for estimating τ_w on regional or global scales involve the determination of "scavenging ratios" or "scavenging efficiencies," in which average precipitation characteristics are utilized. On the local space scale, τ_w must be defined by hydrometeor scavenging assessments, since individual events are of interest. The parameter most difficult to evaluate in such assessments is the "collection efficiency," i.e., the fraction of particles or gases in the path of a hydrometeor that will be captured by it. For particles smaller than about 5 μm, it is generally expected that the retention efficiency (the ratio of particles or gas molecules retained after a collision to those undergoing collision) is unity. The expectation is based on the fact that for these relatively large particles, the relatively small surface to volume ratio permits van der Waals forces to dominate at short distances. In this case, the major uncertainties lie in specifying the collision efficiency and in properly accounting for particle growth by water vapor condensation, interception (especially for ice crystals), phoretic effects, etc. (see Slinn, this volume,for a further discussion of these points, both for particles and gases). Realistic laboratory studies to investigate these processes for a variety of gases and particles are needed, as are field studies that attempt to directly measure collection efficiencies and effects of relative humidity on scavenging processes.

Dry Deposition

Although this paper reviews the role of atmospheric water in cycling atmospheric trace constituents, it is appropriate to consider also some aspects of removal processes in the absence of precipitation, since for many gases dry deposition is the dominant removal process and since dry and wet removal can be of comparable importance for particles under some climatic and surface conditions.

Dry deposition of gases is frequently dictated by their behavior in or on the receptor. Even for reactive gases such as SO_2, surfaces can become saturated with the gas and its reaction products, thereby strongly inhibiting additional deposition. To treat gas transfer to vegetation and to natural water surfaces, it is often useful to identify separate "transfer resistances" in air and in the surface medium. For highly soluble and/or reactive gases, atmospheric resistance dominates; the associated transfer velocities are known reasonably accurately. On the other hand, there remain many uncertainties about gas transfer and reactions in surface media: knowledge of the transfer velocity in natural waters suffers from both theoretical and experimental deficiencies, and gas solubilities and reaction rates in the intercellular fluid of vegetation and in the "surface film" of natural waters are virtually unknown. Almost certainly some of the reactions are fast enough and the products sufficiently reactive to modify the host fluid chemically. In the case of air-sea exchange, chemical-hydrodynamic coupling may be an additional complication.

Measurements of particle dry deposition pose serious difficulties. As yet, we do not know how to relate deposition on artificial surfaces to deposition on natural surfaces. Results from eddy-correlation, concentration-gradient, and concentration-decrement methods, generally utilizing polydisperse aerosol particles, continue to be unsatisfactory. Some satisfying measurements of deposition to snow and vegetation have recently been reported, but progress (using more-nearly monodisperse particles, and for deposition to forests and to natural waters) is needed. It is encouraging that some of the recent studies suggest that wind-tunnel data can be used as first estimates for particle dry deposition to natural canopies of vegetation. In contrast, there is concern that available water/wind-tunnel data for particle deposition cannot be readily extrapolated to the case of particle deposition to natural waters. It is also noted that as an aerosol ages, the mass-average deposition velocity normally decreases (since the larger particles deposit first) and that measurements of actual deposition are strongly weighted according to particle diameter.

CONCLUSIONS

In making assessments as we have done here, an important aspect of the work is the selection of major goals for future work and the identification of major uncertainties. These tasks are not readily accomplished without a clear idea of why they are being done, however, since our topic impinges upon many aspects of modern society and the pursuit of one goal may result in the neglect of another. As our framework, we have taken as our single goal the expression of critical problems in understanding the role of water in the transport, conversion, and removal of tropospheric trace compounds. This perspective, one of general scientific understanding of a series of related chemical systems, then permits us to identify areas in which knowledge is satisfactory and those in which it is deficient. Such identification should then permit those with more applied goals to make use of these conclusions for their purposes as well.

In a field as new as that of the chemistry of atmospheric water, vast seas of uncertainty confront the investigator. These areas are relieved to some extent by small islands of satisfactory (or relatively satisfactory) information. The latter appear to include nucleation scavenging, at least a partial inventory of stable species (but not free radicals) in aerosol particles and raindrops, and some of the needed equilibrium constants and kinetic data. The list of critical deficiencies is longer: it is presented and briefly discussed immediately following the introduction to this paper.

The study of aqueous chemistry in the atmosphere is relatively new, conceptually complex, and experimentally difficult. The information to be gained by such study and the attendant benefits accruing therefrom promise to be great. Progress, however, will require great diligence and the application of techniques and insights from across the full spectrum of chemistry.

GENERAL REFERENCES

(1) Chameides, W.L., and Davis, D.D. 1982. The free radical chemistry of cloud droplets and its impact upon the composition of rain. J. Geophys. Res. 88: in press.

(2) Graedel, T.E., and Weschler, C.J. 1981. Chemistry in aqueous atmospheric aerosols and raindrops. Rev. Geophys. Space Phys. 19: 505-539.

(3) Husar, R.B.; Lodge, J.P., Jr.; and Moore, D.J., eds. 1978. Sulfur in the Atmosphere. Oxford: Pergamon Press (also published Atmos. Env. 12: 1-796).

(4) Lee, Y.-N., and Schwartz, S.E. 1981. Evaluation of the rate of uptake of nitrogen dioxide by atmospheric and surface liquid water. J. Geophys. Res. 86: 11971-11983.

(5) National Academy of Sciences 1978. The Tropospheric Transport of Pollutants and Other Substances to the Oceans. Washington, D.C.

(6) Pruppacher, H.R.; Semonin, R.G.; and Slinn, W.G.N., eds. 1983. Proceedings of the 4th International Conference on Precipitation Scavenging, Dry Deposition, and Resuspension. Amsterdam: Elsevier, in press.

(7) Slinn, W.G.N.; Hasse, L.; Hicks, B.B.; Hogan, A.W.; Lal, D.; Liss, P.S.; Munnich, K.O.; Sehmel, G.A.; and Vittori, O. 1978. Some aspects of the transfer of atmospheric trace constituents past the air-sea interface. Atmos. Env. 12: 2055-2087.

(8) Stumm, W., and Morgan, J.J. 1981. Aquatic Chemistry, 2nd ed. New York: John Wiley.

(9) Twomey, S. 1977. Atmospheric Aerosols. Amsterdam: Elsevier.

(10) Zafiriou, O.C. 1982. Natural water photochemistry. In Chemical Oceanography, eds. J.P. Riley and G. Skirrow, vol. X. New York: Academic Press.

APPENDIX - DEFINITIONS, STANDARDS, AND MEASUREMENTS

Suggested Terminology

Investigations of chemistry in atmospheric water have received less attention than some other areas of atmospheric chemistry, and the terminology is less well established. Although a comprehensive study of terminology and usage has not been attempted, we have identified several specific items upon which recommendations can be made:

Use of accepted nomenclature and symbols

"Shorthand" notation is sometimes practical among specialists, but should not appear in the atmospheric chemistry literature, especially in areas so potentially complex and interdisciplinary. We recommend the chemical nomenclature of the IUPAC (3, 17), the use of radical and ionic symbols as appropriate, and adherence to the kinetic data standards of the CODATA Task Force (4).

The term "acid rain" should be avoided, both because of its linguistic inelegance and because of its chemical inaccuracy. Acidic rain may be used if a reference to a simple pH measurement is intended; the study of the chemical properties of the rain is better designated "the acid/base properties of rain" or "the chemistry of rain."

For the case of the atmospheric water droplet consisting of an aqueous phase surrounding an insoluble core, a catalyst present on the exterior of this core which interacts by catalyzing species diffused to that site is termed a heterogeneous catalyst. Conversely, a catalytic component which is dissolved within the aqueous phase is termed a homogeneous catalyst. Elemental carbon is an example of the former, manganese ion of the latter.

The term rainout may be used as a synonym for "rain scavenging" and washout as a synonym for "precipitation scavenging." The use of these terms to signify in-cloud and sub-cloud precipitation scavenging is obsolete and confusing and is generally being abandoned. In-cloud scavenging and below-cloud scavenging are also satisfactory terms, where appropriate.

The collection of material deposited on an artificial surface by both wet and dry deposition is sometimes referred to as a "total deposition" measurement. Such a term implies that the measured value is representative of the amount received by a natural surface, including the dry deposition of gases. We prefer use of the term bulk deposition for measurements of deposition made with a continuously exposed collector and wet only deposition when the collector is exposed only during rain events.

An "aerosol" is a system consisting of condensed phase particles within a fluid medium. If one is referring only to the condensed phase component, the proper term is aerosol particle.

Reference standards

For many studies of the chemistry of atmospheric water, it is necessary to have a reference point for describing the acid/base properties of an aqueous system. This point has traditionally been taken as pH 5.65 in the two-phase CO_2-H_2O system. It is important to realize that assigning this choice as the reference state for this system is done for convenience, not because it has intrinsic meaning as a global "background level." (Recent studies have demonstrated the wide variability expected in natural acidity.) A reference state that is more useful in a chemical sense is the inflection point of the (pH - total H) titration curve for the CO_2 - H_2O system (see discussion by Morgan, this volume). In the general case, deriving complete acid/base information for a sample requires one to establish the total titration curve and then to assign its features to individual chemical components in the sample.

Cloudwater Sampling

Direct collection of water samples from clouds is a relatively recent achievement, and the acquisition of samples involves techniques not fully perfected. Since clouds contain spatial inhomogeneities in the cloud water distribution, a loss of sample cloud water by evaporation must be expected during the process of cloud water sampling with aircraft. Most cloud water collectors currently in use require relatively long sampling times, and there are questions as to whether the sample collected may be modified by incorporation of gases and aerosols during the in-cloud sampling. An associated problem is that the samplers now in use lead to a fractionation of the droplets of different sizes, with loss of some fraction of the small cloud droplets. New techniques and devices are required to achieve representative sampling with high time resolution.

Rainwater Sampling

Many networks for precipitation chemistry are presently in operation with many different techniques and types of collectors being used. Among the problems encountered are the influence of wind, influence of the surrounding topography, and contamination in automatically opening and closing samplers. Automatic collectors in use show differences with respect to the sensitivity of the sensor: if at low sensitivity, collectors miss samples at small precipitation rates. Also, heating of the sensor during the winter period poses problems. It is important that the sensor-sensitivity of different systems be comparable, and that light rain be included in the samples (since light rain is in most cases more acidic than heavy rain).

Virtually all inorganic and organic substances (and/or their decomposition or dissociation products) found in aerosols and in the gas phase will be found in rain. In addition, other species derived from reactions in aqueous solution will also be found. Except in urban areas, there has been little effort given to separating dissolved and particulate forms of the variour constituents of rainwater. While the relatively high concentrations of many of the potentially important trace metal ions and organic compounds enable one to obtain samples in urban regions with relatively little concern for contamination, the extremely low concentrations of many of these substances expected in remote regions will make contamination a very serious problem. For example, typical iron concentrations in urban rain range from 10^{-4} to 10^{-5}M. However, iron concentrations in some remote marine regions can be in the range of 10^{-9} to 10^{-7}M, i.e., 0.05 to 5 μg/l, while Cu may be as low as 10^{-10}M.

It should be noted that the definition of "particulate" and "dissolved" using conventional techniques is operational only and depends on the specific characteristics of the filter used. Most commonly used filters will pass colloidal size (<0.5 μm) particles.

Increased attention to "event sampling" of individual rain events is recommended, as is differential sampling during individual rainfalls in order to be able to study trends and fluctuations of the concentration of different components occurring during the rainfall. In this case a system can sample over equal time increments or over equal precipitation volumes. It should be mentioned that a detailed meteorological evaluation and interpretation is possible only for a single rain event or group of events, not for samples that are a composite of several different rainfall conditions.

Common, Reliable Methods for Rainwater Analysis

Measurements of pH should be made with addition of KCl to the samples for adjusting ionic strength. The necessary cell calibration can be performed by using standard phosphate buffer or standard strong acid solution with added KCl (14). For conductivity measurements a standard temperature (20°C or 25°C) must be maintained. In order to have the possibility of establishing mass balances, pH and conductivity measurements should be made at the time of analyzing the samples for the principal constituents.

One of the most versatile techniques for rainwater analysis is ion chromatography (20). It is applicable to the analysis of mixtures of cations, such as ammonium, alkali, and earth alkaline metal ions, as well as to anions such as Cl^-, NO_3^-, or

SO_4^{2-}. Besides ion chromatography, a wide variety of single-ion procedures are in use for rainwater analysis. Protons of strong acids are determined by titration (1,13) or by employing a radiochemical tracer technique (12). Several wet chemical photometric procedures have proven to give reliable results. Examples are the thorin method for sulfate (16), the mercury(II)-thiocyanate method for halide ions (7), and the indophenol blue method for ammonium ions (5,6). Nitrate can be determined by a direct UV photometric procedure (19) or after reduction to nitrite and subsequent formation of an azo dye (18). Radiochemical isotrope dilution analysis has been employed for the determination of sulfate and chloride (11,12). Some ions (e.g., NH_4^+) are occasionally measured potentiometrically by using ion selective electrodes (14).

Atomic absorption spectroscopy is still the most frequently used technique in analyzing rainwater for metal ions (Na^+, K^+, Ca^{2+}, Mg^{2+}, Fe^{3+}, Al^{3+}, etc.). Especially for the determination of some trace heavy metals (Zn^{2+}, Cu^{2+}, Cd^{2+}, Pb^{2+}), anodic stripping voltammetry has been shown to be an attractive approach (15). Multielement techniques such as neutron activation or X-ray fluorescence analysis are well suited for investigating airborne particulate matter (9) and have also been used for the determination of trace metals in precipitation.

The investigation of rainwater with respect to dissolved organic material has so far been restricted to a few species. Carboxylic and sulfonic acids can be determined by GC/FID after suitable derivatization (alkyl esters) (2,10). For formaldehyde a photometric procedure has been described (8).

The most important precondition for obtaining reliable results is the avoidance of contamination and change in composition of samples during collection, handling, and storage. Data quality should be verified by the following balances: a) anion equivalents vs. cation equivalents, b) measured conductivity vs. calculated conductivity, and c) pH vs. c_{H^+} (as determined by titration).

APPENDIX REFERENCES

(1) Askne, C., and Brosset, C. 1972. Determination of strong acid in precipitation, lake-water and airborne matter. Atmos. Env. 6: 695.

(2) Barcelona, M.J.; Liljestrand, H.M.; and Morgan, J.J. 1980. Determination of low molecular weight volatile fatty acids in aqueous samples. Anal. Chem. 52: 321-325.

(3) Cahn, R.S., and Dermer, O.C. 1979. Introduction to Chemical Nomenclature, 5th ed. London: Butterworths.

(4) CODATA Task Group on Data for Chemical Kinetics. 1974. CODATA Guidelines on Reporting Data for Chemical Kinetics, NBSIR 74-537, Washington, D.C.: National Bureau of Standards.

(5) Crowther, J., and Evans, J. 1980. Automated distillation-spectrophotometry procedure for determining ammonia in water. Analyst 105: 841-848.

(6) Crowther, J., and Evans, J. 1980. Blanking system for spectrophotometric determination of ammonia in surface waters. Analyst 105: 849-854.

(7) Florence, T.M., and Farrar, Y.J. 1971. Spectrophotometric determination of chloride at the ppb-level by the mercury(II)-thiocyanate method. Anal. Chim. Acta 54: 373-377.

(8) Klippel, W., and Warneck, P. 1978. Formaldehyde in rain water and on the atmospheric aerosol. Geophys. Res. Lett. 5: 177-179.

(9) Klockow, D. 1982. Analytical chemistry of the atmospheric aerosol. In Proceedings of the NATO Advanced Study Institute on Chemistry of the Unpolluted and Polluted Troposphere, Corfu Island, Sept. 28 - Oct. 10, 1981, in press.

(10) Klockow, D.; Bayer, W.; and Faigle, W. 1978. Gas-chromatographic determination of traces of low molecular weight carboxylic and sulfonic acids. Fres. Z. Anal. Chem. 292: 385-390.

(11) Klockow, D.; Denzinger, H.; and Rönicke, G. 1974. Anwendung der substöchiometrischen Isotopenverdünnungsanalyse auf die Bestimmung von atmosphärischem Sulfat und Chlorid in Background-Luft. Chemie-Ing.-Techn. 46: 831.

(12) Klockow, D.; Denzinger, H.; and Rönicke, G. 1978. Sauerstoffhaltige Schwefelverbindungen. In VDI-Berichte, Nr. 314, pp. 21-26. Düsseldorf: VDI-Verlag.

(13) Liberti, A.; Possanzini, M.; and Vicedomini, M. 1972. The determination of the non-volatile acidity of rain water by a coulometric procedure. Analyst 97: 352-356.

(14) Liljestrand, H.M., and Morgan, J.J. 1978. Chemical composition of acid precipitation in Pasadena, California. Environ. Sci. Technol. 12: 1271-1273.

(15) Nquyen, V.D.; Valenta, P.; and Nürnberg, H.W. 1979. Voltammetry in the analysis of atmospheric pollutants. The determination of toxic trace metals in rain water and snow by differential pulse stripping voltammetry. Sci. Total Env. 12: 151-167.

(16) Persson, G.A. 1966. Automatic colorimetric determination of low concentrations of sulfate for measuring sulfur dioxide. Int. J. Air Water Pollut. 10: 845-852.

(17) Rigaudy, J., and Klesney, S.P. 1979. Nomenclature of Organic Chemistry, Section A,B,C,D,E,F, and H. Oxford: Pergamon.

(18) Sawicki, C.R., and Scaringelli, F.P. 1971. Colorimetric determination of nitrate after hydrazine reduction to nitrite. Microchem. J. 16: 657-672.

(19) Slanina, J.; Lingerak, W.A.; and Bergman, L. 1976. A fast determination of nitrate in rain and surface waters by means of UV spectrophotometry. Fres. Z. Anal. Chem. 280: 365-368.

(20) Small, H.; Stevens, T.S.; and Bauman,W.C. 1975. Novel ion exchange chromatographic method using conductimetric detection. Anal. Chem. 47: 1801-1809.

Atmospheric Chemistry, ed. E.D. Goldberg, pp. 119-134. Dahlem Konferenzen 1982.
Berlin, Heidelberg, New York: Springer-Verlag.

The History of Atmospheric Composition As Recorded in Ice Sheets

C. U. Hammer
Geophysical Isotope Laboratory
2200 Copenhagen N, Denmark

Abstract. The polar ice sheets are unique Quaternary deposits which offer a wealth of detailed information on past atmospheric compositions. To what extent and under what assumptions historical records of atmospheric constituents can be established from ice cores will be discussed and exemplified.

INTRODUCTION

Polar ice sheets, in particular the ice sheets of Greenland and Antarctica, play an important role in recent studies of palaeoatmospheric composition and palaeoclimate (2). Their importance stems from the fact that they consist of well layered frozen past precipitations which generally formed between the ice sheet surface and 1-2 km above it, covering perhaps the whole Quaternary era at favorable locations. The special importance of the two major ice sheets relates to their vast extent, high surface elevations, and remoteness to most aerosol sources, which among other advantages, secures a certain uniform mixing ratio of atmospheric trace substances (6) at levels where the precipitation forms.

GENERAL ASPECTS OF ICE CORES AND THEIR INFORMATION ON ATMOSPHERIC COMPOSITION

The bubble filled ice layers carry information on past snow

accumulation, climate, atmospheric gas composition, and aerosols, but the step from ice core information, whatever atmospheric constituent is under study, to historical records of atmospheric composition is not straightforward. It involves a number of important considerations as summarized in the following four questions:

1) Can the ice record be related to air composition?
2) How well does the ice record represent seasonal changes in the air composition, and how does this influence yearly average values?
3) Is the constituent representative of local, regional, hemispheric, or worldwide atmospheric conditions?
4) What is the dating accuracy of the record?

With a little change in the words these questions are also relevant for climatic ice core records.

The second and the third questions are the most difficult to answer. The second is essentially a glaciological question, though related to the meteorology at the drill site, while the third question usually has to be approached with due consideration to the behavior of the general atmospheric circulation and environmental processes. Not surprisingly, the third question can only be given a clear answer in a few cases and is often the subject of much debate in the relevant literature.

A clear answer to the second question is hampered by the lack of data on individual snowfalls, i.e., at what time of the year did they fall and how much snow fell? Usually one can get somewhat around the lack of data by demonstrating that fairly pronounced seasonal changes can be observed in the ice core. A technique which could improve the situation is signal analysis of the observed seasonal changes; if the statistical shape of the observed seasonal variations measured on the ice core is fairly constant with time, it indicates that at least the different seasons of the year are appropriately represented with the same weight. It does not solve the problem in an absolute sense, but it allows relative comparison of ice records.

In areas of high accumulation, i.e., more than approximately 15 gr. of snow/cm^2 year, this kind of reasoning "suffices" for most records.

A case where the answer to question two is a matter of great concern is in the relation between isotopic composition of the ice and average yearly temperatures. As average annual temperatures fluctuate very little from year to year, even slight changes in the seasonal pattern of precipitation may affect the yearly averages of isotopic composition of the ice more than what is inferred by the "expected" annual temperature changes. This is one of the reasons why at present average oxygen-18 concentrations (in the literature given in the relative $\delta(^{18}O)$ scale) of single annual ice layers cannot be related to average air temperatures; the glaciological-meteorological signal/noise ratio is, so to speak, too small. The study of signal/noise ratio should therefore play an important part in the interpretation of $\delta(^{18}O)$ time series, but very little has been published. The problem is, of course, minor when it is a question of demonstrating drastic changes: seasonal changes in $\delta(^{18}O)$ are easily recognized in ice cores as well as climatic changes associated with the last ice age.

$\delta(^{18}O)$ records are just one example of the influence of the seasonal precipitation pattern on the interpretation of ice core records: any record of an atmospheric constituent exhibiting seasonal but small changes in the yearly average values should be interpreted under the scrutiny of question two. It is an important point which, if not answered, may lead to completely erroneous interpretations, especially in the case of data related to ice age conditions.

Questions number 1 and 4 can be answered much more precisely but are more naturally dealt with later.

The data obtained from ice core analysis and their interpretation show some marked differences, with respect to both their general validity and position in a glaciological framework, so that it is practical to divide them into four groups: a) yearly accumulation of snowfall, b) isotopic composition of the snow, c) atmospheric gases, and d) atmospheric impurities.

The first two groups are only of indirect interest here, because yearly accumulations are only of local validity, and over longer time spans both groups are indicators of climatic conditions rather than of atmospheric composition. They do, however, play an important role in the dating of ice cores. Dating is a prerequisite to any historical record from ice sheets, and it is appropriate to touch on the subject before discussing ice sheet records of atmospheric composition.

DATING OF ICE CORES

Dating and Drilling

There are essentially two methods for dating an ice core, if indirect ways are excluded: either to date it stratigraphically or to measure some constituent in the core, which changes with time, e.g., ^{14}C. The latter method suffers from low accuracy but becomes useful where stratigraphic methods fail or are technically difficult, which is often the case for the oldest part of a deep core.

Indirect methods (e.g., flow models, comparison with other records such as pollen in peat bogs or data from sea sediments) are usually not very precise but are for the moment necessary for very old ice. Recent progress in isotope analysis by means of an accelerator technique looks promising for the dating of old ice, but the range and potential of especially the ^{14}C method and the proposed ^{36}Cl-^{10}Be method remain yet to be demonstrated (see discussion in (2)).

The high stratigraphical dating precision which can be obtained for ice cores drilled at favorable locations is in fact one of the fascinating aspects of the polar ice sheets. Even though dating accuracies of ±3 years over 1000 years is the limit at present, it can be inferred from recent experience that in the near future the dating of a 10,000 year old ice layer should be possible without any yearly error at all. Tacitly, however, it is assumed that the ice drilling technique will be able to deliver a perfect ice core, i.e., one which is essentially continuous and of high quality. It may be difficult for "outsiders" to realize the importance of this point, but drilling in ice is a good deal more complicated than drilling for oil (aside from the financial problems). The close link between drilling and ice core analysis, which has developed in order to ensure accurate and comprehensive data, was, e.g., the main reason why US, Swiss, and Danish laboratories moved their equipment into the field during the three summers of 1979-81 that it took to penetrate the 2037 m thick ice sheet at Dye 3 in southern Greenland (Fig. 1).

The deep drilling was a part of the project called GISP (Greenland Ice Sheet Program) and indicates the kind of international

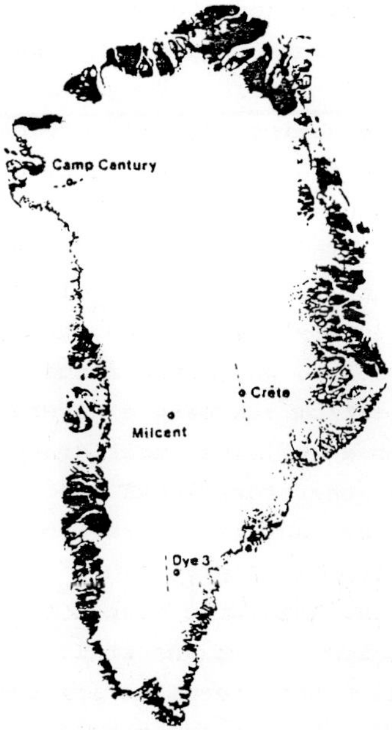

FIG. 1 - Location of four important drill sites in Greenland.

cooperation which is needed in order to obtain all relevant data from the core. In this respect, GISP was just a pilot study, but the drilling itself also gave useful hints for future deep drilling projects: a number of glaciologically oriented scientists entered the drilling "business" and created a computerized drill which recovered 99.97% of nearly excellent core from the 2037 m long drill hole. One outcome of this joint effort will be the first precise dating of the late glacial history, far exceeding the accuracy of the ^{14}C method. It is my hope that these remarks will benefit the international cooperation so that a future deep core can be analyzed in the best possible way.

Stratigraphical Dating Techniques

In areas of high accumulation, seasonal changes in the isotopic composition of the ice serve as a powerful tool in ice core dating (in low accumulation areas diffusive processes obliterate the seasonal $\delta(^{18}O)$ variations). Combined with other stratigraphical dating methods, e.g., methods based on dust, nitrate, or acidity, a very high dating accuracy can be achieved.

The difference in the dating accuracy of ice cores from low and high accumulation areas is important, because in Antarctica, where the accumulation is generally low, the isotopic dating method fails. The low annual precipitation in Antarctica also implies that the previously discussed question 2 becomes a question of utmost importance, especially for certain parts of east Antarctica, where the annual accumulation is a few grams of snow/year/cm^2: a single severe storm could blow a whole annual layer away. The stratigraphic dating of antarctic ice cores is, therefore, less precise than the dating of Greenland ice cores, but it offers considerable advantages for ice older than approximately 10,000 years. Due to its geographical position, Antarctic precipitation is very clean and during the last ice age it was still so clean that the ice remained acidic, while the Greenland ice sheet generally became slightly alkaline (due to larger areas of continents in the northern hemisphere, loss of ice margins, and alkaline dust from previous, sea-covered continental shelves). Irrespective of the low precipitation and perhaps an annual layer missing here or there, the Antarctic Byrd core (see Fig. 2) does show seasonal variations of acidity during the last ice age (see Fig. 3A: yearly snow deposition at present is approximately 13 gr/year cm^2).

Acidities can be measured in great detail, continuously and rather easily (4), while chemical elements and even dust are more difficult to measure in sufficient detail. For the moment, this is in principle only a practical point, but I fear that it will remain so for several years to come, even though dust can also be measured continuously. The seasonal variation in the acid concentration is more pronounced than the corresponding dust variations.

The French Dome C core and the Russian Vostok core (see Fig. 2) also reach back into the ice age. Both cores are from regions of very low precipitation (a few gr/year cm^2) and may be on the limit of stratigraphical methods. An advantage of cores from regions of little precipitation is that they cover many "years" per meter of core.

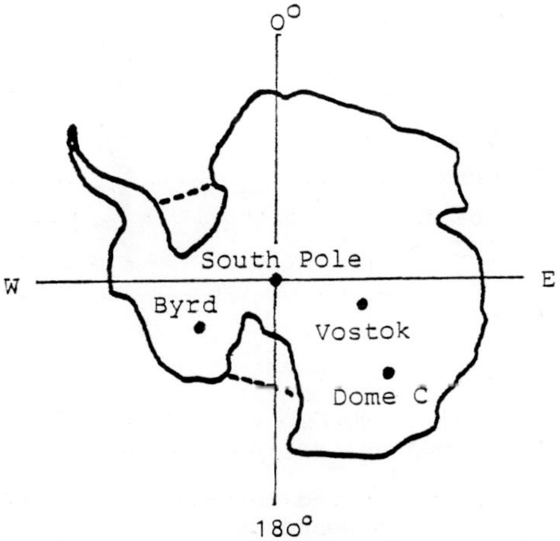

FIG. 2 - Location of four important drill sites in Antarctica.

In Fig. 3 examples of stratigraphical dating are given to illustrate the various methods.

How Can Old Ice Be Recovered?

It is not possible to answer the question in the above subheading in a strict sense. The only way to get a decisive answer would be to drill through the ice sheets in regions where the

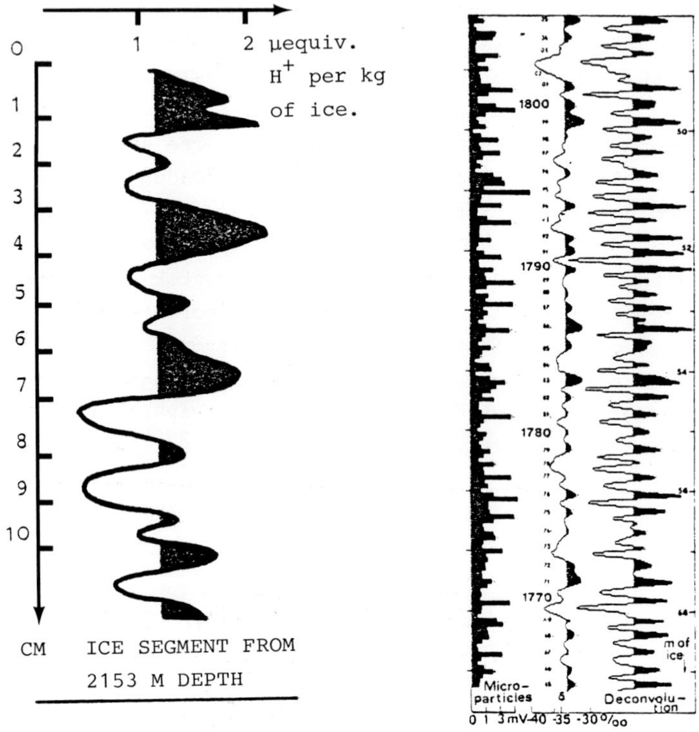

(A) (B)

FIG. 3 - (A) Seasonal variations of acidity in the Byrd core at 2153 meter depth in a 10 cm core segment. Estimated age of segment is approximately 60,000 years B.P. (B) Segment of Crête core showing seasonal variations in dust and $\delta(^{18}O)$ over the period 1765-1805 A.D. The deconvoluted δ profile to the right is first-order corrected for diffusive smoothing in the firn (from (5)). 1 mV corresponds to approximately 50 μgr. dust per kg of ice. (Reproduced from the Journal of Glaciology by permission of the International Glaciological Society.)

oldest ice is expected to be according to available data from deep cores and flow model considerations.

The only three existing ice cores which reach to the bottom of the two major ice sheets were not drilled at favorable locations, but their age-depth relations deduced from measurements or interpretation of data are within the rough expectations deduced from flow models.

In Fig. 4 some climatically interesting ages and their corresponding occurrence with depth for three drill sites in Greenland are deduced from simple flow models to give an impression of the potential age ranges of ice cores: the three cores are Camp Century, Dye 3, and Crête (see Fig. 1 locations). The first two cores reached to the bottom of the ice sheet. Simple flow model ages for the deeper parts of the cores are very inaccurate, but Fig. 4 clearly indicates which of the three drill sites is favorable with respect to recovering old ice.

There seems to be no doubt that the Antarctic ice sheet, at favorable locations, covers ice from several ice ages. To

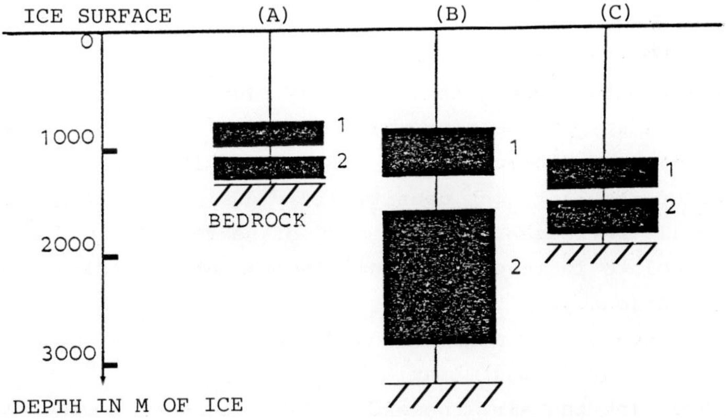

FIG. 4 - Rough age-depth estimates for three Greenland drill sites - (A) Camp Century, (B) Crête, (C) Dye 3. The black areas indicate: 1) Climatic optimum 5000-2000 B.C., and 2) the last ice age, 70,000-10,000 B.P.

what extent the Quaternary is represented in ice sheets is one of the most exciting questions in this field of research. Only future deep drillings can give the answer. The compression of annual layers in the deep ice does, of course, increase the complexity of ice core analysis, but for the moment we simply do not know where the age limit is for detecting single annual layers; data from the Byrd core indicate that 10,000 years is too low a limit for ice cores drilled at - in this respect - favorable locations.

INFORMATION ON GAS COMPOSITION

Ice cores from areas where no summer melting takes place offer the best opportunities for establishing past atmospheric gas composition. As long as melting is not too severe, ice cores from such locations can still be used for reconstructing records of past atmospheric gas composition. Due to an often higher snow accumulation in such regions, a better time resolution can be obtained for the gas composition (see below).

Influence of Summer Melting

Where melting takes place, components such as CO_2 or HNO_3 (if in the gas form) can be highly enriched in the ice compared to their mixing ratios in air. For example, at the Dye 3 core, only a few percent of the summer snow melts. At this location a few percent of the melt layers may even be associated with slight rains. The melting mainly takes place in the upper few centimeters of the ice sheet surface, and leaks down to the bottom of the last high density individual snowfall layer where it usually refreezes. In the case of heavy melting, the water may percolate in favorable channels through several singular precipitations, but it is almost always stopped by the previous winter's dense snow and the lower temperatures at this level of the snowpack. The water or water films on the snow crystals interact with the air, the end results being high concentrations of, e.g., CO_2 and HNO_3.

The effect of melt layers on gas concentrations can be extreme, as has been shown for CO_2 concentrations at Dye 3 (7), and it also holds for acidities (from my own experience). To simplify things, the following discussion is restricted to regions where no melting takes place. This does not mean that one should avoid taking gas measurements in ice cores from such regions; on the contrary, such regions may add to our knowledge on the gas-ice exchange. However, the present trend in gas analysis is to avoid such layers and establish a firm basis for more complex undertakings.

The Gas Enclosing Process - the Firnification Zone

The most important difference - from a glaciological viewpoint - between gases and impurities is related to the physical appearance of the upper 50-100 meters of an ice sheet; in the firnification zone the snow has not yet been transformed into ice, i.e., ice sheet and not solid ice. The firn zone is porous due to the presence of a multitude of air passages and only when these passages have been sealed off are the gas "samples" enclosed in the ice.

In central east Antarctica, the porous zone may cover several thousands of years of precipitation and in central Greenland a few hundred years, thus reflecting the difference in yearly snow accumulation. The existence of the porous zone sets a limit on the time resolution one can expect for records of gas composition. The number of yearly layers covered by the firn zone does not define a priori the resolution, because the open pore space is diminished with depth, and air exchange can be "blocked" even in areas where no melt layers act as "sealing" layers.

The air exchange is driven by gas diffusion in the pore space combined with a "pumping" by barometric pressure changes on the surface. The influence of both effects on the gas exchange may be strongly reduced by a number of very thin (0.5-1 mm), almost icy, layers which are created by strong sunlight (radiation crusts) or strong winds (wind crusts) and seal off the different annual layers in the firn zone. Very little is known about the actual limit of the time resolution and a satisfactory explanation of the strong seasonal CO_2 concentration variation (1) in ice cores is needed.

Total Gas Content (O_2, N_2)

The analysis of the total gas content of ice cores can be used to infer ancient surface elevation of ice sheets (8) because the barometric pressure changes with height. There are still some unsolved questions regarding the precision of the method, but as a first-order indication of surface height it is useful. Only small amounts of O_2 and N_2 diffuse into the ice from the bubbles. Within the error of analysis, the O_2:N_2 ratio has not changed during the last 100,000 years.

Ar, Ne, ^{3}He, and ^{4}He

These noble gases can be used to study diffusion of gases in ice cores, and the helium isotopes may be related to magnetic reversals. Full understanding of their use in ice core analysis is just beginning, but they will probably contribute to our knowledge on the physical processes they undergo during the compression of the ice layers.

CO_2

The burning of fossil fuels has increased the interest in knowing both the preindustrial CO_2 level and its natural fluctuation. The potential of CO_2 analysis of ice cores has recently been clearly demonstrated by Swiss and French groups (3,7). The drop in CO_2 concentrations during the last ice age is extremely interesting and indicates a corresponding drop in the atmospheric air concentration. A great step forward in the CO_2 research of ice cores was stimulated by new and more detailed measuring techniques. It is to be expected that this field of gas research will bring many new and exciting findings.

CH_4

Methane is mentioned here, as it plays an important role in the atmospheric radiation balance. The bulk of the atmospheric methane has a biogenic origin and is removed from the atmosphere by oxidation. Its concentration is generally a few ppm and the atmospheric lifetime on the order of a century. Volcanoes and earthquakes may add to the atmospheric CH_4 inventory, but hardly enough to be traced in ice cores.

A note of caution if CH_4 concentrations are inferred from ice cores: CH_4 suffers oxidation during firnification and to greater depths, thus complicating the interpretation of such records.

There are other gas components of the atmosphere than the above mentioned, but what message they carry is as yet not deciphered. Whatever component one would like to study, the influence of melting, firnification, diffusion, and possible chemical reactions on the concentrations should be given due consideration.

INFORMATION ON IMPURITIES

The number of various impurities which can be studied is as great in ice cores as in the atmosphere. I will only deal with the bulk of them, because we are only beginning to understand the "bulk" components. One of the advantages of bulk components is that they can be related to a source region.

Apart from the blocking effect in the firn, mentioned above (see information on gas composition), the snow apparently acts as a very efficient filter for impurities. In fact, it is so efficient that the different impurity concentrations are almost completely confined to individual precipitations rather than being spread out over the snow pack.

The concentration of impurities in a precipitating air mass over the major ice sheets can be derived from the concentration in the precipitation. Under certain conditions previously discussed by Junge (4), the concentration of atmospheric trace substances could be inferred from ice core concentration. If dry fallout, evaporation, and sublimation only contribute little to the ice concentration, the conditions of Junge would be fulfilled. Today we know that these conditions are almost fulfilled for high accumulation areas. Preindustrial ice sheet impurities consist of marine, continental, volcanic, and extraterrestrial material.

Anthropogenic activities are apparently beginning to add to the impurity concentration of the Greenland ice sheet, but not to Antarctic concentrations (Herron, personal communication). A time series of annual concentrations covering the period from the present back to 1900 A.D. would help to clarify this controversial issue. A typical composition of Holocene ice, e.g., at Dye 3, in a non-volcanic period is: dust - 50 μg/kg of ice; NO_3^-; CL^-; SO_4^{2-}; H^+; NH_4^+; Na^+ and $(Mg^{2+} + Ca^{2+} + K^+)$ - 1.0; 0.5; 0.5; 1.2; 0.3; 0.4; and (0.1) μequiv. per kg of ice, respectively. Comparing these concentrations with available information on atmospheric impurities allows the following conclusion: during non-volcanic periods in the Holocene, the oceans and the continents dominate as source regions, but the stratospheric HNO_3 component is probably an important contributor to the generally acid character of the ice. The input from stratospheric HNO_3 has been estimated from available data on stratospheric HNO_3 concentrations.

In volcanic periods, the continuous acidity profiles reveal several large volcanic eruptions which strongly contribute to the precipitation chemistry up to a few years after the eruptions.

Ice from the Wisconsin glaciation has 3 to 70 times higher dust concentrations than Holocene ice in both of the Greenland deep cores. The concentrations of all major impurities, soluble as well as insoluble, are strongly correlated with $\delta(^{18}O)$. Detection of individual volcanic eruptions by acidity measurements is prevented because Wisconsin ice is generally alkaline, and chemical detection is hampered by the high and variable impurity levels.

However, Byrd core analyses show that Antarctica is better suited for this kind of analysis because the Wisconsin ice is acidic and the impurity level is much lower than in Greenland.

Extraterrestrial material contributes less than 5 μgr/kg of ice. A large part of the black spherules in ice cores are probably extraterrestrial, and it would be interesting to have an ice record on the cosmic "fallout." Unfortunately, the number of large spherules which can safely be identified as extraterrestrial only amounts to a few per kg of ice. Our knowledge of organic impurities is confined to a few publications on, e.g., pollen and wood fibers, a part of ice core analysis which deserves more interest.

CONCLUSION

The history of atmospheric composition during a large part of the Quaternary can be reconstructed from ice cores. At present the gas composition and a few impurity constituents offer the best possibilities for obtaining long records of general validity. Detailed comparative studies of many constituents on the same annual layer and at more locations would add to the number of atmospheric trace substances covered by such records.

REFERENCES

(1) Berner, W.; Stauffer, B.; and Oeschger, H. 1978. Past atmospheric composition and climate, gas parameters measured on ice cores. Nature 276: 53-55.

(2) Dansgaard, W. 1981. Ice core studies: dating the past to find the future. Nature 290: 360-361.

(3) Delmas, R.J.; Ascencio, J.M.; and Legrand, M. 1980. Polar ice evidence that atmospheric CO_2 20,000 yr B.P. was 50% of present. Nature, 284: 155-157.

(4) Hammer, C.U. 1980. Acidity of polar ice cores in relation to absolute dating, past volcanism, and radio-echoes. J. Glaciol. 25 (93): 359-372.

(5) Hammer, C.U.; Clausen, H.B.; Dansgaard, W.; Gundestrup, N.; Johnsen, S.J.; and Reeh, N. 1978. Dating of Greenland ice cores by flow models, isotopes, volcanic debris, and continental dust. J. Glaciol. 20(82): 3-26.

(6) Junge, C.E. 1975. Processes responsible for the trace content in precipitation. IAHS-AISH publication No. 118: 63-77.

(7) Neftel, A.; Oeschger, H.; Schwander, J.; Stauffer, B.; and Zumbrunn, R. 1982. Ice core sample measurements give atmospheric CO_2 content during the past 40,000 yr. Nature 295: 220-223.

(8) Raynaud, D., and Lebel, B. 1979. Total gas content and surface elevation of polar ice sheets. Nature 281: 289-291.

Atmospheric Chemistry, ed. E.D. Goldberg, pp. 135-157. Dahlem Konferenzen 1982.
Berlin, Heidelberg, New York: Springer-Verlag.

Lake and Wetland Sediments As Records of Past Atmospheric Composition

H. E. Wright, Jr.
Limnological Research Center, University of Minnesota
Minneapolis, MN 55455, USA

Abstract. The sediments of a lake contain abundant information about past environments, not only of the lake itself but of the surrounding terrain, especially when their various microfossil and chemical components are analyzed in concert and when stratigraphic sequences are dated by appropriate methods. The peat of raised bogs in some respects is superior as a stratigraphic record of atmospheric composition, for it is not influenced by slope wash from the adjacent terrain. Most of the information from lake or wetland sediments, however, is only indirectly relevant to the subject of changes in atmospheric composition. For example, the pollen sequence in lake and wetland sediments records the history of the vegetation, which is a source for various atmospheric components. The inferred vegetation changes in turn form the basis for reconstruction of past climate, which of course determines the global distribution of atmospheric components. Another constituent of lake sediments with indirect application is charcoal, for fire introduces additional atmospheric components. Several extensive vegetation types are adapted to a relatively high fire frequency, whereas others are not so adapted. Charcoal analysis of lake sediments provides a basis for placing fire frequency in an historical context. The diatom stratigraphy of lake sediments may reflect changes in pH in a lake, and in the proper context with well dated sedimentary sections this may be attributed to the effects of acid precipitation. Similarly, a stratigraphic increase in heavy metals may record industrial air pollution.

INTRODUCTION

The subject of atmospheric composition involves consideration of (a) inputs from a wide variety of terrestrial and oceanic sources, (b) chemical reactions during transport by air masses, and (c) outputs in the form of precipitation and dry deposition. Among the sources of inputs, plants on the landscape provide oxygen, hydrocarbons, and other organic materials by respiration, combustion, and decomposition, and dust is added from areas of sparse plant cover. It is therefore of interest to know the distribution of past vegetation if one wishes to evaluate how inputs from these sources may have changed.

Other inputs can be attributed to human disturbance of the landscape or to direct emissions from modern industrial societies, and it is of interest to know the magnitude of these perturbations to the natural system.

On the output side of the budget, most of the materials that fall out or rain out after stratigraphic transport are carried by slope wash and streams to the sea, where they are mixed and buried beyond recognition. But some are trapped in lake sediments after modest mixing with locally derived terrestrial materials. Thus lake sediments can contain a stratigraphic record of the outputs; they also, through their pollen content, reveal the history of the vegetation and its modification by human influences, and thus they can serve as a guide to some of the inputs.

The stratigraphic record of raised peat bogs can be one step closer to the atmosphere, for in this case little mixing with locally derived materials takes place. In both lake and wetland sediments, the possibility of precise dating of the sedimentary sequence permits the correlation of stratigraphic changes with regional climatic changes as they might affect natural inputs, or with known historical events as they might affect anthropogenic inputs.

THE NATURE OF LAKE SEDIMENTS

Sources

Lake sediments are derived from two sources - internal organic productivity within the lake itself, and the introduction of materials from external sources by inwash from the hill slopes or wind transport from a greater distance (6). Organic production by algal and macrophytic photosynthesis depends largely on nutrient supply. The principal limiting nutrient is phosphorus, which is derived largely from the rocks and soils of the drainage basin, so even this bipartite subdivision between internal and external cannot be rigid. The algae produced may be consumed at higher levels in the food chain, or they may be decomposed directly by bacteria, either in the water column or in the surface sediments, releasing carbon dioxide, methane, and other compounds as well as nutrients for further plant growth. The residue of decomposition is entrapped in the sediments, where further decomposition proceeds at a lessening rate. The role of algae in lake productivity is well established, but the fossil remains of algae in lake sediments are largely restricted to diatoms, which contain siliceous tests that are generally resistant to decomposition, except perhaps when buried in highly acidic or silica-poor sediments. Otherwise, certain algal groups may be identified in the sediments by their distinctive pigments, which are sometimes preserved (10).

The component of lake sediments derived from the drainage basin or beyond is more complex chemically, because it includes both organic and inorganic matter. The organic matter consists in large part of detrital particles such as fragments of leaves and fruits carried into the lake by streams or slope wash, but it also includes pollen grains and charcoal particles that are either washed in or blown in from a greater distance. In the case of lakes bordered by wetlands or by conifer forests with a thick organic soil, the inflowing water may be charged with dissolved humic substances to which iron and manganese may be complexed, giving the lake a characteristic tea color and perhaps a distinctive chemistry.

The particulate inorganic matter entering a lake consists of soil and rock particles eroded from the hill slopes or stream beds, as well as dust particles blown from afar - dust from volcanoes, glacial outwash, deserts, or artificially cleared land. The dissolved major inorganic ions that make up most of the chemical components of lake water are also derived from the drainage basin, except for the portion contributed by atmospheric fall-out or precipitation. The concentration of these ions depends on the susceptibility of the rocks and soil to release the ions, i.e., weathering and ion exchange. It also depends on the residence time of the water on the land surface, in the soil, or in the groundwater, as well as on the proportion of water delivered to the lake via these three pathways, plus direct precipitation on the lake surface itself.

Major Cations

Once in a lake the biologically inactive cations (Na, K, Mg) remain in solution and do not become incorporated into organic matter in significant amounts. Ca plays a special role in carbonate lakes, where it is precipitated in the form of $CaCO_3$ as a result of algal macrophytic photosynthesis. Only in highly saline lakes does the concentration of major cations increase enough to result in strictly chemical precipitation of sulfates and carbonates. Otherwise, stratigraphic cationic analyses of lake sediments usually reveal complacent profiles that merely reflect the composition of mineral particles eroded from the rocks of the drainage basin, rather than chemical processes. Accordingly, they provide little data directly on environmental history. An exception may exist if a volcanic ash with distinctive cationic content intercedes in the stratigraphic succession.

Of course, a gradual increase in the abundance of major cations as a proportion of total sediment may reflect an increase in upland soil erosion in comparison to the production of organic detritus in the drainage basin or in the lake itself (15). Or it may reflect the shallowing of a lake, for in this case mineral sediment can be carried farther out into the basin by inflowing streams.

Biologically Active Elements

The biologically active elements dominate the chemical transformations in a lake, including the nutrient elements (P, N, Si) in addition to C, O, and H. They are incorporated into organic matter and may be recycled many times in the water or at the sediment surface before being buried. Even after burial, bacterial decomposition proceeds and produces methane and other gases. The more mobile constituents, as well as anions such as Cl^- and SO_4^{2-}, may move up or down in the sediment by diffusion along chemical gradients, or by transport with the interstitial water during sediment compaction.

Heavy Metals and Fossils

Heavy metals hold more hope for providing a record of environmental change because they are less mobile and are more likely to be complexed to organic matter or clay minerals (9). But potentially the most useful components of lake sediments are fossils, particularly pollen grains blown into a lake from the vegetation on the surrounding landscape, and diatoms, which are siliceous algae that are sensitive to water chemistry and other environmental characteristics.

Conclusions

It can be concluded that the internally and externally derived chemical inputs to lakes are thoroughly mixed and chemically modified in the lake water and in the surficial sediments so that their sources are not always easily identified, and the contribution of atmospheric gases and many liquid or solid phases may be so thoroughly submerged chemically in the matrix that chemical analysis of the sediments can yield limited direct or indirect information relevant to the history of atmospheric composition. Those particulate materials that resist decomposition may be useful, however, such as volcanic minerals, or pollen grains, diatoms, or other microfossils. Similarly, chemical compounds complexed to organic substances or clay minerals may become fixed in stratigraphic position and not subject to migration after deposition.

RAISED BOGS

Some wetland peats do not hold the same deficiencies as lake sediments if the objective of investigation is the history of atmospheric composition. Specifically, raised bogs are peat accumulations that are not subject to direct inputs from the adjacent terrain, for they receive their moisture, nutrients, and chemical and mineral components solely from the atmosphere. This eliminates the complications involved in the inwash of dissolved or particulate organic or mineral matter derived from the adjacent hill slopes. Because of the acidic nature and limited content of cations in atmospheric inputs, raised bogs are normally acidic (pH about 4.0), and their vegetation is adapted to such a chemical environment.

Raised bogs usually originate on drainage divides, where in-flowing water is minimal, and they involve the rapid build-up of peat under environmental conditions that disfavor bacterial decomposition. A round form may be achieved, resulting in centrifugal flow of surface water and the development of systematic patterns of vegetation growth and in the formation of linear pools and hummocks (7).

Peat does undergo decomposition after deposition, of course, and the migration of interstitial water as well as ionic diffusion may result in the blurring of what may initially have been a stratigraphic record of changes in atmospheric inputs. As with lake sediments, the heavy metals may be complexed to organic matter and preserve the initial stratigraphy, and such particulates as pollen grains and mineral grains (only from the air in this case, not from inwash) can likewise faithfully record past events. Although diatoms may be common in surface peats, they are not preserved at depth, presumably because of the acidic reaction as well as the silica deficiency of peat.

DATING

Radioisotopes

An absolutely essential requirement in the paleoenvironmental study of lake and wetland sediments is the establishment of a satisfactory time scale. This is ordinarily provided by radiocarbon analysis within the time range of the last 40,000 years, or in exceptional cases to 65,000 years. The statistical accuracy of the dates is usually about 1%. In many cases, however, the sample may be contaminated by carbon deficient in ^{14}C. For example, if a lake is located in an area of carbonate rocks, groundwater or surface water may introduce old carbon to the lake, and the organic matter that derives from algal detritus may include this carbon. Such a problem does not exist with the terrestrial plant material of peat. Contamination by deep plant roots may be a problem in peat, however, resulting in dates younger than the matrix of the peat.

The only radioisotope suitable for dating in the time range of a century or so is ^{210}Pb, although for lakes in areas where the vegetation has been disturbed during this period the inflow of lead complexed to organic detritus may distort the stratigraphic record.

Non-isotopic methods for the longer time range depend indirectly on radiocarbon dating. Thus layers of volcanic ash provide excellent time markers, for many of them can be identified precisely by mineral or chemical analysis, and once satisfactorily dated in a few localities they can be used elsewhere for correlation. The same procedure perhaps can apply to the abrupt reduction in hemlock pollen in the northeastern United States and adjacent parts of Canada, attributed to the apparent decimation of the hemlock population resulting from a pathogen of some type (5).

Stratigraphic Time Markers

Apart from these two dating methods - directly by radioactive isotopes or indirectly by correlation of unique catastrophic events that themselves are well dated isotopically - the most practical technique for establishing a time scale is the identification of a stratigraphic feature that can be correlated with an historical event of known date.

Thus in central and eastern North America the increase in pollen of ragweed and other plants of agricultural land clearance or cultivation is a reliable marker for the time as early as 1700 in New England to 1890 in the Dakotas. Other recent pollen markers in America can include the decrease in chestnut pollen in 1925-1950 in the northeastern states, related to the catastrophic chestnut blight, or the decrease in pine pollen in the Great Lakes region, related to timber cutting.

For western Europe the identification of time markers by pollen analysis is generally less precise, because agricultural land clearance started with the Neolithic farmers more than 5000 years ago, and the basis for the time designation is archaeological, in turn controlled by radiocarbon dates, at least for prehistoric time. More recent discrete events may be useful locally, such as close correlation of the Cannabis pollen curve with the required planting of hemp in East Anglia during a particular interval (8), or the planting of pine and beech in eastern Ireland after 1700 (17).

The chemical stratigraphy of lake sediments can provide other time markers. Of particular use are those markers recording global events, such as the products of atmospheric bomb tests in 1954-1963, especially ^{137}Cs. The introduction of lead in gasoline in the 1950s is another discrete event, although the effects may be more pronounced in urban districts or in those areas where automobile traffic is heavy than in areas more remote from sources. The same goes for the development of DDT, PCB, or other hydrocarbons now identifiable in sediments. Local disruptions within the drainage basin may be recorded by clear stratigraphic changes, e.g., introduction of mine tailings, silt inflow from road construction, application of $CuSO_4$ as an algicide, or diatom blooms resulting from phosphate pollution. In most cases the stratigraphic record of the termination of the event is blurred by the fact that the marker in question may be deposited in shallow as well as deep water, and during the years following the termination it may be resuspended in shallow water and then redeposited in deep water.

Furthermore, both the beginning and the end of the event may be blurred by the mixing of bottom sediments by currents and especially by bioturbation.

Varves

Far superior to either isotope dating or to fossil or chemical markers are annually laminated lake sediments (varves), which can simply be counted from the top down (19). Unfortunately these are not common, for they usually depend on a small, deep lake well protected from wind-driven currents by steep sides and a high forest cover. Such conditions inhibit physical movement of the bottom water by currents, as well as deep circulation of the oxygen necessary for populations of most of the bottom fauna. High content of dissolved solids in such lakes may result in chemical stratification, further stabilizing the bottom water and inhibiting oxygen penetration.

The varves themselves result from seasonal cycles of organic activity or perhaps stream inflow. Conditions are most favorable in lakes with a winter ice cover. A cycle might involve photosynthetic carbonate precipitation or diatom blooms in the summer, followed by slow deposition of fine-grained organic detritus during the quiet season of ice cover. Varves provide the ideal situation for dating, for such sediments not only can be correctly dated but they are not subject to blurring of the record by bottom mixing. They even can be used to correct the radiocarbon time scale (20).

POLLEN STRATIGRAPHY

Pollen grains are produced in large quantities by flowering plants, are commonly broadcast widely by the wind, and are preserved in perennially wet sediments, where anoxic conditions can inhibit bacterial decomposition. The assemblage of pollen grains at any stratigraphic level in the sediments thus is a rough reflection of the regional vegetation. The method is tried and true in most temperate regions, but in tropical areas the pollen of many flowering plants is dispersed by insects rather than by the wind, so it is not cast so broadly nor represented so abundantly in wet sediments.

Pollen stratigraphy provides a means of reconstructing regional vegetation history and, through this, climatic history. Its relation to the history of atmospheric composition is indirect in two respects. First, atmospheric composition is affected by the general circulation of the atmosphere, i.e., by the movement of the air masses and storm tracks that distribute atmospheric materials of all sorts from their sources to their sinks, and variations in climatic zones particularly during the last glacial period can be documented by pollen studies in different regions. Second, certain vegetation types provide discrete sources for various hydrocarbons and other atmospheric components, and a shift from tropical rain forest to relatively dry savanna over wide areas could reduce source emissions. In the following sections a few examples are used to indicate the magnitude of vegetational and inferred climatic changes during the last glacial period and subsequent time. In the process, some of the problems in interpretation are pointed out.

North America

On a continental scale, climatic zones are reflected by vegetation. In North America, for example (Fig. 1), the latitudinal vegetation bands of tundra, boreal conifer forest, hardwood forest, and southern pine forest result from the southward gradient of increasing temperature, and the longitudinal bands of prairie and steppe in the western plains reflect increasing distance from the Caribbean moisture source. More specifically, the northern limit of the boreal forest in Canada coincides with the mean frontal position of the arctic air mass in summer, and the southern limit with its position in winter (4). The relative sharpness of these vegetational boundaries reflects the fact that the frontal positions are fairly constant for several months at a time, before shifting to other positions. The wedge of grassland extending eastward from the western plains (the prairie peninsula) records the high incidence of summer drought resulting from the dry Pacific air that loses its moisture in crossing the western mountains.

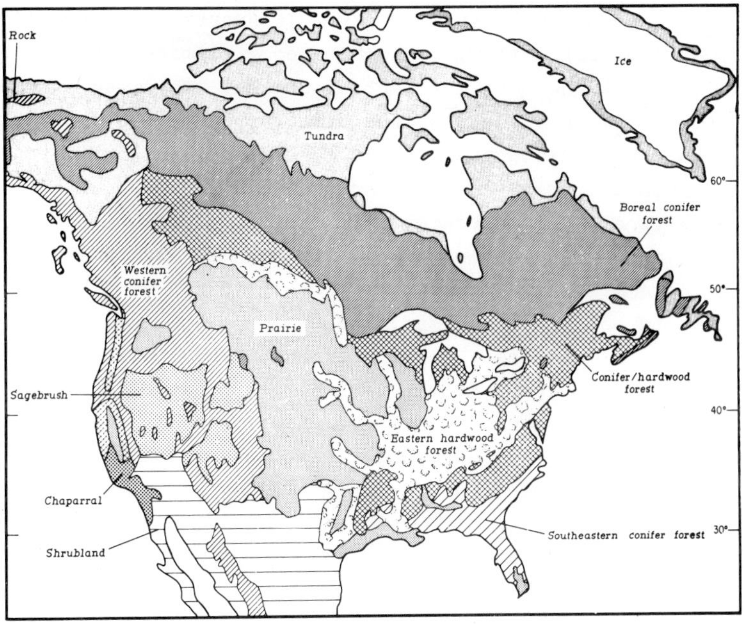

FIG. 1 - Vegetation map of North America.

During the maximum of the last glacial period about 18,000 years ago, these vegetation zones were shifted to lower latitudes and, in the western mountains, to lower altitudes. But in the process they were substantially changed in composition, presumably because some critical climatic parameters such as seasonality were different from those of today. Thus few completely satisfactory modern analogues exist for ice age vegetation. A fringe of tundra-like vegetation existed close to the ice sheet in some sectors, and boreal forest extended to the southern states (25). Extensive prairie apparently did not exist. In the western cordillera, alpine vegetation and forest zones reached far down the mountains, and steppe was restricted to basins to the south.

With the climatic warming and the retreat of the ice sheets, rapid vegetational change ensued (2). The postglacial sequence is illustrated by a pollen diagram from the present pine forest in northwestern Minnesota (Fig. 2). It starts with a dominance of Picea (spruce) pollen, implying a cool climate such as that found today in the boreal forest of Canada. This was succeeded abruptly by Pinus (pine), as the climate became too warm for regeneration of Picea. About 8500 years ago, according to radiocarbon dating, Pinus was replaced by Quercus (oak) as the dominant tree, with enough open landscape to produce the abundant pollen representation of prairie plants. By this time the climate must have been drier than it is today. This climatic trend was then reversed, and by about 4000 years ago Betula (birch) and then Pinus returned and have since dominated the upland landscape. At the same time the extensive plains of poorly drained glacial lakes north of this area became converted to peatlands, and boreal trees such as spruce expanded on the uplands, indicating a trend to cooler climatic conditions.

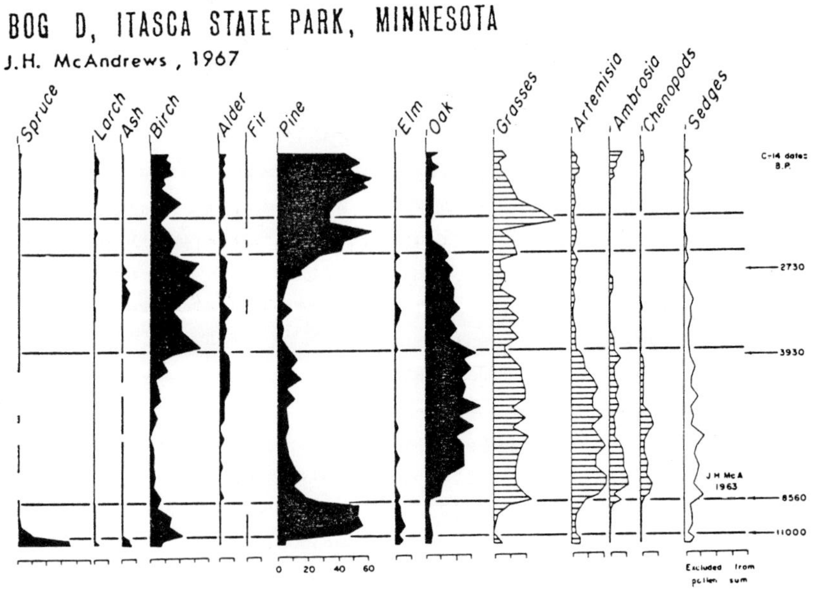

FIG. 2 - Abbreviated pollen diagram for Bog D in northwestern Minnesota. Modified from McAndrews (16).

This sequence of vegetational changes provides about the best documentation for the mid-postglacial period of maximum warmth and dryness that characterized much of North America, culminating there about 7000 years ago (24). The prairie/forest border shifted about 100 km to the east during this interval (23), representing a substantially greater incidence of summer drought at the prairie border, probably accompanied by a mean annual temperature 1-2°C higher than today. Confirmation of this interpretation comes from the evidence for lower lake levels provided by the macrofossils of shallow-water plants from the same stratigraphic levels. These warm, dry climatic conditions are considered by many to constitute the best analogue for the conditions of 2050 A.D., if the predicted doubling of atmospheric CO_2 has a substantial effect on world climate. The cooling trend has continued to today, and in fact there is pollen evidence near the prairie/forest border that the trend accelerated about 400 years ago, correlative with the expansion of glaciers in the western mountains in the Little Ice Age (11).

Evidence for this warm, dry interval elsewhere in North America is somewhat less clear, and the timing is uncertain. In the western mountains the vegetational belts were slightly higher during the first half of postglacial time, implying conditions warmer than today's. In the arctic regions fossil logs dating to this interval have been found in the tundra far north of the present tree line. In Florida most of the abundant lakes of today were dry basins until 8000-5000 years ago, and the contemporaneous vegetation contained many open prairie-like areas because of low rainfall. In the northeastern United States, where the vegetation has been forest throughout postglacial time, the warm, dry interval is not so easily recognized in the pollen sequence, although the increasing percentages of spruce pollen during the last few thousand years imply a subsequent trend to cooler conditions, as does a downward shift in the vegetation belts in the mountains of northern New England.

Europe

In western Europe the paleoclimatic interpretation of pollen diagrams is more difficult in some respects. The early postglacial pollen sequence and thus the vegetation history is complicated by the delayed immigration of different dominant tree types from distant glacial refuges (probably the Balkan peninsula) into the tundra landscape that had prevailed during the glacial period. The generalized pollen diagram from Denmark illustrates the sequence (Fig. 3). Early alternations of herb pollen with birch and pine pollen are closely correlated with ice margin fluctuations in southern Norway, Sweden, and Finland and are believed to represent a temporary reversal in the late-glacial warming trend, lasting about 500 years before 10,500 years ago. Following this event, however, the warming trend resumed, and various deciduous trees arrived in Denmark from the south during the next few thousand years, with uncertain influence by contemporaneous climate. Interpretation of the pollen sequence becomes complicated after about 5000 years ago,

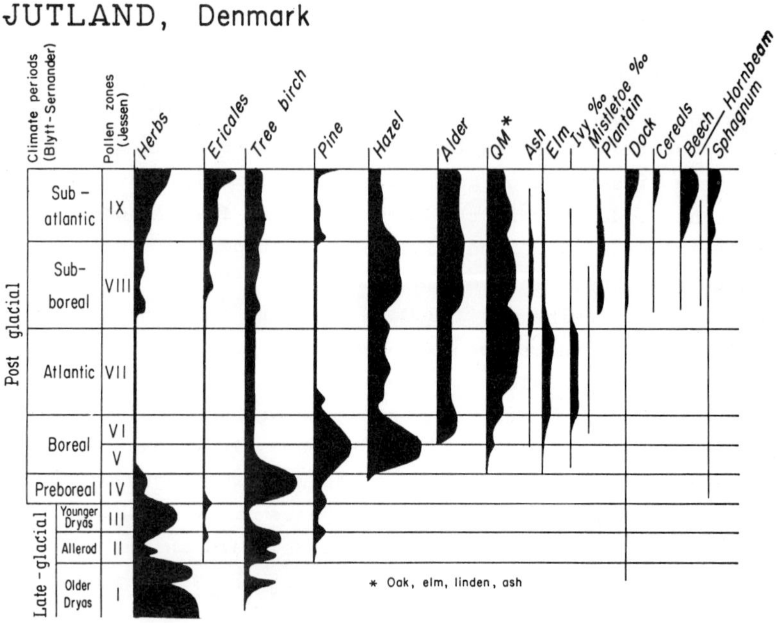

FIG. 3 - Generalized pollen diagram for the late-glacial and postglacial times of western Denmark. From Iversen (13).

for at this time the first Neolithic farmers affected the vegetation, as they cleared some of the forest for grazing and cultivation of grain crops introduced from the Mediterranean region. In some areas of poor soils, particularly in the western part of the British Isles and other humid areas, the forest never recovered, and the landscape is now treeless blanket bog. A cooler, moister climate during the last few thousand years may have contributed to this trend, however.

Other Areas

In the Mediterranean latitudes of southern Europe and the Near East, the vegetation of the last glacial period was characterized largely by shrub steppe rather than forest, and the climate was dry and relatively cool compared to today. Summer as well as winter rains occurred, unlike today, when summers are dry. Following the climatic change at the end of the glacial period about 11,000 years ago, deciduous oak and pine expanded in the area, and the typical Mediterranean vegetation developed by about 5000 years ago, only to be severely degraded later as a result of agricultural disturbance.

In what is now the tropical rain forest of South America, pollen sites are extremely rare, so that little is known of the vegetation cover during the maximum of the last glacial period. Lake Valencia in Venezuela, at an elevation of 400 m, extends back to 13,000 years ago, when the lake was dry. The early vegetation following the dry interval was grassland, and the tropical rain forest apparently did not develop until about 10,000 years ago (3).

The Role of Fire

The gross changes in vegetation recounted above for the glacial period and subsequent time for different parts of the world were clearly controlled primarily by climatic change. Other environmental factors, however, have a significant influence on the local distribution of individual plants and of communities and even major vegetational formations.

Soils, topography, aspect, drainage conditions, nutrient supply, competition, and the action of herbivorous animals are some of the factors. Fire as an ecological factor, however, is of especial interest in the present context, because it potentially leaves a record in lake sediments in the form of charcoal concentrations, thus permitting the determination of fire frequency over a long span of time. Furthermore, combustion of biomass yields an abundant supply of gases and other substances to the atmosphere, and the fire frequency is thus of direct interest in the study of atmospheric composition.

Fire plays a dominant role in the long-range stability of many ecosystems. In the boreal forest of Canada many tree species are adapted to frequent fires, which hold back the competition and permit the regeneration of well adapted species (12). Pinus banksiana (jack pine) is readily consumed by fire, but the cones persisting on the tree are opened by the heat of a fire, and the seeds are broadcast to the burned ground, where the lack of competition permits rapid regrowth of a new pine forest. Pinus resinosa (red pine) is adapted in a different way, for it has thick fire-resistant bark and few low branches, so that it is usually only scarred rather than killed by a fire which cleans out the ground competition and at the same time prepares a clean seedbed for germination of pine seeds. Betula papyrifera (white birch) and Populus tremuloides (aspen) are fire-adapted, because they readily sprout from the roots after fire even though the tree may be killed. Between fires Picea (spruce) and Abies (fir) tend to invade areas of maturing forest stands and gradually succeed them, but they are killed when a fire occurs. The result in the boreal forest is a mosaic of fire-adapted, fire-susceptible, and mixed forest stands, with the distribution of individual components determined by the topography, fire breaks, vagaries of local fuel accumulation, the time since the last fire, and the winds and other weather factors at the time of the fire. The same patterns exist in the mountains of the American West, where certain trees are well adapted to fire, e.g., Pinus contorta (lodgepole pine), Pinus ponderosa, Pseudotsuga menziesii (Douglas fir), and Sequoiadendron giganteum (giant redwood).

Fire frequency in the boreal forest or the conifer forests of the western mountains can be determined from fire scars on surviving trees and from the age classes of trees regenerated after fires. Results in the conifer forest of northern Minnesota show that fire burns the entire area (in segments, of course) once every 100 years or less, depending on how calculations are made, and this figure is probably representative of the frequency in the boreal forest of central and western Canada, where summers are relatively dry and the forests are subject to many lightning strikes. In the boreal forest of Labrador, however, the interval between complete fire coverage may be as long as 500 years, because of the more humid climate.

Stratigraphic counts of charcoal particles in short cores of annually laminated lake sediments in northern Minnesota show a fair correlation of charcoal maxima with fire years as determined from the fire-scar chronology (21). Additional maxima downward through older sediments imply a comparable fire frequency during earlier centuries. Although some of the charcoal may come from direct deposition from smoke, the bluntness of the maxima implies that much of the charcoal may be washed into the lake from hill slopes for some years after a fire, or it may be resuspended from shallow water and redeposited at greater depths. Correlation of thick annual layers with charcoal maxima also suggests greater slope erosion after fires. Evidence is lacking for significant nutrient enrichment in streams and lakes after a forest fire, however (26).

Regardless of the mechanism of charcoal deposition in lake sediments, it is apparent from pollen diagrams in these conifer forests that the mosaic of fire-adapted and fire-susceptible trees has persisted for many millennia under the influence of repeated fires. It has not yet been demonstrated from charcoal stratigraphy that fire frequency was greater or less when the regional vegetation was substantially different (e.g., during the time of the late-glacial Picea forest), although such a presumption is reasonable.

The evidence for frequent natural fires is good in most of the boreal forests of North America, the conifer forests of the western mountains, the southern pine forests of the southeastern United States, and the prairies of the Great Plains, and charcoal in lake sediments in these areas is relatively common. On the other hand, the fire frequency in the deciduous forests of the eastern United States is probably quite low, and accordingly the charcoal content of lake sediments is also low. The dominant trees in these forests are not in any way adapted to fire, and the regeneration of forest stands results from other types of disturbance (e.g., windthrow). The same applies to the deciduous forests of western Europe, and even the Scandinavian forests of Pinus, Picea, and Betula may have a low fire frequency.

For the role of fire in other forest ecosystems, most is known about the eucalypt forests and shrubland of eastern Australia. Many species of Eucalyptus are precisely adapted to fire in different ways - persistence of fruits on the tree until heated, sprouting of twigs directly from burned trunks after fire, or sprouting of new stems from the root collar. Abundance of charcoal in lake sediments has been used there to show the long-range association of fire with Eucalyptus forests, compared to earlier times when tropical rain forest dominated in the absence of fire.

Interest in recent changes in the CO_2 content of the atmosphere and its implications for the global carbon budget have prompted speculation on the role of fire. In many areas of fire-sensitive vegetation, effective policies of fire suppression have greatly reduced the contributions of fire to atmospheric composition, as documented by a decrease in the influx of charcoal to lake sediments, for example, in the conifer forest in northern

Minnesota (21). In the tropical rain forest of Brazil, however, accelerated land clearance in recent decades is said to have the reverse effect, although no stratigraphic studies have been made to support the statement. In a study of varved lake sediments in southern Finland, however, charcoal first became abundant at levels correlated with pollen and archaeological evidence for land clearance, presumably by slash-and-burn practices (22).

CHEMICAL STRATIGRAPHY

Chemical analyses of lake sediments have been used to reconstruct past conditions not only in lakes but in the drainage basins as well, in conjunction with the pollen record of vegetation (15,18). Complication in interpretations arise from the varied pathways of different ions from atmospheric precipitation and rock weathering to lake and sediment interaction (6), and some of these have been mentioned in the early pages of this review. A maximum in stratigraphic distribution of halogens and boron in lake sediments in the English Lake District was interpreted by Mackereth (15) as marking a time of more oceanic climate, for these elements are more abundant in precipitation in areas close to the sea. Otherwise, virtually the only elements that may have direct source in the atmosphere and may leave a stratigraphic record with few complications are heavy metals, for they tend to be complexed to organic matter and clay minerals and do not move stratigraphically once deposited. Thus an upward increase in the influx of heavy metals in a dated stratigraphic sequence can be used to document the local history of industrial air pollution. Such investigations have been made for lake sediments in Lake Erie (14) and for peat in Denmark (1). An extensive investigation specifically of raised bogs, in which drainage-basin inputs are not involved, is being undertaken by Gorham and others (University of Minnesota) in a transect from Labrador and Newfoundland on the east to Minnesota in the mid-continent, to examine recent changes in the flux of heavy metals as well as synthetic organic compounds and other pollutants. Dating will be by ^{210}Pb, ^{137}Cs, and pollen analysis. Some long cores will be analyzed to see if the natural influx of chemical components has changed in concert with vegetational and climatic changes as inferred from the pollen stratigraphy.

DIATOM STUDIES

Stratigraphic diatom studies have been useful in documenting the occurrence of phosphate pollution in lakes in instances where the appearance of indicators of high nutrient requirements, such as Asterionella formosa or Stephanodiscus hantzschii, comes at a time of nutrient enrichment in a lake. Of more direct relevance to the history of atmospheric composition is the increase in Tabellaria binalis in a lake in southern Norway in which the historically documented increase in acidity is attributed to acidic precipitation during the last few decades. Dating in this case was accomplished by ^{137}Cs analysis.

CONCLUSIONS

Of all the components of lake sediments subject to analytical discrimination, pollen grains are the most extensively studied, primarily for the purpose of reconstructing the regional vegetational and climatic history. The relevance of pollen analysis to changes in the composition of the atmosphere, however, is only indirect - first, in that vegetation of different types releases certain chemicals to the atmosphere, and second, in that the inferred climates control the global distribution of atmospheric constituents from local or regional sources. Charcoal in sediments has further indirect significance, for fires introduce other substances into the atmosphere. Of other microfossil constituents in lake sediments, diatoms have potentially the greatest application, for many species are sensitive to pH, and in poorly buffered lakes they can record stratigraphically a progressive increase in pH related to acidic precipitation.

Of the chemical constituents in lake sediments, heavy metals that are industrial pollutants may be of greatest interest for atmospheric composition, for the contribution from non-atmospheric sources (i.e., from the rocks of the catchment) is ordinarily minimal. Raised peat bogs have the clearest records, for they are supplied only from the atmosphere, not from the hill slopes.

Acknowledgements. Research on which this paper is based is currently supported by the US National Science Foundation (Grants ATM81-12840, DPP-81-00124) and US Department of Energy (Contract DE-AC02-79EV). Contribution 248, Limnological Research Center, University of Minnesota.

REFERENCES

(1) Aaby, B.; Jacobsen, J.; and Jacobsen, O.S. 1978. Pb-210 dating and lead deposition in the ombrotrophic peat bog, Draved Mose, Denmark. Danmarks Geologiske Undersøgelse, Arbog 1978: 45-68.

(2) Amundson, D.C., and Wright, Jr., H.E. 1979. Forest changes in Minnesota at the end of the Pleistocene. Ecol. Monog. 49: 1-16.

(3) Bradbury, J.P.; Leyden, B.; Salgado-Labouriau, M.; Lewis, Jr., W.M.; Schubert, C.; Binford, M.W.; Frey, D.G.; Whitehead, D.R.; and Weibezahn, F.H. 1981. Late Quaternary environmental history of Lake Valencia, Venezuela. Science 214: 1299-1305.

(4) Bryson, R.A. 1966. Air masses, streamlines, and the boreal forest. Geogr. Bull. 8: 228-269.

(5) Davis, M.B. 1981. Outbreaks of forest pathogens in Quaternary history. IV International Palynology Conference, Lucknow (1976-1977), vol. 3, pp. 216-227.

(6) Engstrom, E.R., and Wright, Jr., H.E. 1982. Chemical stratigraphy of lake sediments as a record of environmental change. In Studies in Palaeolimnology and Palaeoecology. Essays in Honour of Winifred Pennington, eds. H.J.B. Birks and E.Y. Haworth. Leicester: University of Leicester Press, in press.

(7) Glaser, P.H.; Wheeler, G.A.; Gorham, E.; and Wright, Jr., H.E. 1981. The patterned mires of Red Lake peatland, northern Minnesota: vegetation, water chemistry and landforms. J. Ecol. 69: 575-599.

(8) Godwin, H. 1967. Pollen-analytic evidence for the cultivation of Cannabis in England. Rev. Palaeobot. Palynol. 4: 71-80.

(9) Goldberg, E.D.; Hodge, V.F.; Griffin, J.J.; and Koide, M. 1981. Impact of fossil fuel combustion on the sediments of Lake Michigan. Env. Sci. Tech. 15: 466-471.

(10) Gorham, E., and Sanger, J.E. 1976. Fossilized pigments as stratigraphic indicators of cultural eutrophication in Shagawa Lake, northeastern Minnesota. Geol. Soc. Am. Bull. 87: 1638-1642.

(11) Grimm, E.C. 1981. An ecological and paleoecological study of the vegetation of the Big Woods region of Minnesota. Ph.D. Thesis, University of Minnesota, Minneapolis.

(12) Heinselman, M.L. 1981. Fire and succession in the conifer forests of northern North America. In Forest Succession: Concepts and Applications, eds. D.C. West, H.H. Shugart, and D.B. Botkin, pp. 374-405. New York: Springer-Verlag.

(13) Iversen, J. 1941. Land occupation in Denmark's Stone Age. Danmarks Geologiske Undersøgelse II 66: 20-68.

(14) Kemp, A.L.W.; Thomas, R.L.; Dell, C.I.; and Jaquet, J.-M. 1976. Cultural impact on the geochemistry of sediments in Lake Erie. J. Fish. Res. Board Can. 33: 440-462.

(15) Mackereth, F.J.H. 1966. Some chemical observations on post-glacial lake sediments. Phil. Trans. Roy. Soc. Lond. B 250: 165-213.

(16) McAndrews, J.H. 1967. Pollen analysis and vegetational history of the Itasca region, Minnesota. In Quaternary Paleoecology, eds. E.J. Cushing and H.E. Wright, Jr., pp. 218-236. Minneapolis, MN: University of Minnesota Press.

(17) Mitchell, G.F. 1965. Littleton Bog, Tipperary: an Irish agricultural record. J. Roy. Soc. Antiq. Ire. 95: 121-132.

(18) Pennington, W.; Haworth, E.Y.; Bonny, A.P.; and Lishman, J.P. 1972. Lake sediments in northern Scotland. Phil. Trans. Roy. Soc. Lond. B 264: 191-294.

(19) Saarnisto, M. 1979. Applications of annually laminated lake sediments: a review. Acta Universitatis Ouluensis, Series A, 82, Geologica 3: 97-108.

(20) Stuiver, M. 1971. Evidence for the variation of atmospheric C^{14} content in the late Quaternary. In The Late Cenozoic Glacial Ages, ed. K.K. Turekian, pp. 57-70. New Haven, CT: Yale University Press.

(21) Swain, A.M. 1973. A history of fire and vegetation in northwestern Minnesota as recorded in lake sediments. Quatern. Res. 3: 383-396.

(22) Tolonen, M. 1978. Paleoecology of annually laminated sediments in Lake Ahuenainem, S. Finland. I. Pollen and charcoal analyses, and their relation to human impact. Ann. Botan. Fenn. 15: 209-222.

(23) Webb, III, T.; Cushing, E.J.; and Wright, Jr., H.E. 1982. Holocene changes in the vegetation of the Midwest. In Late Quaternary Environments of the United States: The Holocene, ed. H.E. Wright, Jr., vol. 1. Minneapolis, MN: University of Minnesota Press, in press.

(24) Wright, Jr., H.E. 1976. The dynamic nature of Holocene vegetation. A problem in paleoclimatology, biogeography, and stratigraphic nomenclature. Quatern. Res. 6: 581-596.

(25) Wright, Jr., H.E. 1981. Vegetation east of the Rocky Mountains 18,000 years ago. Quatern. Res. 15: 113-125.

(26) Wright, R.F. 1976. The impact of forest fire on the nutrient influxes to small lakes in northeastern Minnesota. Ecology 57: 649-663.

Atmospheric Chemistry, ed. E.D. Goldberg, pp. 159-179. Dahlem Konferenzen 1982.
Berlin, Heidelberg, New York: Springer-Verlag.

The History of the Atmosphere As Recorded by Carbon Isotopes

M. Stuiver
Quaternary Isotope Laboratory, University of Washington
Seattle, WA 98195, USA

Abstract. A 9000-year-long record of atmospheric $^{14}C/^{12}C$ isotope ratio change is given and interpreted in terms of solar modulation and geomagnetic field intensity changes. Upper limits for the $^{13}C/^{12}C$ isotope ratio changes in atmospheric CO_2 of the A.D. era are derived and related to changes in exchange rate and size of carbon reservoirs. The ^{13}C variability added by the trees' photosynthetic cycle is discussed, and an attempt is made to separate the global ^{13}C signal from the internal ^{13}C variability through normalization on constant assimilation rates.

INTRODUCTION

The current measurements of the concentration and composition of carbon compounds in the various terrestrial carbon reservoirs yield information on the redistribution of carbon in these reservoirs by man during the 20th century. For atmospheric CO_2 there is now a detailed record available showing CO_2 increases from 315 ppm to 339 ppm over the 1958-1980 interval (1). From the CO_2 concentration of the air occluded in bubbles of ice core samples, paleo-CO_2 concentrations of 271 ± 9 ppm are inferred for the later part of the Holocene, with CO_2 concentrations dropping to the 200 to 250 ppm level during the last ice age (17). These paleoatmospheric samples are derived from a mixture of air accumulated during many years, perhaps even centuries. A detailed history of yearly or decadal atmospheric CO_2 change cannot be obtained in this manner.

The history of the $^{14}C/^{12}C$ and $^{13}C/^{12}C$ "labels" of atmospheric CO_2 provides additional information on past changes in atmospheric CO_2 content and carbon exchange between the carbon reservoirs. Different information is derived from both isotopes because of their different production mode. Whereas the stable ^{13}C and ^{12}C are of primordial origin, any ^{14}C accreted during the time of formation of the earth has long since disappeared due to the relatively short half-life of 5730 years. The natural ^{14}C currently encountered in our reservoirs is produced by cosmic ray-generated neutrons that interact with atmospheric nitrogen. The ^{14}C, after oxidation to $^{14}CO_2$, mixes with atmospheric CO_2 and labels this compound with a $^{14}C/^{12}C$ ratio that is approximately 10^{-12}. The oceanic and biospheric carbon reservoirs obtain similar $^{14}C/^{12}C$ ratios through carbon exchange with atmospheric CO_2.

The history of the $^{13}C/^{12}C$ isotope ratio of atmospheric carbon relates to changes in the size of, and exchange rate between, the various terrestrial carbon reservoirs. The $^{14}C/^{12}C$ history is tied to these aspects of the carbon cycle in a similar manner, but it also gives information on geophysical and solar parameters that influence the cosmic ray fluxes interacting with the earth's atmosphere.

RECORDERS OF ISOTOPE CHANGE

Paleoatmospheric samples from ice cores undisturbed by melting can provide the most direct information on atmospheric CO_2 isotope ratios. However, accurate dating of the samples used for the determination of an isotope record is of critical importance. The precise dating of ice core samples appears to be rather complicated. The use of radiocarbon dating has been limited because large quantities of ice are needed for a conventional radiocarbon date (18). Mass spectrometric ^{14}C dating, with its milligram carbon samples, appears to have much promise (25). In regions with sufficient annual snow accumulation other dating methods, such as the counting of seasonal changes in oxygen isotope ratios and/or acidity, come into play (20). Once the difficulties associated with the isotope ratio measurement of small gas samples have been resolved, ice cores from regions with large annual accumulation rates may provide a good historical record of isotope changes of atmospheric CO_2.

Plants are potential recorders of atmospheric CO_2 isotope change because their organic structures acquire $^{14}C/^{12}C$ and $^{13}C/^{12}C$ ratios that reflect the carbon isotope ratios of atmospheric CO_2 during the time of photosynthesis. Trees provide the best history because dendrochronological dating of the rings often yields a record that is accurate to a single year. In addition to reflecting the global change in atmospheric CO_2 isotope ratios, the tree's isotope ratio is also influenced by other variables, such as changes in isotopic composition caused by recycled biospheric CO_2 in the local microenvironment (13). The tree also has a variable discrimination against the heavier carbon isotopes during photosynthesis.

More complex systems of accumulated organic materials, such as lake and marine sediments (21,23), indirectly provide information on the history of isotopic change in the atmosphere. However, for the purposes of this paper it appears advisable to restrict the discussion to the isotope record derived from trees because this record is tied to the isotope changes in atmospheric CO_2 in a less complicated manner.

ANALYTICAL METHODS

The cellulose portion of wood is the preferred material for isotope analysis because other compounds, such as resins, may have been added to the ring after the year of growth. Analysis of tree rings for $^{13}C/^{12}C$ ratios is accomplished by oxidation of carbon in cellulose to carbon dioxide, which is subsequently analyzed on a mass spectrometer. Three basic types of combustion are possible: a) flow-through, b) circulation, and c) closed-tube. Experience with both the flow-through (16) and closed combustion methods (5) in the Quaternary Isotope Laboratory leaves us with a preference for the closed combustion method because appreciably higher precision can be obtained. The precision was 0.12 and 0.11 per mil for closed tube CO_2 prepared from, respectively, Sigma Chemical Company cellulose (number of samples, n = 122) and Sitka Spruce Cellulose (n = 43) vs. 0.20 per mil (n = 44) for Sigma Chemical Cellulose combusted in a flow-through system.

The $^{13}C/^{12}C$ isotope ratios are expressed as $\delta^{13}C$, which is the relative deviation in ^{13}C content from the PDB standard according to

$$\delta^{13}C = \left[\frac{(^{13}C/^{12}C)_{sample} - (^{13}C/^{12}C)_{PDB}}{(^{13}C/^{12}C)_{PDB}}\right] 1000 \text{ per mil.}$$

For the determination of the ^{14}C activity, the sample is combusted to CO_2 in a flow-through or closed-tube system, and the radioactivity of the CO_2 gas, or any other compound derived from the CO_2 gas such as methane or benzene is counted. The measured ^{14}C activity is corrected back to 1950 A.D. for the decay that occurred since the cellulose was incorporated in the tree's cell walls (^{14}C half-life of 5730 years). Isotope fractionation is accounted for by normalizing on a fixed $\delta^{13}C$ ratio of -25 per mil (29). After normalization, the relative difference between age-corrected sample activity (A_s) and the National Bureau of Standards oxalic acid activity (0.95 times this activity = A_{ox}) is expressed as a $\Delta^{14}C$ value (28):

$$\Delta^{14}C = \left[\frac{A_s - A_{ox}}{A_{ox}}\right] 1000 \text{ per mil.}$$

The precision of a ^{14}C determination depends on the counting time and amount of sample counted. Whereas a typical mass spectrometric ^{13}C determination on a 10 mg cellulose sample can be done with an 0.1 per mil precision (see above), the precision obtained for ^{14}C counting is much less. The highest precision obtained in the Quaternary Isotope Laboratory for four days of counting is about 1.5 per mil for 6 gram carbon samples. Smaller samples can also be used but at a loss in precision.

ENVIRONMENTAL FACTORS INFLUENCING THE ^{13}C RECORD OF TREES

For the interpretation of the ^{13}C record an understanding of the mechanisms that induce ^{13}C variability is required. This understanding is of critical importance because substantial disparity exists in published ^{13}C trends of tree cellulose. Many variables, such as precipitation, temperature, light intensity, humidity, and availability of nutrients appear to play a role.

Urey (34) first suggested that the ^{13}C content of plant material may be related to temperature. Despite substantial research, the link between $^{13}C/^{12}C$ ratio and temperature is uncertain and there appears to be conflicting evidence even on the sign of the temperature coefficient (for instance (12,33,35)). Factors such as precipitation and humidity also play a role in the correlations (15).

Farquhar has recently put forward an attractive theory of carbon isotope discrimination in plants (9). He relates the isotope discrimination to the ratio of intercellular leaf CO_2 concentration and atmospheric CO_2 concentration and shows that the effects of changes in irradiance and temperature on this ratio are in reasonable accord with the $\delta^{13}C$ changes observed. His calculations lead to the following expressions:

$$\delta^{13}C_{leaf} = -11.1 - 27.6\ C_i/C_a, \text{ and} \qquad (1)$$

$$A = g(C_a - C_i), \qquad (2)$$

where C_a and C_i are the mole fractions of CO_2, respectively, in the atmosphere and the intercellular spaces, g the leaf conductance to the diffusion of CO_2, and A is the rate of assimilation. Together the two expressions result in:

$$\delta^{13}C_{leaf} = -a_1 + a_2\ A/gC_a, \qquad (3)$$

where a_1 and a_2 are constants.

The major variables are the rate of assimilation and leaf conductance. As pointed out by Farquhar, factors that influence

A by influencing stomatal aperture often have a similar relative effect on g. On the other hand, g would not change much for conditions where CO_2 assimilation rate is drastically reduced by some factor directly limiting the assimilatory metabolism. In this instance, $\delta^{13}C$ would be proportional to the rate of assimilation and least negative under optimum growth conditions. When the C_i/C_a ratio is independent of C_a, little change in $\delta^{13}C$ can be expected (Eq. 1) due to changes in atmospheric CO_2 content (9). A more detailed discussion of the possible influence of a change in atmospheric CO_2 concentration on $\delta^{13}C$ is given elsewhere (Stuiver, Burk, and Quay, in preparation).

The above "single leaf" equations can also be used to describe the change in $\delta^{13}C$ ratios in more complex plants, such as trees (Stuiver, Burk, and Quay, in preparation). The average annual rate of assimilation per leaf (or needle) times the number of leaves (n) equals the amount of wood produced each year, and thus as a first-order approximation

$$A\ n = h\rho R_a, \qquad (4)$$

where h is the height of the tree, ρ the density, and R_a the surface area of the tree rings in a cross section.

Combining Eqs. 4 and 3 gives

$$\delta = -a_1 + a_2 h\rho R_a/gn\ C_a. \qquad (5)$$

For a mature tree with approximately constant height and limited variation in wood density with aging, the above expression is of the form

$$\delta = -a_1 + a_3\ R_a/gn. \qquad (6)$$

In addition to leaf conductance, g, and the assimilation rate (for which ring area, R_a, is a substitute), we now have to contend with changes in the number of leaves over the lifetime of the tree.

The dependence on n of the relationship between δ and R_a can be evaluated in a simple manner for constant leaf conductance g. In this instance,

$$\frac{d\delta}{dR_a} = \frac{a_3}{g}\left[\frac{1}{n} - \frac{R_a}{n^2}\frac{dn}{dR_a}\right]. \qquad (7)$$

For a constant number of leaves ($dn/dR_a = 0$), the tree behaves as a "single leaf" with the slope equal to a_3/ng. The derivative is positive, and $\delta^{13}C$ values increase with ring area.

A negative slope is obtained when

$$\frac{1}{n} - \frac{R_a}{n^2}\frac{dn}{dR_a} < 0$$

or

$$dn/n > dR_a/R_a.$$

Accordingly, when the relative change in leaf number is larger than the relative change in ring area, the $\delta^{13}C$ values decrease with respect to ring area. A typical example of this mode is a tree that loses half of its leaves due to frost but compensates for this loss by letting each leaf assimilate more so that the ring area is not reduced by the same factor of two.

A particular strength of these theoretical considerations is that specific correlations with environmental parameters (temperature, humidity, light intensity, nutrients) are not needed as long as a record of change of the rate of assimilation (ring area) is available. The rate of assimilation, stomatal conductance, and the number of leaves are critical factors.

The following examples of tree $\delta^{13}C$ change can be visualized:

A. If stomatal aperture causes the rate of assimilation to change, the CO_2 conductance, g, will change also, and little correlation between ring area, R_a, and $\delta^{13}C$ values is expected.

If stomatal conductance for CO_2 is constant, two possibilities exist:

B. For a tree with a constant number of leaves, $\delta^{13}C$ will increase linearly with ring area, R_a (provided the height and density of the tree are constant).

C. For a tree with a variable number of leaves, $\delta^{13}C$ may vary linearly with R_a, but the slope may be either positive or negative. In the latter case, $\delta^{13}C$ values decrease with respect to ring area. The slope is determined by the relative change in leaf number versus the relative change in ring area. A zero slope is also possible, and this case would be indistinguishable from mode A.

Examples of all three modes of ^{13}C fractionation exist in the tree records (Stuiver, Burk, and Quay, in preparation).

The ^{13}C variability can also result from the influence of ^{13}C deficient biospheric CO_2 on the $\delta^{13}C$ ratios of the local atmosphere (13). The respiration of land plants induces an annual $\delta^{13}C$ fluctuation of about 0.5 per mil in atmospheric CO_2, with the atmosphere heavier in the autumn (14). A change in the amount of recycled biospheric CO_2 during early growth (juvenile effect) when the tree increases in height, or when the influence of the canopy on recycled CO_2 varies (canopy effect) may result in $\delta^{13}C$ changes.

THE ^{13}C RECORD

The large-scale combustion of fossil fuel since 1850 A.D. has appreciably influenced the isotope ratios of atmospheric carbon. Release of biospheric CO_2 by man may also have influenced the pre-1850 $\delta^{13}C$ record to some extent. These pre-1850 man-induced changes probably have been small, and a distinction will be made between the pre-1850 "natural" record and the post-1850 anthropogenic record. The natural record is important because only by comparison with this record is it possible to evaluate the anthropogenic component.

An early study of natural variability was made by Craig on whole wood of Sequoia rings (6). Thirty-one samples were measured over the interval from 1072 B.C. to A.D. 1649. These samples indicated constant ^{13}C isotopic composition of atmospheric CO_2 to at least 1 per mil during the 900 B.C. to A.D. 1600 interval.

Freyer and Belacy (11), Francey (10), and Grinsted et al. (12) provide $\delta^{13}C$ records on tree cellulose that extend beyond the anthropogenic interval. These records are given in Fig. 1 relative to the natural average over the pre-1850 time range. There are insufficient data in the original publications to attempt correlations with ring area, and the data presented are not corrected for possible isotope ratio variability induced by assimilation rate changes. The $\delta^{13}C$ records were obtained for, respectively, Scots Pine (Pinus sylvestris) from Northern Sweden, King Billy Pine (Athrotaxis selaginoides) from Tasmania, and Bristlecone Pine (Pinus aristata) from California.

Stuiver, Burk, and Quay (in preparation) give $\delta^{13}C$ records for two Californian Sequoia (Sequoiadendron giganteum) trees (CS and RC), one Californian Bristlecone Pine, and one Alaskan Spruce tree. The $\delta^{13}C$ values of the Sequoia trees correlate strongly

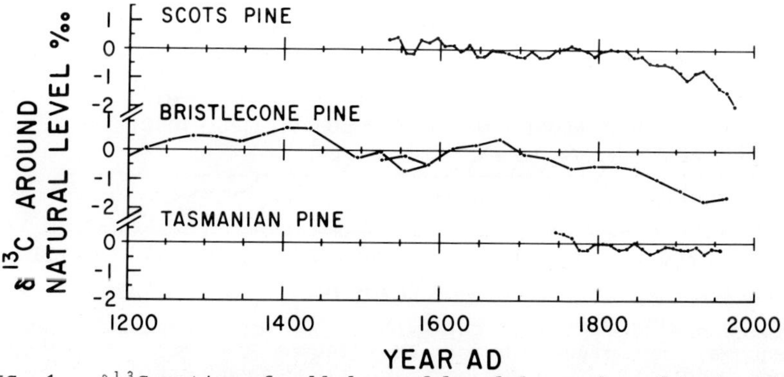

FIG. 1 - $\delta^{13}C$ ratios of cellulose of decadal samples of Scots Pine (11), averaged for five trees between 1780-1969 A.D., of Bristlecone Pine (12) (each sample containing 30 rings), and of decadal samples of King Billy Pine (10), averaged over three trees. The $\delta^{13}C$ ratios are relative deviations from the pre-1850 averages.

(r = 0.8) with ring area according to mode B discussed in the previous section. For Bristlecone Pine, circumferential growth is very irregular. However, ring-width correlation with $\delta^{13}C$ yields a negative slope, as discussed under mode C. The Spruce tree from Sitka does not show any significant correlation with ring area.

The $\delta^{13}C$ values of the above four trees, normalized on constant ring area for the Sequoias and on constant ring width for the Bristlecone Pine, are plotted in Fig. 2 relative to the natural (pre-1850) average $\delta^{13}C$ value. Figure 3 gives the $\delta^{13}C$ deviation from the natural baseline averaged for these trees over the A.D. 200 to 1970 interval.

The variability in the natural $\delta^{13}C$ record is fairly small. The standard deviation, σ, equals 0.31 per mil for the $\delta^{13}C$

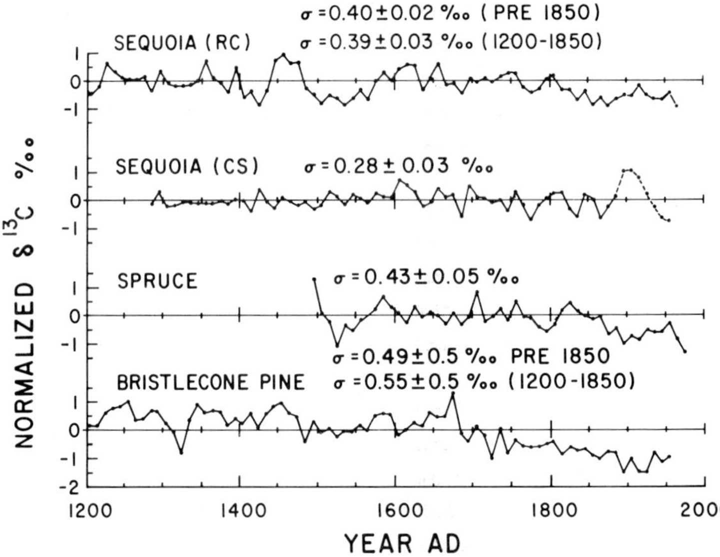

FIG. 2 - $\delta^{13}C$ ratios of cellulose of decadal samples of two Sequoia, one Spruce, and one Bristlecone Pine tree (Stuiver, Burk, and Quay, in preparation). The dashed portion of the CS curve is most likely influenced by nearby log cabin construction activities. Data are residuals obtained after normalizing to constant ring area. The $\delta^{13}C$ ratios are relative deviations from the pre-1850 averages.

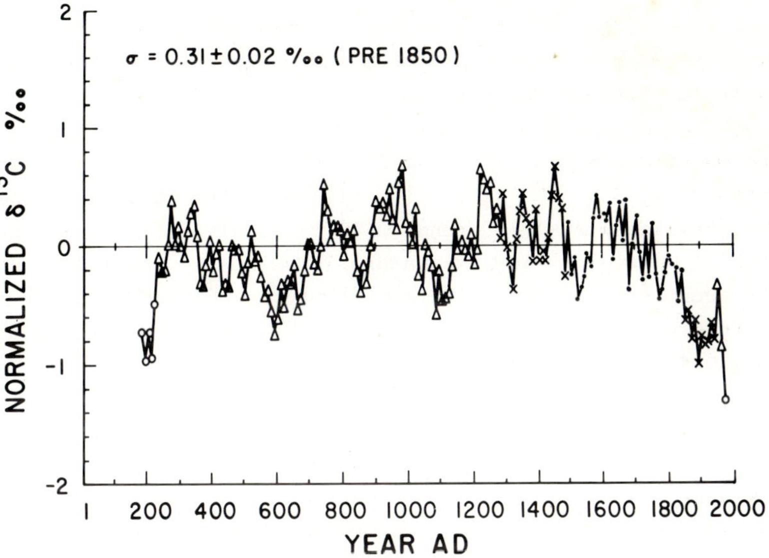

FIG. 3 - Average normalized $\delta^{13}C$ record, relative to the pre-1850 average, for the trees given in Fig. 2. Solid dots give $\delta^{13}C$ values obtained by averaging for four trees, crosses for three trees, triangles for two trees, and open circles for one tree.

values covering a 1700 year interval in Fig. 3. For the A.D. 1300 to 1600 interval of the CS tree this variability is even smaller, with $\sigma = 0.15$ per mil (Fig. 2). This level of variability for decadal samples is close to the measuring precision.

The Scots Pine data in Fig. 1 have a standard deviation, σ, of 0.21 ± 0.03 per mil around the pre-1850 average, whereas the Tasmanian King Billy Pine record has a σ of 0.22 ± 0.5 per mil.

The Sequoia trees (Fig. 2) grew about two kilometers apart, and the $\delta^{13}C$ values of these trees correlate significantly with each other over the 1500 to 1850 interval ($r = 0.57$, $n = 35$). However, the records as a whole do not correlate well with each other (Stuiver, Burk, and Quay, in preparation). Evidently a major component of the $\delta^{13}C$ variability (Fig. 3) is of local origin, and not related to global atmospheric CO_2 changes. The smaller 0.15 per mil variability encountered for the CS Sequoia between A.D. 1300 and 1600 may be indicative of an upper limit for such global changes.

The magnitude of the 19th and 20th century anthropogenic disturbance of the $\delta^{13}C$ record is still a matter of controversy. The detailed post-1750 trends in $\delta^{13}C$, as given in Fig. 4, all show $\delta^{13}C$ levels below the natural baseline by the middle of the twentieth century. The extent of the $\delta^{13}C$ change, however, differs and is in Fig. 4 from top to bottom, respectively, 0.2, 0.3, 0.4, 0.9, 1.3, and 1.7 per mil by 1955. Model calculations show a $\delta^{13}C$ depression due to fossil fuel CO_2 release alone of about 0.4 per mil by the mid-fifties (Stuiver, Burk, and Quay, in preparation). The larger tabulated values would indicate substantial deforestation CO_2 fluxes during the 19th and 20th century, resulting in preindustrial atmospheric CO_2 levels as low as 235 ppm, whereas the smaller values would point towards a twentieth century biospheric sink of CO_2 (24).

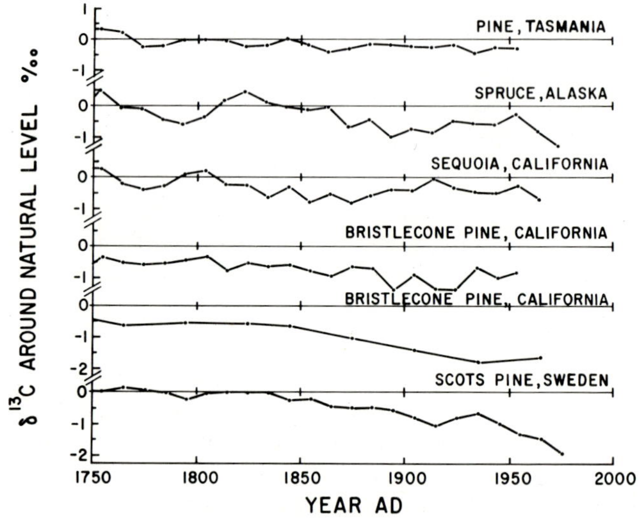

FIG. 4 - Post-1750 A.D. $\delta^{13}C$ changes, relative to the "natural" pre-1850 averages, of the trees listed in Figs. 1 and 2.

The disagreement in recent $\delta^{13}C$ trends perhaps can be reduced if more attention is paid to growth rates (ring area). For instance, there is the possibility that the often large anthropogenic $\delta^{13}C$ reduction in European trees is tied to changes in growth rate that are caused by an environmental response to increased acidity of precipitation.

The cause of the natural ^{13}C variability, for which an upper limit was given previously, is not clear. Global ocean temperature changes may play a role, but probably to a minor extent only.

The equilibrium fractionation of carbon isotopes between gas phase (atmosphere) and dissolved HCO_3^- (ocean) is temperature dependent, with a temperature coefficient of about 0.10 per mil per ^{o}C (16). The 1300-1600 portion of the $\delta^{13}C$ record with $\sigma = 0.15$ per mil would be equivalent to changes in mean ocean temperature of $\pm$ 1.5^{o}C. Such changes are fairly large and unlikely to have occurred during the last millennium. The entire glacial-interglacial mean global temperature change amounts to only 6°C.

A change in the size of the biospheric reservoir also introduces changes in the atmospheric $\delta^{13}C$ record because organic materials ($\delta^{13}C$ ca. - 25 per mil) are deficient in ^{13}C relative to the atmosphere ($\delta^{13}C$ atmosphere ca. - 7 per mil). Tentative results obtained from a box-diffusion carbon reservoir model (19), with slightly modified parameters (30), shows the A.D. 1300-1600 Sequoia CS record to be equivalent to changes in the size of the biosphere of a few percent, resulting in atmospheric CO_2 ppm changes of $\pm$ 5 ppm (Stuiver, Burk, and Quay, in preparation).

THE ^{14}C RECORD

A global record of atmospheric ^{14}C change dating back to nearly 9000 years before the present has been derived from dendrochronologically dated wood. Discrepancies in measurements are negligible within the measuring precision of about 2 per mil when different trees are used (27,30). The good agreement is partially due to the elimination of tree isotope fractionation effects which is achieved by normalizing on a fixed $\delta^{13}C$ value of -25 per mil. For such a normalization, the tree's ^{14}C isotope fractionation is taken as twice the ^{13}C fractionation (28).

^{14}C measurements of dendrochronologically dated wood are available for Bristlecone Pine samples dating back to about 6500 B.C. Results obtained by the La Jolla laboratory (32), smoothed by drawing a spline function through the experimental points, are given in Fig. 5 together with a best-fit sine curve. Quaternary Isotope Laboratory results (Fig. 6) show the more detailed ^{14}C patterns of decadal samples of the current millennium (29).

In the preceding section ^{13}C variability was tentatively attributed to changes in the size of the biosphere (i.e., carbon transfer from biosphere to atmosphere lowers the atmospheric $\delta^{13}C$ ratios). Changes in carbon reservoir size and exchange rates also influence $\Delta^{14}C$ levels. However, the observed magnitude of the ^{14}C variations can only be produced by very large

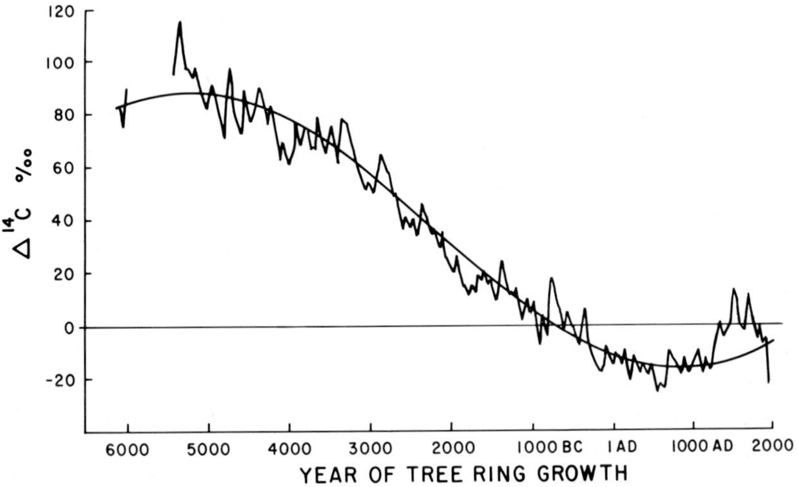

FIG. 5 - $\Delta^{14}C$ changes during the last 8000 years. Figure from Suess (32).

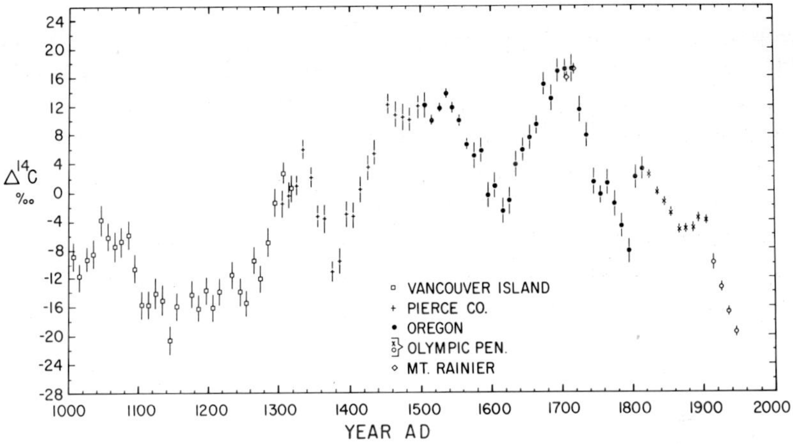

FIG. 6 - $\Delta^{14}C$ decadal changes for the current millennium (29). Part of the post-1850 A.D. decline is caused by fossil fuel CO_2 release. Error bars represent one standard deviation.

changes in reservoir parameters. For instance, the 1700 A.D. perturbation in ^{14}C levels would require a 40-year-long reduction in global oceanic eddy diffusivity (or atmosphere-ocean exchange rate) of 40 percent (29). Such large global changes, with the exception of glacial-interglacial changes, are not very likely. Cosmic ray flux variability appears to be a more likely cause for inducing atmospheric ^{14}C variability.

Part of the incoming galactic cosmic radiation is deflected by the earth's geomagnetic field. The ^{14}C production rate (Q in at/sec cm^2 earth surface) is proportional to $M^{-1/2}$, where M is the magnetic dipole moment. The known variations in Earth geomagnetic intensity (2) suggest that a major portion of the long-term sinusoidal trend (Fig. 5) is caused by geomagnetic field changes.

The modulation of the cosmic ray flux by the sun appears to be responsible for most of the "wiggles" (Fig. 6) that last a few centuries or less (29). The modulation is due to changes in the intensity and distribution of the magnetic field carried by the solar wind in the interplanetary space. Direct observations of neutron fluxes in the atmosphere (available since 1937) yield relationships between the ^{14}C production rates Q and the 11 year sunspot numbers. During intervals of low sunspot numbers, ^{14}C production is high and vice versa. It is possible to calibrate the ^{14}C production rate, as derived from carbon reservoir modeling, with the historical record of sunspot observations (29). Such a comparison shows contemporaneity of the ^{14}C production maximum between 1654 and 1714 and the extended period of low sunspot numbers (the Maunder Minimum) between 1645 and 1715 (8,29).

Application of the relationship between sunspot numbers and ^{14}C production rates leads to a record of solar change that reveals several sunspot minima similar to the Maunder Minimum (29,31).

The substantial 11 year modulation of the cosmic ray flux does not result in measurable 11-year atmospheric ^{14}C variability (Fig. 7) because the attenuation of these rather short cycles is very strong. Spectral analysis of the Fig. 7 annual variability demonstrates the lack of an 11 year cycle in these data.

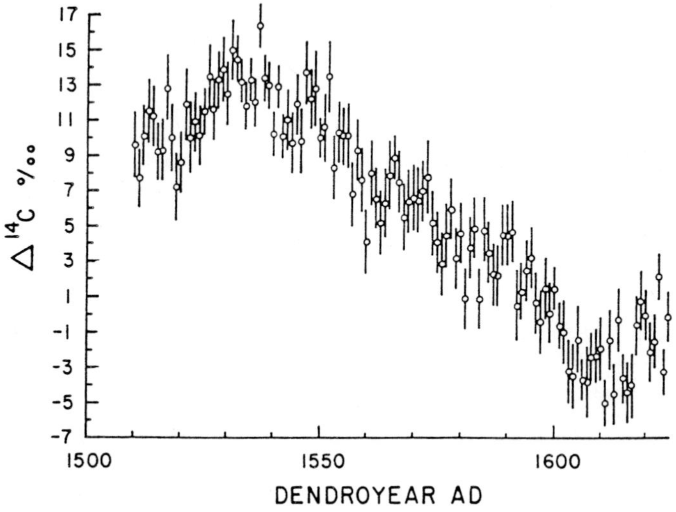

FIG. 7 - Single year $\Delta^{14}C$ variability of a Washington State Douglas Fir. Error bars denote one standard deviation (27).

Through detailed carbon reservoir modeling that takes into account solar modulation, it is possible to calculate the natural ^{14}C levels of the twentieth century (30). Addition of fossil fuel CO_2 fluxes to the model results in a decrease in calculated $\Delta^{14}C$ levels for the first half of the twentieth century (solid line in Fig. 8). The actual $\Delta^{14}C$ change (Suess effect) measured precisely for trees from the state of Washington (30) and from The Netherlands (7) is in excellent agreement with the calculated values (Fig. 8) and confirms the validity of predicting atmospheric CO_2 isotopic trends with the box-diffusion carbon reservoir model.

The long-term record of solar change derived from the atmospheric ^{14}C variations can be compared with records of climate change. Such a study of the ^{14}C production rate record and regional climate changes given by ice core oxygen isotope ratios, winter severity indices, and tree-ring thicknesses shows

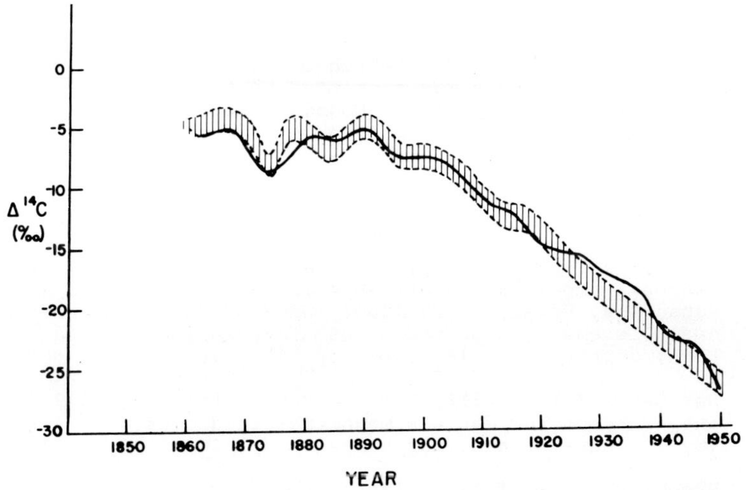

FIG. 8 - Box diffusion carbon reservoir model calculated $\Delta^{14}C$ levels when fossil fuel CO_2 release is included (single line (30)) compared with measured $\Delta^{14}C$ levels of Dutch and German oaks (7). The width of the shaded area represents ± one standard deviation; the experimental data were smoothed with a spline function (7).

little correlation between these variables for the current millenium (26).

Most of the Holocene atmospheric ^{14}C variability can be attributed to solar modulation and geomagnetic change. Holocene climate changes evidently influenced the ^{14}C record to a limited extent only. Such is not the case when approaching the glacial-interglacial boundary. Recent tree-ring work in our laboratory and others (4), and also previously published $\Delta^{14}C$ variations derived from a varve chronology of Lake of the Clouds, Minnesota (22), show atmospheric ^{14}C levels to be 8 to 10 percent above our NBS oxalic acid baseline around 9000 to 10,000 years ago. Part of this increase may be related to changes in oceanic properties that produce changes in atmospheric CO_2 content during glacial episodes (3,17). However, earth geomagnetic field changes also may be responsible for these early changes in ^{14}C levels. A choice between both mechanisms is difficult because global geomagnetic field intensity is not well-defined for the pre-8000 B.P. Period.

Acknowledgements. The ^{14}C research of the Quaternary Isotope Laboratory reported here was supported by the Climate Dynamics Program of the National Science Foundation grant ATM8022240. The ^{13}C research was supported by a Department of Energy contract 79EV10206.

REFERENCES

(1) Bacastow, R.B.; Keeling, C.D.; and Whorf, T.P. 1981. Seasonal amplitude in atmospheric CO_2 concentration at Mauna Loa, Hawaii, 1959-1980. WMO/ICSU/UNEP Abstracts, Analysis and Interpretation of Atmospheric CO_2 Data, Bern, September 14-18, 1981, pp. 169-176.

(2) Barton, C.E.; Merrill, R.T.; and Barbetti, M. 1979. Intensity of the earth's magnetic field over the last 10,000 years. Phys. Earth Plan. Int. 20: 96-110.

(3) Broecker, W.S. 1981. Glacial to interglacial changes in ocean and atmosphere chemistry. In Climatic Variations and Variability: Facts and Theories, ed. A. Berger, pp. 111-121. Dordrecht, The Netherlands: Reidel Publishing Company.

(4) Bruns, M.; Linick, T.W.; and Suess, H.E. 1981. The atmospheric carbon-14 level in the 7th and 8th millenia B.C. Proceedings of a Conference on Carbon-14 and Archaeology, Groningen, The Netherlands, August 1981.

(5) Buchanan, D.L., and Corcoran, B.J. 1959. Sealed tube combustions for the determination of carbon-14 and total carbon. Analyt. Chem. 31: 1635-1638.

(6) Craig, H. 1954. Carbon-13 variations in Sequoia rings and the atmosphere. Science 119: 141-143.

(7) De Jong, A.F.M. 1981. Natural ^{14}C variations. Thesis, University of Groningen, Groningen, The Netherlands.

(8) Eddy, J.A. 1976. The Maunder minimum. Science 192: 1189-1202.

(9) Farquhar, G.D. 1980. Carbon isotope discrimination by plants: effects of carbon dioxide concentration and temperature via the ratio of intercellular and atmospheric CO_2 concentrations. In Carbon Dioxide and Climate: Australian Research, ed. G.I. Pearman, pp. 105-110. Canberra: Australian Academy of Science.

(10) Francey, R.J. 1981. Tasmanian tree rings belie suggested anthropogenic $^{13}C/^{12}C$ trends. Nature 290: 232-235.

(11) Freyer, H.D., and Belacy, N. 1981. $^{13}C/^{12}C$ record in northern hemispheric trees during the past half millennium - anthropogenic impact and climate superpositions. WMO/ISCU/UNEP Abstracts, Analysis and Interpretation of Atmospheric CO_2 Data, Bern, September 14-18, 1981, pp. 209-215.

(12) Grinsted, M.J.; Wilson, A.T.; and Ferguson, C.W. 1979. $^{13}C/^{12}C$ ratio variations in Pinus Longaeva (Bristlecone Pine) cellulose during the last millennium. Earth Plan. Sci. Lett. 42: 251-253.

(13) Keeling, C.D. 1958. The concentration and isotopic abundances of atmospheric carbon dioxide in rural areas. Geochem. Cosmochem. Acta 13: 322-334.

(14) Keeling, C.D.; Mook, W.G.; and Tans, P.P. 1979. Recent trends in the $^{13}C/^{12}C$ ratio of atmospheric carbon dioxide. Nature 277: 121-123.

(15) Mazany, T.; Lerman, J.C.; and Long, A. 1980. Carbon-13 in tree-ring cellulose as an indicator of past climates. Nature 287: 432-435.

(16) Mook, W.G. 1968. Geochemistry of the stable carbon and oxygen isotopes of natural waters in the Netherlands. Thesis, University of Groningen, Groningen, The Netherlands.

(17) Neftel, A.; Oeschger, H.; Schwander, J.; Stauffer, B.; and Zumbrunn, R. 1982. Ice core sample measurements give atmospheric CO_2 content during the past 40,000 years. Nature 295: 220-223.

(18) Oeschger, H.; Alder, B.; and Langway, C.C. 1967. An in situ gas extraction system to radiocarbon date glacier ice. J. Glaciol. 6: 939.

(19) Oeschger, H.; Siegenthaler, U.; Schotterer, U.; and Gugelmann, A. 1975. A box-diffusion model to study the carbon dioxide exchange in nature. Tellus 27: 168-192.

(20) Risbo, T.; Clausen, H.B.; and Rasmussen, K.L. 1981. Supernovae and nitrate in the Greenland Ice Sheet. Nature 294: 637-639.

(21) Schackleton, N.J. 1977. Carbon-13 in Uvigerina: tropical rainforest history and the equatorial Pacific carbonate dissolution cycles. In The Fate of Fossil Fuel CO_2 in the Oceans, eds. N.R. Andersen and A. Malahoff, pp. 401-428. New York: Plenum Press.

(22) Stuiver, M. 1971. Evidence for the variation of atmospheric ^{14}C content in the late Quaternary. In The Late Cenozoic Glacial Ages, ed. K.K. Turekian, pp. 267-306. New Haven: Yale University Press.

(23) Stuiver, M. 1975. Climate versus changes in ^{13}C content of the organic component of lake sediments during the late Quaternary. Quatern. Res. 5: 251-262.

(24) Stuiver, M. 1978. Atmospheric carbon dioxide and carbon reservoir changes. Science 199: 253-258.

(25) Stuiver, M. 1978. Carbon-14 dating: a comparison of beta and ion counting. Science 202: 881-883.

(26) Stuiver, M. 1980. Solar variability and climatic change during the current millennium. Nature 286: 868-871.

(27) Stuiver, M. 1982. A high-precision calibration of the AD radiocarbon timescale. Radiocarb. 24-1: 1-26.

(28) Stuiver, M., and Polach, H.A. 1977. Discussion - reporting of ^{14}C data. Radiocarb. 19: 355-363.

(29) Stuiver, M., and Quay, P.D. 1980. Changes in atmospheric carbon-14 attributed to a variable sun. Science 207: 11-19.

(30) Stuiver, M., and Quay, P.D. 1981. Atmospheric ^{14}C changes resulting from fossil fuel CO_2 release and cosmic ray flux variability. Earth Plan. Sci. Lett. 53: 349-362.

(31) Stuiver, M., and Quay, P.D. 1981. A 1600-year-long record of solar change derived from atmospheric ^{14}C levels. Solar Phys. 74: 479-481.

(32) Suess, H.E. 1980. Radiocarbon geophysics. Endeav. (New Series) 4: 113-117.

(33) Troughton, J.H., and Card, K.A. 1975. Temperature effects on the carbon-isotope ratio of C_3, C_4 and Crassulacean-acid-metabolism (CAM) plants. Planta (Berl.) 123: 185-190.

(34) Urey, H.C. 1947. The thermodynamic properties of isotopic substances. J. Chem. Soc.: 562-581.

(35) Wilson, A.T., and Grinsted, M.J. 1977. $^{12}C/^{13}C$ in cellulose and lignin as palaeothermometers. Nature 265: 133-135.

Group on
Changes in Atmospheric Composition

Standing, left to right:
Claus Hammer, Claude Lorius, Wolfgang Seiler, Detmar Wagenbach, Herb Wright, Jr., Lothar Schütz, Jochen Rudolph, Minze Stuiver, and Bob Garrels.

Seated, left to right:
Ed Goldberg, Tom Wigley, Jim Walker, Steve Schneider, and Jürgen Hahn.

Atmospheric Chemistry, ed. E.D. Goldberg, pp. 181-198. Dahlem Konferenzen 1982.
Berlin, Heidelberg, New York: Springer-Verlag.

Changes in Atmospheric Composition Group Report

J. H. Hahn, Rapporteur
R. M. Garrels, E. D. Goldberg, C. U. Hammer, C. Lorius, J. Rudolph,
S. H. Schneider, L. Schütz, W. Seiler, M. Stuiver, D. Wagenbach,
J. C. G. Walker, T. M. L. Wigley, H. E. Wright, Jr.

INTRODUCTION

There is little doubt that the atmosphere, in the course of about 4.5 billion years of earth history, has been influenced by changes in the lithosphere, hydrosphere, and in the biota. Geological time, i.e., the time span of the geological record, is commonly divided into various intervals as listed in Table 1. Very little is known about the time prior to the beginning of the rock record at about 3.8 billion years BP. As indicated by the 3.8 billion-year-old Isua metasediments in West Greenland, the atmosphere-ocean system was definitely established by that time. The compounds outgassed from the interior of the earth were distributed within this system depending on their solubilities, saturation vapor pressures, and their reaction rates with the material of the earth's crust.

The main constituents of the Archean atmosphere were very likely N_2, CO_2, and water vapor. In addition to the noble gases, there was probably a suite of minor constituents such as H_2, CH_4, CO,

TABLE 1 - Chief divisions of geological time (in millions of years).

Era	Period	Age	
Cenozoic	Quaternary	Holocene	present - 0.010 BP
		Pleistocene	0.010 - 2.0 BP
	Tertiary	Pliocene	2 - 7 BP
		Miocene	7 - 26 BP
		Oligocene	26 - 33 BP
		Eocene	33 - 54 BP
		Paleocene	54 - 65.5 BP
Mesozoic	Cretaceous	65.5 - 137 BP	
	Jurassic	137 - 195 BP	
	Triassic	195 - 225 BP	
Paleozoic	Permian	225 - 280 BP	
	Carboniferous	280 - 345 BP	
	Devonian	345 - 395 BP	
	Silurian	395 - 435 BP	
	Ordovician	435 - 500 BP	
	Cambrian	500 - 600 BP	
Proterozoic	600 - 2,500 BP		
Archean	2,500 - 3,800 BP		

SO_2, H_2S, NH_3, HCl, HF, and others, suggesting that the Archean atmosphere was very likely anaerobic, but with probably only a small reducing capacity (e.g., (20)). At this point a definition of the terms aerobic and anaerobic and oxidizing and reducing with regard to the atmosphere is in order. An atmosphere is aerobic if it contains enough free molecular oxygen to support aerobic respiration as an energy source for organisms. A reasonable estimate for the O_2 mixing ratio in such an atmosphere would be > 1 per mille by volume. An anaerobic atmosphere (O_2 mixing ratios of less than 1 per mille) does not support aerobic life. An oxidizing atmosphere oxidizes iron minerals exposed to it during the course of weathering. This can give rise to the formation of red beds on the continents. Red beds cannot be formed in a reducing atmosphere. Kasting and Walker (11) interpreted an oxidizing atmosphere as containing more oxidizing free radicals than reducing free radicals (e.g., more atomic oxygen, OH, and HO_2 than H). This interpretation assumed that weathering by free radicals dominates in anaerobic atmospheres. Photochemical models suggest that the transition from reducing to oxidizing occurs at O_2 mixing ratios of about 100 parts per trillion by volume (pptv) (11). A highly reducing atmosphere, as preferred by chemical evolutionists, contains more CH_4 than CO_2, more NH_3 than N_2, and more H_2 than H_2O, or, at least, contains H_2, CH_4, and NH_3 in more than trace amounts, i.e., mixing ratios in excess of 100 parts per million by volume (ppmv).

The time period of frequent catastrophic events, i.e., of major bombardment by meteorites, probably ended around 3.9 billion years BP. From then on, major changes in atmospheric composition seem, with few exceptions, to have been primarily the result of long-lasting trends rather than of catastrophies, i.e., such changes occurred on a fairly large time scale (thousands to millions of years) and were thus the more pronounced the longer the trends lasted.

The resolution of time sequences can be a year or less at any time in the geological record where annually layered sediments (varves) are found, even in the Precambrian (e.g., banded iron formations). Other annually laminated materials are the large polar ice sheets and smaller ice caps in more temperate latitudes and rings of trees known for at least the last 9,000 years. Otherwise, time resolution depends largely on the technique of dating. For the last few decades or even centuries, correlation of the stratigraphic record with known historical events (e.g., pollen changes as a consequence of land clearance for agriculture, the introduction of artificial radionuclides from weapons testing) provides local, regional, or even global time markers. Resolution in these cases may be better than

10 years. Dating by the ^{210}Pb method provides a resolution to plus or minus one or so years. For the last 40,000 years, the ^{14}C method usually provides a resolution of about 1%. For the rest of the Quaternary, resolution depends on other isotope dating (K/Ar for volcanic ashes or paleomagnetic stratigraphy, U-series for marine stratigraphy) and may be accurate to 5,000-20,000 years. For the Tertiary and earlier periods, various isotope dating methods are possible with a resolution of perhaps 2% (e.g., 10^5-10^6 years for the Tertiary, 10^6-10^7 for the Cambrian, and 10^7-10^8 for the Precambrian).

Except for the Holocene and perhaps for the last glacial period, the evidence for changes in atmospheric composition which may be extracted from existing records is largely indirect (proxy records) and requires model assumptions for interpretation. Both climate and atmospheric composition appear to have varied greatly over the geological past. These changes are closely coupled to variations of the biota. Because of the strong interactions between atmospheric composition, climate, and life, knowledge of the changes of all three may be important to the process of reconstructing any one of them. Although our emphasis here is on paleoatmospheric compositions, we recognize the comparable importance of concurrent changes in climate and in the biota. The following discussion begins with potential changes in atmospheric composition during the Holocene and goes from the present backwards in time on an increasingly larger time scale. It is mainly concerned with the geological record.

RECORDS OF CHANGES IN ATMOSPHERIC COMPOSITION DURING THE QUATERNARY

For the time period of the Quaternary, especially for the late Quaternary, there are a number of sources of information on potential changes in atmospheric composition. In the following, we will discuss three principal ones, namely, a) ice cores obtained by drilling into the polar ice sheets and into ice caps at high elevations in more temperate latitudes, b) marine and lake sediments, and c) rings of trees.

a) Ice Cores

Ice cores from the polar ice sheets in both hemispheres are most valuable records for potential changes in atmospheric composition. The information contained in these ice cores is more or less direct, as atmospheric air is trapped in the form of gas bubbles when the deposited snow turns into ice. Ice traps not only gas-phase compounds but also airborne particles. The pathways of transport and the alteration of the various materials found in ice cores are by no means clear. Except for a few surrounding sources, there are usually transport distances of several thousand kilometers between major source areas of crustal and anthropogenic aerosol particles and the polar ice sheet regions. For this reason and because of the vertical extension of the ice sheets of up to several thousand meters, only a relatively small fraction of the bulk of the airborne particulate matter is deposited in the ice. Large volcanic eruptions reaching the stratosphere (e.g., the Laki eruption on Iceland in 1783 or the Krakatoa eruption in 1883) provide more favorable conditions for long-range transport of airborne particulate matter. Therefore, such eruptions are well documented in the polar ice sheets. Since Hammer's paper (this volume) deals with the polar ice sheets and the problems of their analysis, only a few important points will be highlighted here.

It looks as if, at favorable locations, the Greenland ice record covers at least the last 100,000 years. The Antarctic ice sheet might even contain ice from several glacial and interglacial periods and thus reach much further back in time. Due to horizontal viscous flow and the associated horizontal movement in the ice sheets, the individual annual ice layers grow increasingly thinner with depth. In South Greenland, an annual layer equivalent to 50 cm of water on the surface is, for example, thinned out to approximately 2 cm after 10,000 years of flow. As a result of this process, the resolution of time sequences in the ice record diminishes with depth. The decrease in time resolution depends to some extent on the substance under consideration. The annual deposition rate of snow in Greenland is generally higher (equivalent to 10-100 cm of water per year) than in arid Antarctica (equivalent to 2-20 cm of water per year). From ice flow considerations as well as from experimental data, it appears that, for the time span of the Holocene, the time resolution of the Greenland ice record is higher, while for ice layers older than about 10,000 years, the time resolution of the Antarctic ice record is higher (except for the ice in Central Greenland). At present, a resolution of time sequences of better than 1 year is possible for the upper part of Greenland ice cores covering the Holocene (accuracy: ± 3 years for dating a 1,000 years period).

Contrary to Antarctica, the Greenland ice turned slightly alkaline during the last ice age as a result of increased amounts of carbonates in the dust incorporated in the ice layers. (In Antarctica, the amounts of continentally-derived dust incorporated in the ice layers are generally lower and have thus less influence on the acidity of the ice.) Therefore, any dating method based on the annual variation of acidity is of no use for the ice layers formed during the last ice age in Greenland. $^{18}O/^{16}O$ dating of ice layers laid down during the last ice age is also not practicable, because the annual layers are too thin in both Greenland and Antarctica. There are other stratigraphic methods for dating such thin ice layers, but the potential of these methods has not been fully evaluated. Presently, the experimental resolution is still insufficient.

Ice cores from areas without any summer melting give the best information on past atmospheric gas compositions. Core samples from locations with some melting during the summer season are still useful, as the often higher rate of snow accumulation at those locations minimizes the error due to partial melting (such as the enrichment of gases like CO_2 in the ice relative to N_2 and O_2) and allows a higher time resolution.

There is an important difference between the resolution of time sequences in the record of gases and in the record of airborne particles in ice cores. While the time resolution of the particulate record is better than one year in Greenland ice cores for more than the last 10,000 years, the time resolution for changes in atmospheric gas composition is, in the worst case, a few hundred years in Greenland and not better than several thousand years in regions such as Central East Antarctica. The cause of this difference is the porosity of the upper 50-100 meters of the ice sheets where the snow has not yet been transformed into ice. In this firnification zone, gases may diffuse more or less freely and equilibrate with the atmosphere as long as there are not too many of the continuous, very thin icy layers which form in intense sunlight (radiation crusts) or during periods of strong winds (wind crusts). It is conceivable that the crusts enhance time resolution, but to what extent is still unknown.

The O_2/N_2 ratio in the gas bubbles found in the ice cores covering the last 100,000 years is, within the error of the analytical method used (accuracy better than 1%), the same as in today's atmosphere. It is thus reasonably safe to conclude that the atmospheric O_2/N_2 ratio has not changed, at least not more than 1%, during the last 100,000 years.

Carbon dioxide is the only trace gas which has been measured so far in gas samples extracted from ice cores (100 cm^3 of ice yield an average of about 10 cm^3 of gas). The CO_2 data obtained (5,14) suggest that during most of the Holocene the atmospheric CO_2 level was somewhere between 260 and 310 ppm by volume. During the last glacial period, however, the atmospheric CO_2 mixing ratio decreased to values between 190 and 230 ppmv close to the end of the ice age. The cause of the drop in atmospheric CO_2 is still a matter of debate. Explanations have been offered by Broecker (3) and by Berger (2). The arguments go as follows: according to Broecker, the falling sea level exposed sediments in shelf areas so that increased weathering and leaching occurred. This, in turn, led to higher phosphate concentrations in seawater. The elevated phosphate levels encouraged primary production of marine biomass and a greater flux of biogenic carbon and carbonate to the deeper ocean. The vertical profile of total carbonate in seawater changed with a lowered CO_2 partial pressure in the surface layer and, as a consequence, the atmospheric CO_2 mixing ratio decreased. According to Berger, when the sea level rose after the last glaciation, reef-building activity increased and more oceanic bicarbonate was converted into solid carbonate, releasing CO_2 to the atmosphere and thus increasing atmospheric CO_2 levels.

These processes might have played a role, but more measurements must be performed to determine more accurately the magnitude of the changes in atmospheric CO_2 during and at the end of the last glaciation. It then remains to be seen whether the processes sketched above can, in fact, account for the magnitude of the changes.

As was mentioned above, the record of changes in atmospheric composition as far as particulate matter is concerned shows much better time resolution than the record of atmospheric gas composition. Snow appears to act as such an efficient scavenger and trap for airborne particles that even individual precipitation events are discernible in the ice records. In areas with high snow accumulation rates, the influence of dry deposition and evaporation or sublimation of snow on the concentration of particulate matter observed in the ice layers is small. Furthermore, there are indications that atmospheric particulate concentrations are to some extent reflected in ice cores. Preindustrial ice layers contain particulate matter from marine and continental sources (e.g., soil dust, volcanic material, pollen, plant debris) and particulate matter formed by gas-to-particle conversion processes in the atmosphere. There is also a small fraction of extraterrestrial material. For the Greenland ice sheet, typical particulate concentrations in Holocene ice from a non-volcanic period are 50 μg dust per kg of ice and 0.3 to 1.2 micro-equivalents per kg of ice for each of the anions NO_3^-, Cl^-, and SO_4^{2-}, and for each of the cations H^+, NH_4^+, and Na^+. The sum of Mg^{2+}, Ca^{2+}, and K^+ accounts for not more than 0.1 micro-equivalents per kg of ice. In Central Antarctic ice, the concentrations are generally lower due to the longer distances to the source areas.

During a long period in the last ice age, the amounts of dust trapped in the layers of the Greenland ice sheet were 3 to 70 times higher than during the Holocene. A significant increase (5-10 times the Holocene concentrations) has also been observed in Antarctica. This increase has been interpreted as an effect of climatic changes and changes in atmospheric circulation patterns. Indeed, in both ice sheets, the concentrations of all particulates, both soluble and insoluble in water, show strong correlations with $^{18}O/^{16}O$ ratios in the ice. In ice cores from periods during which the amount of alkaline material (carbonates) was not too large in the dust scavenged, volcanic eruptions can be detected by measuring acidity profiles. Many major volcanic eruptions which noticeably perturbed the chemical composition of precipitation for up to a few years after the eruptions can be identified in ice cores from the polar ice sheets.

Some of the man-made changes in the composition of atmospheric particulate matter are also recorded in the polar ice sheets. Artificial radioactive fallout from weapons testing is well documented in both ice sheets. Other anthropogenic effects on particulate compositions appear to be obscured by natural fluctuations in Antarctica. In Greenland, man-made changes may be more apparent. But there is a need for measurements of continuous time series covering the last century to clarify this point.

In summary, one may say that the polar ice sheets have great potential for the investigation of past atmospheric compositions. This potential is far from being exhausted by the studies carried out thus far. The value of smaller ice caps at lower latitudes lies in the usually shorter distances to continental, particularly anthropogenic, sources of airborne material. However, their value depends to a large degree on their location (elevation). Major leaching and melting, which are typical for temperate glaciers, can severely cut down on the usefulness of the small ice caps for the study of past atmospheric compositions.

For future activities in the field of ice core studies, one would like to see

1) measurements of other trace gases in addition to CO_2;
2) an examination of the relevance of gas concentrations measured in the gas bubbles trapped in ice for coeval atmospheric concentrations;
3) an examination of the scavenging efficiency of snow for various vapor phase compounds and particles as a function of crystal form and shape of snowflakes (see also Graedel et al., this volume);
4) a determination of the chemical composition of the trapped dust;
5) more detailed measurements of ionic species such as Cl^-, NO_3^-, and SO_4^{2-};
6) an examination of individual particles isolated from ice cores for size, morphology, chemical composition, etc.; and
7) an investigation of the pathways of transport of particulate material from more distant continental areas to Greenland and to Antarctica.

The increase in atmospheric CO_2 which apparently occurred between 15,000 and 10,000 years BP still awaits a satisfactory explanation. The "seasonal" variation of CO_2 in some ice cores cannot be caused by atmospheric fluctuations, since the time resolution in the ice record of atmospheric gas compositions cannot reveal seasonal variations for reasons explained above (see also Hammer, this volume).

In view of the large quantities of sample material obtained from drillings in the ice sheets, analytical techniques should be developed for continuous measurements over large sections of an ice core. Also, improved dating techniques (which could make use of the minute amounts of ^{14}C, ^{10}Be, or ^{36}Cl) are desirable. The linear accelerator technique for ^{14}C dating (17) is one possibility.

b) Lake and Marine Sediments

Lake sediments may contain dust particles, pollen grains, charcoal, trace metals, or other atmospheric components, all transported through the atmosphere either directly to a site or indirectly after a brief residence on the slopes of the drainage basin. These materials occur in a matrix composed largely of unidentified organic and inorganic matter washed in from adjacent hill slopes, as well as particulate organic matter originating within the lake from photosynthetic production and food-chain processes.

The dust from volcanoes may be identified by its distinctive mineralogy or chemistry, but it provides only local information about contemporaneous atmospheric composition.

The pollen grains found in lake sediments reflect local and regional vegetation, and a network of pollen diagrams permits the mapping of past vegetation over broad areas. These may be interpreted in terms of distributions of air masses and storm tracks which in turn control the pathways of transport of atmospheric components from source to sink. Furthermore, specific vegetation types may be sources of certain atmospheric compounds (e.g., hydrocarbons from heavy forests). Also, it is known from both studies of fire ecology and the abundance of charcoal fragments in lake sediments that natural fires are more common in certain vegetation types (e.g., savannas, pine forests, Eucalyptus forests) than in others (e.g., temperate deciduous forests, tropical rain forests), and fires may release gases and aerosol particles to the atmosphere. Stratigraphic analysis of charcoal abundance, when combined with pollen analysis and archaeological data, can suggest the degree to which fire frequency was altered as a result of agricultural practices or other human activities (Wright, this volume).

The stratigraphic record in lake sediments cannot be fully interpreted without a time scale. The main dating method for lake sediments is the ^{14}C method, but varves are also useful. For the time span of the last 100-200 years, ^{210}Pb dating can be utilized, or some marker can be identified that can be firmly correlated with a known historical event. In this category as a global marker is the spike of products of atmospheric tests of nuclear bombs, especially ^{137}Cs and Pu (A.D. 1953-1964). Regionally in central North America, the increase in ragweed pollen is useful (generally 1790 to 1890, depending on the area). Local markers can be identified in many other cases, e.g., mine tailings, $CuSO_4$ algicides, silt inflows from road construction, diatom blooms from phosphate pollution. Once a time scale is established, other stratigraphic changes more closely related to atmospheric composition can be dated and related to anthropogenic events, after the long-term pre-cultural background is defined.

In some cases it may be difficult to determine whether the source of atmospheric pollutants is local or regional. For example, an increase in lead may reflect the development of leaded gasoline by the automobile industry or the increased highway traffic in the local area. Likewise, DDT, PCBs, and other synthetic organic compounds may have either regional or local increases. An effective reflection of pH change attributable to acid precipitation, at least for lakes in poorly buffered areas of quartzitic or granitic bedrock, can come from stratigraphic analyses of the abundance of diatoms which are sensitive to pH.

Complications in lake-sediment stratigraphy that come from components derived ultimately from the rocks of the drainage basin (e.g., trace metals) instead of from the atmosphere can be overcome in the case of raised peat bogs (Hochmoor) which are built up solely under the influence of atmospheric precipitation and the nutrients which come with it.

Stratigraphic analysis of marine sediments has included micropaleontological, isotopic, chemical, and mineralogical studies, but interpretations generally concern variations in ocean temperature and/or salinity, ice-sheet volume, or the magnitude of continental erosion rather than the composition of the atmosphere. Silt influx during the last glacial interval off West Africa has been attributed to dust from a latitudinally-shifted desert zone; although dust is an atmospheric component, the relevance here is mostly to climatic change. The marine sedimentation rate is ordinarily not fast enough for the resolution necessary to detect changes in the atmospheric level of global pollutants such as DDT or the PCBs, largely because of mixing by the bottom fauna. Exceptional opportunities arise in the case of annually laminated sediments at anoxic sites such as in the Santa Barbara basin off southern California (e.g., (9)). Anoxic layers of marine sediments at some other locations, for example, in Buzzards Bay, Massachusetts (e.g., (8)) or in the Baltic Sea (e.g., (18)), have also been found to be useful for the study of the history of air pollution.

c) Tree Rings

Tree rings provide a record which extends to about 9,000 years BP. Dating of tree rings is generally accurate to the year. With the aid of a model that incorporates the carbon exchange between atmosphere, hydrosphere, and the biota, tree rings may be used to estimate the global ^{13}C record and thus past atmospheric CO_2 concentrations. The derivation of the global ^{13}C record is not simple, because the photosynthetic processes do not always provide uniform $^{13}C/^{12}C$ isotope ratios (Stuiver, this volume).

A precise signal for atmospheric ^{14}C can be derived from tree rings and direct information on changes in the relative proportions of atmospheric $^{14}CO_2$, $^{13}CO_2$, and $^{12}CO_2$ with time.

Moisture Records

Variations in atmospheric moisture content in time and space can be inferred from isotope measurements in ice cores, speleothems, and tree rings. Simultaneous measurements of δD and $\delta^{18}O$ values in ice cores show variations in the deuterium excess (i.e., the intercept in a plot of δD against $\delta^{18}O$) which can be interpreted in terms of 1,000 year time scale changes in the mean relative humidity of tropical to mid-latitude regions (13). Speleothem data can be used in the same way. The use of simultaneous δD and $\delta^{18}O$ data from tree ring measurements to reconstruct past variations in relative humidity on a year-to-year time scale also has potential (4).

RECORDS OF CHANGES IN ATMOSPHERIC COMPOSITION DURING EARLIER PERIODS OF EARTH HISTORY

The geological record shows long-term fluctuations of the organic carbon content and the abundances of stable isotopes in the sedimentary rocks of the Phanerozoic (last 600 million years). Furthermore, we have information on changes in tectonic activity, distribution of continents and their position relative to the poles, lithology of sediments, and, last but not least, abundance of life forms. However, we have no direct evidence for major changes in atmospheric composition during the Phanerozoic.

Paleoatmospheric compositions cannot be derived from the geological record without model assumptions, i.e., paleoatmospheric compositions derived from the geological record for a given time period are largely model-dependent. As one goes back in time, the geological record becomes scanty and less certain, so that constraints on paleoatmospheric compositions or climate reconstructions become less firm. Hence, widely varying interpretations of the record are possible. It is imperative, therefore, to recognize the often speculative nature of most inferences of paleoatmospheric composition or climate. However, there are some constraints which are widely accepted. For example, most C-3 plants today could hardly survive if CO_2 mixing ratios in the troposphere dropped below 150 ppmv, since Calvin cycle photosynthesis is then no longer possible (12); mammals and even invertebrates could not exist if atmospheric O_2 levels dropped below 1% by volume; fire frequency would generally increase if atmospheric O_2 increased significantly above the present atmospheric level; and most grasses and many other plants would die, even if they escaped the then frequent fires, if atmospheric O_2 increased to mixing ratios of more than 40% by volume.

Biogeochemical cycle studies and inferences from the geological record (e.g., (6,10)) suggest the possibility that atmospheric CO_2 mixing ratios were different in the past. In addition, climate modeling experiments for mid-Cretaceous times suggest that it is very difficult to explain the warm climates inferred from the geological record for this time period without recourse to a large heating source such as the greenhouse effect of atmospheric CO_2 several times today's level (1). The paleontological record of the Phanerozoic contains several major extinction events. A most spectacular event of mass extinctions occurred at the Cretaceous/Tertiary boundary about 65 million years ago. The effects of this event were obviously most severe among marine organisms. Although the extinctions did not all occur suddenly and simultaneously, nearly half of all genera of the floating, swimming, and bottom-dwelling marine organisms, only 14 percent of freshwater genera, and 20 percent of terrestrial genera became extinct at that time. Among the marine organisms, the calcareous nannoplankton was hit hardest (e.g., (16)). The $\delta^{13}C$ data for the calcareous component of the pelagic sediments above the Cretaceous/Tertiary boundary show a shift towards less ^{13}C on an apparently global scale. This shift can be precisely correlated using the planktic-foraminiferal zones (15). A number of explanations has been offered covering the range from a rather undramatic depletion of nutrients for the phytoplankton (leading to extinctions gradually moving up the food chain) to spectacular catastrophies such as a nearby supernova explosion or the impact of a comet. The discovery of a distinctive geochemical anomaly, an enrichment of the element iridium, at the precise Cretaceous/Tertiary boundary in marine sediments exposed in Italy, Denmark, Spain, the South Atlantic, the North Pacific, Texas, and New Zealand supports the hypothesis of an impact by an extraterrestrial body. No matter what caused the mass extinctions, there must have been significant changes in atmospheric composition which were certainly most dramatic in the case of an impact of a large body. The evidence was recently discussed during a multidisciplinary conference in Snowbird, Utah. The proceedings of this conference will be published by the Geological Society of America (7).

A major change in atmospheric composition of which we can be reasonably sure is the change of the atmosphere from anaerobic to aerobic, which occurred around 2 billion years BP (e.g., (20)). There was very likely a release of free molecular oxygen by photosynthesizing microorganisms long before this time. However, this oxygen was probably consumed in the oxidation of reduced species such as ferrous iron in the oceans. Atmospheric oxygen could not build up until the reservoirs of the reduced species in the oceans and on the continents (minerals such as siderite ($FeCO_3$), greenalite ($(Fe,Mg)_6Si_4O_{10}(OH)_8$), and pyrite ($FeS_2$) exposed to the atmosphere by weathering) were exhausted. It is not clear when the present atmospheric level of oxygen was reached. The buildup can be modelled, and the results are dependent on the model applied.

There are several lines of geochemical evidence suggesting that there has been an irregular but continuous trend for decreasing atmospheric CO_2 levels from the Archean to the present. In addition to the inorganic climate-weathering-atmospheric CO_2 feedback processes (e.g., (19)), life is a significant component in the loop (12). Outstanding problems do, of course, still remain. In this context, it is important to

1) resolve the question of how weathering was related to temperature in the Archean if the earth surface was largely water;
2) perform climate model experiments to estimate how much CO_2 is needed to maintain equable climate for the existence of life under varying assumptions of surface conditions, cloud cover, and solar luminosity;
3) obtain more detailed information on the stable isotope record, particularly for the time period of the Phanerozoic

(the stable isotopes are currently our major tool in acquiring more information on changes in the biogeochemical cycles); and

4) improve the models for the Phanerozoic by a thorough study of the coupling of the different biogeochemical cycles, especially the coupling of the carbon and the phosphorus cycle.

Proxy records - this holds for most of the geological record, but also for tree rings, pollen assemblages, charcoal abundance, etc. - yield information on paleoecological or paleoclimatic changes rather than on paleoatmospheric composition. Therefore, a great deal of imaginative work is needed to find ways to infer atmospheric composition from analyses of the various proxy records.

REFERENCES

(1) Barron, E.J.; Thompson, S.L.; and Schneider, S.H. 1981. An ice-free Cretaceous? Results from climate model simulations. Science 212: 501-508.

(2) Berger, W.H. 1982. Increase of carbon dioxide in the atmosphere during deglaciation: the coral reef hypothesis. Naturwissenschaften 69: 87-88.

(3) Broecker, W.S. 1981. Glacial to interglacial changes in ocean and atmosphere chemistry. In Climatic Variations and Variability: Facts and Theories, ed. A. Berger, pp. 111-121. Dordrecht, The Netherlands: D. Reidel Publishing Co.

(4) Burk, R.L., and Stuiver, M. 1981. Oxygen isotope ratios in trees reflect mean annual temperature and humidity. Science 211: 1417-1419.

(5) Delmas, R.J.; Ascencio, J.-M.; and Legrand, M. 1980. Polar ice evidence that atmospheric CO_2 20,000 years BP was 50% of present. Nature 284: 155-157.

(6) Garrells, R.M., and Lerman, A. 1981. Phanerozoic cycles of sedimentary carbon and sulfur. Proc. Natl. Acad. Sci. USA 78: 4652-4656.

(7) Geological Society of America. 1981. Proceedings of the Conference on Large Body Impacts and Terrestrial Evolution: Geological, Climatological, and Biological Implications, Snowbird, Utah, October 1981. Geological Society of America, in press.

(8) Hites, R.A.; Laflamme, R.E.; and Farrington, J.W. 1977. Sedimentary polycyclic aromatic hydrocarbons: The historical record. Science 198: 829-831.

(9) Hom, W.; Risebrough, R.W.; Soutar, A.; and Young, D. 1974. Deposition of DDE and polychlorinated biphenyls in dated sediments of the Santa Barbara basin. Science 184: 1197-1199.

(10) Junge, C.E.; Schidlowski, M.; and Pietrek, H. 1975. Model calculations for the terrestrial carbon cycle: Carbon isotope geochemistry and evolution of photosynthetic oxygen. J. Geophys. Res. 80: 4542-4552.

(11) Kasting, J.F., and Walker, J.C.G. 1981. Limits on oxygen concentration in the prebiological atmosphere and the rate of abiotic fixation of nitrogen. J. Geophys. Res. 86: 1147-1158.

(12) Lovelock, J.E., and Whitfield, M. 1982. Life span of the biosphere. Nature 296: 561-563.

(13) Merlivat, L., and Jouzel, J. 1979. Global climatic interpretation of the deuterium-oxygen 18 relationship for precipitation. J. Geophys. Res. 84: 5029-5033.

(14) Neftel, A.; Oeschger, H.; Schwander, J.; Stauffer, B.; and Zumbrunn, R. 1982. Ice core sample measurements give atmospheric CO_2 content during the past 40,000 yr. Nature 295: 220-223.

(15) Perch-Nielsen, K.; McKenzie, J.; and Qixiang, H. 1982. Precision bio- and isotope-stratigraphy and the "catastrophic" extinction of calcareous nannoplankton at the Cretaceous/Tertiary boundary. Proceedings of the Second Symposium Kreide, München, June 1-7, 1982. Zitteliana, in press.

(16) Russell, D.A. 1979. The enigma of the extinction of the Dinosaurs. Ann. Rev. Earth Planet. Sci. 7: 163-182.

(17) Stuiver, M. 1978. Carbon-14 dating: a comparison of beta and ion counting. Science 202: 881-883.

(18) Suess, E., and Erlenkeuser, H. 1975. History of metal pollution and carbon input in the Baltic Sea sediments. Meyniana 27: 63-75.

(19) Walker, J.C.G.; Hays, P.B.; and Kasting, J.F. 1981. A negative feedback mechanism for the long-term stabilization of Earth's surface temperature. J. Geophys. Res. 86: 9776-9782.

(20) Walker, J.C.G.; Klein, C.; Schidlowski, M.; Schopf, J.W.; Stevenson, D.J.; and Walter, M.R. 1982. Environmental evolution of the Archean-Early Proterozoic Earth. In Origin and Evolution of Earth's Earliest Biosphere: An Interdisciplinary Study, ed. J.W. Schopf. Princeton, NJ: Princeton University Press, in press.

Atmospheric Chemistry, ed. E.D. Goldberg, pp. 199-213. Dahlem Konferenzen 1982.
Berlin, Heidelberg, New York: Springer-Verlag.

The Production and Fate of Reduced Volatile Species from Oxic Environments

J. E. Lovelock
Coombe Mill Experimental Station, St. Giles on the Heath
Launceston, Cornwall PL 15 9RY, England

Abstract. The oxic part of the biosphere includes the surface, the top soil, and the upper regions of lakes and oceans. It is the site of a surprisingly large proportion of the total biosynthesis of reduced molecular species. The production under oxic conditions of reduced compounds of the principal groups of elements used by the biota is discussed.

INTRODUCTION

When the topic of this background paper was chosen I do not think that the planning committee intended to hand me a cockatrice egg. But taken literally the assignment could demand a description of the complete past history and future of life itself, a monstrous task to encompass in a brief background paper. I shall therefore limit it to a short account of the present-day production and the fate of reduced molecular species in the fully oxidized environment. An example of this activity is the production of ethylene by apples and other fruit. Should we also consider processes, such as that by which we as individuals produce up to 30 liters of methane daily? Unfortunately, interesting though it may be, this process is not a true example of the production of a reduced species in an oxic environment. The methane we exhale is produced by intestinal microflora existing in a local anaerobic niche in our guts and cannot therefore be the concern of this paper.

The elements which are known to have volatile compounds biologically produced under oxidizing conditions are: carbon, nitrogen, sulfur, selenium, tellurium, arsenic, mercury, chlorine, bromine, and iodine. Less well established is the production of volatile reduced compounds of phosphorus, tin, zinc, and antimony. We still do not know even approximately the gross annual production of many of the compounds; with others such as dimethyl sulphide, methyl chloride, and methyl iodide we can only use that word "substantial," favored by patent agents who would avoid a numerical commitment, to express the proportion of the total flux contributed by life in the oxic zone.

The structure of the biosphere is such that the oxic environments overlie the anoxic zone which in turn overlies the crustal rocks. The detritus from plant life and consumers in the photosynthesis layer at the interface with the atmosphere is moved downwards by the burial process of sedimentation. A counterflow of gases, nitrogen, nitrous oxides, and methane moves upwards carrying with it methylated derivatives of the elements listed above and also those produced in the anoxic zone. The organic matter which survives to the base of the anoxic zone is buried deeper to become part of the crustal rock. Some of it is recycled to the atmosphere through volcanism.

The upper mixed zone of the ocean is a fully oxidized environment, and it is rich in biochemical transactions leading to the production of reduced molecular species. Compounds of elements other than those listed above are produced here. Lack of space and information prevent any detailed discussion of this otherwise interesting oxic region.

CARBON-, HYDROGEN-, AND OXYGEN-BEARING COMPOUNDS

By far the largest emission of hydrocarbons comes from the metazoan flora, especially from trees. The total emission is estimated by Zimmerman et al. (22) to be between 350 and 480 Mtons/year of carbon. This quantity is almost the same as that of the total methane production, most of which comes from the anoxic sector. The hydrocarbons emitted by the vegetation are mostly isoprene and terpenes. These are very reactive and their atmospheric residence time is only a few hours. The detailed pathways of the oxidation of terpenes in the air are not known, but the first step may well involve reaction with tropospheric ozone which is always present in considerable excess at 30 to 50 ppb. The reaction of unsaturated hydrocarbons with ozone is chemiluminescent, and light emission is often used in the analysis of the aerial concentration of either the hydrocarbons or of ozone. It is interesting to speculate that the otherwise obscure phenomenon of the "Will of the Wisp," a wavering glow seen at night in marsh and woodland areas, is in fact the visible manifestation of this chemiluminescent reaction. It is usually ascribed to the oxidation of methane, but this gas is far too stable kinetically to react rapidly enough at ambient temperatures to give a visible glow.

It is interesting to consider what is the competitive advantage conferred upon coniferous and eucalyptus trees by the production of these vast quantities of reactive hydrocarbons. In this connection it must be borne in mind that the biota are normally most efficient in conserving organic compounds. Even mammals, which are more or less open reaction systems involving extensive organic chemistry at above ambient temperatures, do not exhale any significant quantity of organic vapor. Pheromones excluded, the presence of volatile compounds in the breath is pathological and, indeed, can be used to confirm the diagnosis of a malady such as diabetes mellitus in which the acetone is exhaled in detectable quantities.

Watson (19) has commented that it may be significant that the genera of tree most active in producing terpenes are those which benefit from fires. Natural fires are more destructive to tree species which compete with conifers, such as oaks, than to the conifers themselves. It is suggested that the forest floor detritus shed by the conifers and eucalyptus trees has evolved for maximum flammability and the frequent occurrence of forest fires. The presence of terpenes and their polymers in the detritus would certainly assist the development and spread of a fire.

The greatest area of forest in the world is in the tropics. These are rapidly being cleared for agricultural use; if we need to know the emissions of hydrocarbons and other organic chemical compounds in these regions, the measurements should soon be made. The tropical rain forests in their normal state maintain under their canopies a more or less permanent stable environment. The production and fate of reactive hydrocarbons in such a protected environment is likely to be very different from that of the wholly or partially farmed forest.

We are also in great ignorance about the numerous chemical processes in the oxic surface layers of the oceans. During a voyage of the German research vessel Meteor in 1973, Rasmussen observed the presence of saturated and unsaturated hydrocarbons in the surface waters of the Atlantic and in the air above the water. On the same voyage I observed the copious production of PAN and other photochemical end products of the reaction of these hydrocarbons in sunlight with oxides of nitrogen also coming from the sea. These observations, now nearly ten years old, have not been repeated or extended.

Biochemistry makes available a vast catalogue of chemical compounds which may be used when appropriate to enhance the fecundity of some entrepreneurial organism. Thus the cockroach produces normal octane which is used as a solvent for the waxes of its cuticle. The cockroach life-style requires slithering over rough surfaces which render it liable to dehydration through loss of the integrity of its hydrophobic waxy coating. The hydrocarbon solution of waxes is a convenient toilet preparation for the cockroach and enables it to maintain its cuticle intact. In this process the hydrocarbon evaporates to the environment. We do not know if there are similar uses of hydrocarbons by other insect species, nor what the total emission of hydrocarbons from cockroaches is. Before this is dismissed as a trivial curiosity, it is worth considering the recent observation by Zimmerman (22) that termites, through the anaerobic flora of their hind guts, emit methane. Their total production is claimed to be a sizeable proportion of all the methane made in the biosphere. The emissions of a single invertebrate may be small but the quantity of hydrocarbons from all insects and invertebrates may be significant on a global scale.

The direct use of the matter and energy involved in the production of volatile hormones, pheromones, perfumes, and stenches is not large. The importance of these products lies in the information transactions they enable. Without their use the stability in a cybernetic sense and the efficiency of the planetary ecosystem would be diminished. E.O. Wilson (20) in his book on sociobiology gives a lucid description of some of the information transfer systems of the biota. In it the concept of residence time as applied to a chemical compound released into the environment takes on a new significance. The persistence of a pheromone has an optimum. Confusion can come from too long a lifetime just as it can come from too short a lifetime.

The principal fate of hydrocarbons and other organic chemicals in the oxic environment is to be oxidized through reaction with OH and possibly also ozone. Apart from methane this compound class does not persist long enough in the troposphere for a detectable quantity to reach the stratosphere. A few compounds, as for example, diacetyl, absorb visible or near ultraviolet radiation, are photolyzed directly, and will have residence times of seconds to hours. A similarly short life is to be expected of those compounds which react rapidly with OH which is always present in the troposphere. Saturated hydrocarbons persist longer, for their rate constant is lower (10). Methane has a residence time of seven years and terpenes only ten hours (9). The oxidation of all hydrocarbons in the atmosphere is a potential source of CO, H_2, and of ozone.

The fate of volatile organic chemicals is different in the natural environment from that in the polluted atmosphere. Combustion processes wherever they occur release oxides of nitrogen to the atmosphere. These have a profound effect on the tropospheric chemistry of the region of their emission, particularly on the photochemistry of the oxidation of the natural hydrocarbon emissions. This topic is reviewed by Seiler and Crutzen (15).

NITROGEN AND PHOSPHORUS COMPOUNDS

To a chemist the gases NH_3, N_2, N_2O, and NO are all reduced species when referred to our current oxygen-rich atmosphere. They also originate in the oxic or partially oxic sector of the biosphere, although some of the dinitrogen gas comes from the anoxic sector.

Ammonia is important in the chemistry of the troposphere through its effect on the chemistry of rainwater (17). As with so many other important gases, its flux through the atmosphere and its abundance are very poorly understood. Ammonia is present over those land areas where measurements have been made, almost always in temperate agricultural regions, at between 1 and 10 ppbv. Tjepkema et al. (18) observed that particulate NH_4^+ was over ten times more abundant in the atmosphere over forest regions in Massachusetts than was gaseous ammonia. Crutzen (9) has wisely commented that the role of cycling of ammonia in the atmosphere is complex; plants can both secrete and adsorb ammonia. Consequently, the ammonia cycling within a natural forest economy or even an ungrazed field can be very different from that over intensively cultivated land. Factors such as these make it very difficult to more than guess at the gross turnover of ammonia through the atmosphere.

Ammonia comes from the bacterial decomposition of animal urine and feces. The total production from this source, including the contribution of both wild and domestic animals, is between 20 and 30 M tons N/year (4, 16). The output/input of NH_3 from the oxic regions of the oceans is unknown, although from a few measurements (2), it is thought to be much less than from the land. Crutzen also commented that biomass burning is potentially a large ammonia source, perhaps as great as 60 M tons/year. Some of this is from natural wildfires. Fires also release HCN and methyl nitrite. PAN is a product of the photochemical processing of combustion emissions and has a net reducing tendency.

The fate of ammonia is to be absorbed, rained out, or deposited as ammonium salts. Nitrogen, nitrous, and nitric oxides are all substantial products of biosynthesis in the oxic environment. They are returned to the surface by the effects of high energy processes which include lightning, corona discharges, ionizing radiation, and combustion. These all recombine oxygen and nitrogen, and the immediate nitrogen oxide products then react further to produce nitric acid. This then dissolves in rainwater and if not for the biota would become a stable solution of nitrate ions dissolved in the oceans. The biological process of denitrification returns nitrogen to the atmosphere mostly as dinitrogen gas, but also as nitrous and nitric oxides. The flux of each of these gases is constantly revised often by as much as 100%. Very approximately, the total flux of nitrogen is near 300 M tons/year and of nitrous oxide, 10 M tons/year.

SULFUR COMPOUNDS

H_2S, CS_2, COS, $(CH_3)_2S$ and CH_3SH all are biologically produced and emitted to the atmosphere along with a miscellaneous collection of minor sulfur emissions of compounds such as diallyl sulfide (garlic), ally isothiocyanate (mustard) and dicrotyl sulfide and butanethiol as flavors and as stench agents from skunks and other animals. The fact that these last mentioned compounds are quantitatively insignificant belies their importance in the quality of life for us and in survival for the skunk. H_2S and C_2S are almost certainly produced from anoxic sources as is COS, although the possibility of the oxic production of COS cannot yet be excluded. Our concern in this paper is therefore principally with the production and fate of dimethyl sulfide and methane thiol, both products of the oxic sector. Dimethyl sulfide (DMS) was observed by Challenger (6) to be emitted at high concentration by certain marine macroalgae, for example, polysiphonia fastigiata. Fifteen percent or more of the sulfur of this organism is in the form of propylthetin:

$(CH_3)\ 2S^{+} \cdot CH_2CH_2COOH.$

This is decomposed enzymatically or by alkali to yield DMS.

Polysiphonia are ubiquitous in temperate coastal waters as an epiphyte on large wracks, such as Fucus, and could serve as a major source of DMS in these regions. Limited investigations over several years in western Ireland suggested that the emission is very seasonal, being most marked in the early spring, February and March, and also in October and November. An expedition to measure the DMS emissions from these coastal waters could all too easily find only the low background rate associated with many species of marine algae. The total DMS flux is now thought to be 30 - 50 x 10^{12} g/yr^{-1} (3, 14). Andreae (1) has also commented that in the open sea, blooms of dinoflagellates are often characterized by a strong odor of DMS.

Professor Challenger (6) drew my attention to the work of Ishida (11), who investigated the evolution of sulfur compounds from various unicellular algae. Ishida showed that DMS production was common among algae of marine or saline habitat but only a minor activity of fresh water algae. He also confirmed that the precursor of DMS from the most active of the producers among the unicellular marine algae was dimethyl-propiothetin. Perhaps the most interesting of Ishida's findings was the discovery that the enzymic cleavage of the propiothetin was strongly dependent upon the ionic strength of its medium. The rate of production of DMS rose from near zero at 0.05M NaCl to a plateau at 0.5M and above.

Ishida also recognized that the Japanese phrase "iso no kaori," for the unique smell of the sea, was indeed that of DMS. His paper was published in a Japanese journal not well-known in the West which may account for the otherwise undeserved infrequency of its citation.

Methane thiol is the product of numerous biological systems. The pure compound at high aerial concentrations has an intolerable stench, but when diluted and especially when mixed with the odors of other compounds, it becomes a key component of many desirable flavors. Allyl caproate, for example, is the essential smell of pineapple, but alone is unconvincing. The addition of a small amount of methane thiol brings verisilmilitude. The total flux of sulfur from this and the other thiols and thioethers made in the oxic sector is quite unknown although probably small.

The fate of the thioethers and thiols in the atmosphere is to be oxidized through the attack of OH radicals. Cox (8) estimated a residence time of a few weeks for DMS. The products of the oxidation of DMS and of methane thiol include (along with such obvious end products as SO_2 and H_2SO_4) the compounds dimethylsulfoxide (DMSO) and methansulfonic acid. The latter compound is not easily distinguished from sulfuric acid in the analysis of the atmospheric aerosol. Its presence in the aerosol might be a useful indicator of the proportion of sulfur cycled as DMS and methane thiol.

DMSO is not very volatile and is hygroscopic, it would not long persist in the atmosphere, and the equilibrium concentration would be near or below the limits of detection. It is, however, abundant in the surface waters of the sea. Andreae (1) observed concentrations of DMSO in the surface waters of 20 to 40 nanomoles per liter. This was about 100 times larger than the concentration of DMS I found in comparable coastal regions. For the oxic marine environment DMSO appears to be a sulfur carrier of major importance; like DMS, its production in fresh water is small by comparison.

REDUCED COMPOUNDS BEARING THE HALOGEN ELEMENTS

There is a similarity between the production of methyl iodide and that of DMS by marine algae. Polysiphonia fastigiata was found by Challenger (6) to specialize in the biosynthesis of DMS, and similarly the large wracks of the inshore waters include the genus Lamminaria which specialize in the production and release of methyl iodide. Just as with the production of sulfur gases from marine algae, methyl iodide production is a common property, and the sea almost everywhere has been found to bear methyl iodide. Another similarity is that production is very variable, particularly seasonally. It is maximal for the northern Atlantic waters in the early spring and late autumn.

Methyl iodide is a strong alkylating agent and not surprisingly therefore a mutagen and carcinogen. It is interesting to contemplate the political problems that might have arisen had its ubiquitous presence in the inshore waters been man-made rather than natural. The sequestration of the rare element iodine from seawater by the large kelps has long been known, and these were once harvested and burned and the ash extracted as a commercial source of iodine. The local ecological advantages which might be associated with methyl iodide production are not known, although it could have antibiotic possibilities.

There are numerous iodine-bearing compounds biologically synthesized. The family of iodine-bearing metabolites of the thyroid hormones is common to most mammals, including humans. Many iodine-bearing compounds are also present in marine algae. None of these iodine compounds, however, appears to enter the environment in significant quantities, and methyl iodide appears to be the dominant carrier of iodine from the living cell to the environment.

The fate of methyl iodide after it has been released in the oxic marine environment has been investigated by Zafiriou (21). He observed that it reacts readily with the abundant chloride ion of seawater to yield methyl chloride. The same reaction may also yield some methyl bromide. The possible fate of methyl iodide and iodine in the atmosphere is well reviewed by Chameides and Davis (7). Methyl iodide is readily photolyzed in the troposphere and would have a residence time measured in hours or days according to the intensity of sunlight. The iodine released from methyl iodide could have a much more significant role in tropospheric chemistry than its low aerial concentration might otherwise suggest. Calculations by Liss and Slater (12) suggest that the source strength of the ocean-to-atmosphere transfer of methyl iodide is 0.3 M tons/year. Others have proposed a source strength of 1 to 2 M tons/year. It should be recalled that, like DMS, the emission varies from place to place and with the seasons, hence some disagreement at this stage is to be expected.

METHYL CHLORIDE AND METHYL BROMIDE

The principal natural halocarbon of the oxic environment is methyl chloride (13), present at about 1.0 ppbv. Methyl bromide is also present but at an abundance thirty times less (.025 ppbv). These two halocarbons are both products of the marine environment and most probably arise by abiological processes through the reaction of methyl iodide with the chloride and bromide ions of seawater (21). The possibility of the direct biosynthesis of these compounds by the marine biota has not yet been excluded. Methyl chloride has also been observed as a product of the bacterial and fungal decomposition of wood (13); the size of this source is quite unknown. The other possible large source of methyl chloride and bromide is biomass burning. The smoke from the smouldering combustion of vegetable detritus can carry as much as 100 ppmv of methyl chloride. Watson, on an expedition to equatorial Africa during the grass burning season, found concentrations of this gas in the region of 2 ppbv in the lower troposphere (19).

The fate of methyl chloride and bromide in the atmosphere is to be oxidized by hydroxyl radicals. This takes place mostly in the troposphere, with the end products being CO_2, H_2O, and HCl. The HCl would rapidly be rained out. The residence time of methyl chloride is several years, and some of it penetrates to the stratosphere and contributes a natural component to the chlorine burden of that region of the atmosphere. The total natural production of methyl chloride is large and in the range of 7 to 30 M tons/year.

ORGANOMETALS AND ORGANOMETALLOIDS

Brinckman and Bellama (5) have gathered in one volume much of our contemporary knowledge of this compound class as it is related to their occurrence and fate in the environment. The volume starts with a historical review paper by Challenger (6), who did much to pioneer this topic and who coined the term biological methylation.

The subject has a long history, and there is a whiff of Victoriana as well as of garlic in the exhalations of Gosio gas which so stealthily pervaded those 19th century bedrooms and killed the occupants as they slept. The lethal emanation was trimethyl arsine, released by fungi growing on wallpaper patterned in arsenial pigment. The vigor of this process is impressive. The concentration of trimethyl arsine needed to kill humans in eight hours is at least 1.0 ppmv. Assuming that a typical bedroom has a ventilation rate of 2 air changes per hour and a volume of 50 cubic meters, the production of trimethyl arsine needed to sustain a concentration of 1.0 ppmv is 100 milliliters per hour or about a liter for a night.

The natural methylation of arsenic is not limited to the rather artifical conditions of Gosio gas generation. Andreae (1) has found dimethyl arsinic acid to be widely distributed in the oxic marine environment at a concentration of about 2 nanomoles/liter. This is about one tenth the concentration of dimethyl sulfoxide which is a normal component of the same environment. Indeed, so common and so well tolerated is arsenic in the marine photosynthetic environment, that it is interesting to consider to what extent it is a surrogate for phosphorus.

There is no unequivocal evidence for the methylation in the oxic zone of antimony, bismuth, mercury, lead, or thallium leading to the release of volatile compounds. These elements may be involved in methylation in the anaerobic sector.

The aerosol deposited in regions remote from pollution sources has a distribution of elements substantially different from that of the same elements in crustal rocks. Among the elements that are in excess in the aerosol are those which are or might be biologically methylated. Could it be that this process transfers the element to the atmosphere where, no matter how rapid the oxidation, the resulting aerosol is able to travel and subsequently deposit in remote places? Goldberg (private communication) offered the alternative explanation that the anomalous elements in the aerosol have volatile compounds and may be dispersed to the air through high temperature processes such as combustion, natural or otherwise, and by volcanism.

Selenium and tellurium both are methylated under oxic conditions. It would be interesting to know if the natural emissions follow the same pathways as sulfur. Marine algae are known to have an arsenic analogue of choline; do they also possess a selenium or tellurium analogue of propylthetin?

Perhaps the most interesting of all organometallic compounds is the methyl cobalt compound which is a variant of vitamin B12. This substance is able to transfer its methyl group to other elements and to perform biological methylation extracellularly. The role of this compound in the biochemistry of the oxic marine sector is an open question.

REFERENCES

(1) Andreae, M.O. 1980. Dimethylsulfoxide in marine and freshwaters. Limnol. Oceanogr. 25: 1054-1063.

(2) Ayers, G.P., and Gras, J.L. 1980. Ammonia gas concentrations over the Southern Ocean. Nature 284: 539-540.

(3) Barnard, W.R.; Andreae, M.O.; Watkins, W.E.; Bingemer, H.; and Georgii, H.-W. 1982. The flux of dimethylsulfide from the oceans to the atmosphere. J. Geophys. Res., in press.

(4) Bottger, A.; Ehhalt, D.H.; and Gravenhorst, G. 1980. In Atmosphärische Kreisläufe von Stickoxyden und Ammoniak, Report July-1558, Institut für Chemie 3, Kernforschungsanlage Jülich, FRG.

(5) Brinckman, F.E., and Bellama, J.M. 1978. In Organometals and Organometalloids Occurrence and Fate in the Environment. Washington, DC: American Chemical Society.

(6) Challenger, F. 1951. Biological methlyation. Adv. Enzymol. 12: 429-491.

(7) Chameides, W.L., and Davis, D.D. 1980. Iodine: its possible role in tropospheric photochemistry. J. Geophys. Res. 85: 2-25.

(8) Cox, R.A., and Sandalls, F.J. 1974. The photo-oxidation of hydrogen sulphide and dimethyl sulphide in air. Atmos. Env. 8: 1269-1281.

(9) Crutzen, P.J. 1982. In The Interaction of Biogeochemical Cycles, SCOPE, in press.

(10) Ehhalt, D.H., and Schmidt, U. 1978. Sources and sinks of atmospheric methane. Pageoph. 116: 452-464.

(11) Ishida, Y. 1969. Memoirs of the College of Agriculture, Kyoto University, vol. 94, pp. 48-82.

(12) Liss, P.S., and Slater, P.G. 1974. Flux of gases across the air-sea interface. Nature 247: 181-184.

(13) Lovelock, J.E. 1975. Natural halocarbons in the air and in the sea. Nature 256: 193-194.

(14) Nguyen, B.C.; Gaudry, A.; Bonsang, B.; and Lambert, G. 1978. Reevaluation of the role of dimethyl sulphide to the sulphur budget. Nature 275: 637-639.

(15) Seiler, W., and Crutzen, P.J. 1980. Estimates of gross and net fluxes of carbon between the biosphere and the atmosphere from biomass burning. Climate Change 2: 207-247.

(16) Söderland, R., and Svensson, B.H. 1976. The global nitrogen cycle. Ecol. Bull. (Stockholm) 22: 23-73.

(17) Taylor, G.S.; Baker, M.B.; and Charlson, R.J. 1982. In The Interaction of Biogeochemical Cycles, SCOPE, in press.

(18) Tjepkema, J.D.; Cartica, R.J.; and Hemond, H.F. 1981. Atmospheric concentration of ammonia in Massachusetts and deposition on vegetation. Nature 294: 445-446.

(19) Watson, A.J. 1978. Consequences for the Biosphere of Grassland and Forest Fires. Ph.D. Thesis, University of Reading (UK).

(20) Wilson, E.O. 1974. Sociobiology. Cambridge, MA: Harvard University Press.

(21) Zafiriou, O.C. 1975. Reaction of methyl halides with seawater and marine aerosols. J. Mar. Res. 33: 75-81.

(22) Zimmerman, P.R.; Chatfield, R.B.; Fishman, J.; Crutzen, P.J.; and Hanst, P.L. 1978. Estimates on the production of Co and H_2 from the oxidation of hydrocarbon emissions from vegetation.

Atmospheric Chemistry, ed. E.D. Goldberg, pp. 215-229. Dahlem Konferenzen 1982. Berlin, Heidelberg, New York: Springer-Verlag.

The Production and Fate of Reduced C, N, and S Gases from Oxygen-deficient Environments

B. B. Jørgensen
Institute of Ecology and Genetics, University of Aarhus
8000 Aarhus C, Denmark

Abstract. The oxygen-deficient environments comprise the subsurface parts of soils, lake sediments, and the sea bed, and they penetrate into the bottom layers of some stratified water bodies. Organic matter is here decomposed mainly by microorganisms which release metabolic products containing carbon, nitrogen, and sulfur. Among the gaseous products considered here are CH_4, NH_3, N_2O, NO, H_2S, $(CH_3)SH$, $(CH_3)_2S$, CS_2, OCS, and H_2. As most of these trace gases are very useful as nutrients for the biosphere, their emission into the atmosphere is regulated both by biological and by physical mechanisms. The actual emission rates are discussed in view of quantitative flux data as well as regulating factors.

INTRODUCTION

Reduced gases of carbon, nitrogen, or sulfur are produced from biological processes within the anoxic world of soils, stagnant waters, and sediments. For each element, the anaerobic food chains have a few dominant, gaseous end products, such as CH_4, NH_3, and H_2S. There is no biogenic form of gaseous phosphorus from anoxic environments which is known to play a role in the phosphorus cycle, and there are no significant gas-phase reactions of phosphorus in the atmosphere.

The reduced gases carry much of the chemical energy which was introduced into the anoxic world in the form of organic matter.

This chemical energy can be exploited by organisms living in the oxic world. Evolution has adapted microbial communities to assimilate and oxidize the reduced gases very efficiently at the oxic-anoxic interface. Special bacteria may even carry out the oxidation within the anoxic environment and, for example, oxidize CH_4 with sulfate or H_2S with nitrate. Especially for H_2S, there is also a rapid chemical oxidation taking place with O_2, which efficiently competes with the biologically catalyzed process (2).

The high efficiency of the biological communities strongly restricts the possibility for the gases to escape into the atmosphere. The amounts of C, N, or S gas that do escape to the air are mostly trivial relative to the cycling of the elements that takes place within the soil or water. Yet, the small fraction which is emitted becomes important for the mass balance of the terrestrial or aquatic environment over long time scales. The biological (and chemical) gas oxidation tends to shift the spectrum of gases which are finally emitted towards more stable and inert forms. The composition of reduced gases which reach the atmosphere therefore reflects their reactivities just as much as their relative production rates by the anaerobic biosphere. This discrimination between emission fractions for the different gas species is important in the understanding of how natural emission rates are regulated and may be perturbed by our present and future activities.

REGULATING MECHANISMS OF GAS EMISSION

The efficiency of natural ecosystems to retain specific reduced gases is well illustrated in the case of H_2S. The major, global zone in which this gas is produced by the process of bacterial sulfate reduction is the sediments of the continental shelves. Based upon literature data on mineralization rates combined with radiotracer measurements of sulfate reduction rates, a global H_2S production by shelf sediments of 1000 Tg H_2S yr^{-1} can be calculated (10,11). The global emission of biogenic H_2S is only a few tens of Tg, of which coastal sediments may contribute a very small fraction. Instead of H_2S, DMS ($(CH_3)_2S$) is the major sulfur source to the atmosphere from shelf waters, although this compound is produced in trivial amounts relative to H_2S on an areal basis. The biological oxidation of H_2S at the sediment surface is so efficient that even a thin film of bacteria can completely cut off a high diffusive flux of H_2S. In the Black Sea water column, for example, H_2S oxidation is a relatively slow, chemical process with a 40-50 m deep oxidation zone where H_2S has a retention time of a day. In contrast, a film of sulfur bacteria on a sediment with a high H_2S production rate may carry out exactly the same process in a zone of 0.1 mm thickness where H_2S has a retention time of one second. It takes, however, only a slight mechanical disturbance to break up such a film after which the H_2S is suddenly released.

The balance between the decomposition of reduced gases, which retains them within the soil or water, and their emission into the atmosphere is very complex. In most cases, we have only a very crude understanding of how the emission is regulated. There is a general trend of decreasing fluxes when going from the tropical regions towards higher latitudes, from coastal zones and wetlands towards dry mineral soils, or from summer to winter. Such trends can be important for the modelling of the present atmospheric chemistry and to construct budgets of the global mass balance of the elements. The black box approach to the study of emission rates from terrestrial and aquatic environments is, however, insufficient to evaluate local patterns of emission or even to predict future changes in these patterns. Among the important questions with respect to anaerobic gas production are the effects of eutrophication of lakes and coastal waters, increased fertilizer application in agriculture, development of paddy soils, increase in artificial irrigation, drainage of wetlands, and development of tropical agriculture at the expense of natural rain forests.

A high resolution in the measurement of local emission rates has been obtained by the use of flux chambers. In these chambers, the biogenic gases, which are released, can be directly collected and their composition analyzed. This technique has shown that emission rates vary over several orders of magnitude within short distances. The number of measurements required to determine average, local or regional emissions is therefore great and may even seem prohibitive. Because such data are strongly needed in the atmospheric budgets, however, there has been a tendency to make global extrapolations of even the first scattered flux measurements. This has led to overestimates of the contribution from specific types of environments since the flux chambers were frequently applied in locations selected for their high natural emission rates. As the number of measurements increases, a more balanced application of such data is being reached. The flux chamber techniques will also become more useful if combined with simultaneous atmospheric measurements of vertical diffusion gradients and horizontal air trajectories of the relevant gases. An important application of the flux chambers, which has hardly been exploited yet, is the determination of emission fractions of gases produced in the soil or sediments. This requires direct measurements also of the biogenic gas production rates and will yield information on how emission fractions are regulated by environmental conditions. Information on such regulation mechanisms should improve predictions of large scale and future source strengths.

It is mainly due to heterogeneity and instability of natural environments that the biological barrier against reduced gas emission is leaky. As an example, even well drained soils can contain anoxic microenvironments within soil particles of a few millimeters size. Reduced gases can diffuse from their center through their oxic surface and into the soil air within seconds and further into the atmosphere by the slow ventilation of the soil. Such reduced microenvironments may be the diffuse source of H_2S or N_2O released from many soils. Among other mechanisms which rapidly drive gases from anoxic environments into the air and thus bypass the biological barrier are:

a) ebullition of methane from lake sediments and swamps, occasionally triggered by atmospheric depressions; b) periodic exposure of anoxic porewater from coastal mud or sand flats by low tides and by tidal pumping; c) respiratory ventilation of tubes and burrows in anoxic sediments by benthic animals; d) gas transport through air channels in marsh grasses, water lilies, and other plants growing in waterlogged soils; e) diurnal variations between light and dark, regulating the gas emission from shallow coastal sediments (benthic microalgae and photosynthetic bacteria may efficiently oxidize H_2S at the sediment surface during the day while at night this biological filter does not function and H_2S is released into the air); and f) seasonal overturn of thermally stratified lakes in which CH_4, H_2S, and other gases have accumulated in the anoxic bottom water during the warm season and are suddenly exposed to the air.

The overturn of lakes is an example of local, episodic events which may be difficult to quantify relative to more persistent and diffuse sources. Storms, which are often avoided by field scientists, are another example that may cause an episodic peak emission of biogenic gases by stirring up anoxic sediments and waters. It is difficult at the present stage to evaluate the quantitative significance of such effects.

CARBON

The only reduced, carbon-containing gas produced in anoxic environments, which is presently known to contribute significantly to the atmospheric chemistry, is methane. Formaldehyde and CO are primarily produced within the atmosphere during the gas-phase reaction of CH_4 with OH radicals, and they are therefore only secondarily dependent upon natural emissions. Other carbon gases, such as ethane or methylated nitrogen and sulfur compounds, are produced only in small amounts relative to methane. Terpenes and isoprenes are released from oxic environments.

Methane is a main product of biological decomposition in the anoxic world, provided that no electron acceptor such as nitrate or sulfate is available for bacterial respiration. Methane is produced only by strictly anaerobic bacteria which live in waterlogged soils and sediments or in the rumen or gut of livestock and other animals. In swamps and eutrophic lakes, the bacteria often produce such high methane pressures that the gas bubbles rise to the water surface. This rarely happens in sulfate-rich marine environments where the methanogenic bacteria are outcompeted by sulfate-reducing bacteria and where methanogenesis therefore is a less important process of anaerobic decomposition relative to H_2S production.

Swamps and paddy fields are environments of high source strength for CH_4. They have been estimated to contribute 45% and 25%, respectively, to the global CH_4 emission (4). The numbers are, however, highly variable, especially for the wetlands where few measurements have been made. It is difficult from the atmospheric chemistry to point out the source areas for CH_4 emission due to its high background concentration (1600 ppbv CH_4) and long residence time (about 7 years). Flux chamber and other direct measurements have confirmed that marshes and swamps often have extremely high emissions of 1-10 mmol CH_4 m^{-2} day^{-1}. This accounts for a significant fraction of the carbon flow of these ecosystems. Thus, a salt marsh may lose 5% of its annual productivity as atmospheric CH_4 (12), and eutrophic lake sediments may lose up to 30-50% of the carbon input as methane (16). During the fall overturn of a stratified lake, 40% of the large CH_4 pool that had accumulated in the bottom water was lost to the atmosphere (19).

Although terrestrial environments function as net sources of methane to the atmosphere, mainly due to natural wetlands or flooded soils, the role of well drained soils is not well-known. The ocean contributes only a few percent to the atmospheric load. Based on source estimates and on calculations of the reaction rate between CH_4 and OH in the atmosphere, global budgets of atmospheric methane flux range between 330 and over 1000 Tg CH_4 yr^{-1} (4,5,24). The lower estimate implies that drained soils are not significant sources of CH_4.

The methane release from symbiontic bacteria in ruminants and other, mainly cellulose-digesting, animals is around 60 Tg yr^{-1} (4). Symbionts in termite guts were also recently suggested to contribute significantly to this atmospheric methane input (Zimmerman, personal communication).

Taken together, all the biogenic sources account for 80-90% of the present atmospheric methane flux. Due to the current eutrophication of soils, lakes, and coastal waters, the natural emissions will expectedly increase considerably in the coming years. The number of livestock as well as the area of rice paddy fields have both increased by 1-2% per year during the seventies (4). The use of nitrogen fertilizers has further increased the methane production in paddy fields. This enhancement of biogenic methane production is to some extent neutralized by the drainage of wetlands for agricultural use. The increasing load of sulfate via acid precipitation to inland areas of high industrial SO_2 emission may also limit the natural methane production due to competitive inhibition from sulfate-reducing bacteria. Yet, a current increase in the atmospheric methane concentration of 2% per year has been suggested from intensive air monitoring at the US west coast (17). Further measurements are needed to verify the magnitude of this increase which could lead to a doubling of the atmospheric methane pool in 40-50 years.

The main sink for atmospheric methane is its destruction in the troposphere by reaction with OH radicals: $CH_4 + OH \rightarrow CH_3 + H_2O$. This process removes about 90% of the atmospheric methane flux within the troposphere with an additional 10% being removed in the stratosphere. The destruction of methane at the earth's surface by, e.g., uptake in soil microorganisms seems to be trivial relative to the atmospheric reaction (5).

NITROGEN

The following nitrogen gases are produced in anoxic environments, mainly through bacterial processes:

1) NH_3 is the immediate product of almost all organic nitrogen during the anaerobic bacterial hydrolysis and metabolism of proteins. The amino acids are deaminated before the fermentative breakdown of the carbon structure and the free ammonia gradually diffuses back into oxic environments. The fate of ammonia is discussed by Lovelock (this volume).

2) Methylated amines are a minor product from the breakdown of, e.g., choline or creatine in organic matter. They are especially notable from the odor of spoiled fish. Their significance as atmospheric trace gases is not known.

3) N_2O and NO both are produced during the reduction of NO_3^- by denitrification to N_2, by dissimilatory reduction of NO_3^- to NH_4^+, by NO_3^- assimilation into organic nitrogen, and during the oxidation of ammonium by nitrification to NO_3^-. In the former processes, however, N_2O and NO are obligate intermediates, while in nitrification they are minor by-products:

Denitrification: $NO_3^- \rightarrow NO_2^- \rightarrow NO \rightarrow N_2O \rightarrow N_2$

Nitrification: $NH_4^+ \xrightarrow[\searrow N_2O \quad \searrow NO]{} NO_2^- \rightarrow NO_3^-$.

The production of N_2O and NO from denitrification in nature should therefore be much greater than the production from nitrification. It is generally less than one percent of the total nitrogen flow that goes to N_2O or NO during nitrification under conditions of high oxygen and neutral pH, which are favorable for the bacteria. Under adverse conditions, it may occasionally reach 10%. In spite of the small fraction, nitrification may be equally as important as denitrification for the accumulation of these gases in the environment. This is because N_2O and NO are readily assimilated again and reduced by denitrifying bacteria; but they are not further metabolized by nitrifyers and are therefore biologically more stable in oxic environments. The preferential accumulation of N_2O in the oxic region of the water column in lakes or in the sea suggests that this gas was produced by nitrification (13). In coastal sediments, however, the peaks of N_2O and NO may coincide with the zone of denitrification (23). There is here a rapid production and consumption of the gases, and only an insignificant fraction may escape to the water or air above. NO is also produced in the surface layers of the ocean by photolysis of NO_2^-, but in the presence of O_2 it is oxidized to NO_2 within seconds.

The emission of N_2O from soils into the atmosphere is well documented from flux chamber measurements. Release of NO has also been detected, but its role is uncertain due to the limited number of studies. N_2O production is stimulated by ammonia fertilization and to a lesser extent by nitrate fertilization. The net emission does not exceed a few percent of the fertilizer added (21). When soils transiently become oxygen deficient, e.g., after a rainfall, there is a burst of N_2O emission. Under low oxygen conditions, the nitrifying bacteria become O_2-limited, ammonia oxidation is less efficient, and relatively more N_2O is produced. Sulfate acidification of soils due to acidic rain has the same effect. The N_2O production from denitrification is stimulated as well when the balance between production and consumption of N_2O is disturbed after a rainfall. This N_2O is produced in reduced microenvironments surrounded by the aerated soil and can therefore escape from the anoxic zone.

Estimates of the terrestrial N_2O emission have decreased to about 5 Tg yr^{-1} in recent years as a consequence of new atmospheric measurements as well as direct flux measurements. This is also the case with estimates of the flux from the sea surface, about 2 Tg yr^{-1}, after more data have been obtained for N_2O concentrations in the oceans (14). The large pool of N_2O in the troposphere, which may presently be increasing at 0.2% annually, is about 300 ppbv and has an estimated residence time of 100 years (4). N_2O is therefore completely mixed in the atmosphere and its distribution cannot be used to identify the sources. The N_2O is mainly of biogenic origin while NO and NO_2 are mainly anthropogenic (7,22).

SULFUR

Biogenic sulfur gases are produced by two different processes in anoxic environments. One is the dissimilatory reduction of sulfate by a special group of bacteria which yield H_2S as the only product. The bacteria oxidize about 2 mol of organic carbon to CO_2 for each mol of SO_4^{2-} reduced. The other process is the biological decomposition of sulfur-containing organic matter which yields a wider range of sulfur gases: H_2S, $(CH_3)SH$, $(CH_3)_2S$, $(CH_3)_2S_2$, OCS, and CS_2 (3). $(CH_3)SH$ and $(CH_3)_2S$ are intermediates during decomposition of methionine while OCS and CS_2 are products from cysteine and cystine. $(CH_3)_2S_2$ is probably a secondary product formed by the aerobic oxidation of $(CH_3)SH$. The methylated gases can be further metabolized by anaerobic bacteria to form CH_4 and H_2S (25).

The bacterial sulfate reduction produces 30- to 100-fold more sulfur gas per amount of organic matter mineralized than does the release of organic sulfur. Marine tidal flats and other coastal areas, which have connections to sulfate-rich seawater, consequently have extremely high productions of H_2S which completely dominate over those of the other sulfur gases. Their source strength for atmospheric sulfur is, however, limited by the high chemical and biological reactivity of H_2S in the presence of oxygen. H_2S has a lifetime of one or a few hours in oxic water (2) and therefore cannot normally escape through a water column of a few meters. DMS, OCS, and CS_2 are more stable. The H_2S is oxidized to sulfate in the oxic water or sediment with elemental sulfur or thiosulfate as common intermediates. The relative contribution of H_2S to the atmospheric sulfur cycle on a global scale may therefore not be greater than that of other biogenic sulfur gases in spite of its much greater rate of formation.

After the discovery that DMS is present as a trace gas at 2-3 times supersaturation in oceanic surface waters (15), it was suggested that this might be a major biogenic source of atmospheric sulfur. Current estimates of DMS emission from the ocean are 20-30 Tg yr^{-1}. In comparison, the total emission of biogenic sulfur gases to the atmosphere as calculated by different authors is within the range of 30-100 Tg yr^{-1}.

Coastal wetlands have the highest natural source strength for the sulfur gases. The concentration of H_2S in the air here reaches 0.1-10 ppbv in comparison to a level of <0.01-0.1 ppbv in inland or oceanic air. Due to the short residence time of a day or less for atmospheric H_2S, its site and intensity of emission can sometimes be inferred from concentration changes which follow air mass trajectories, vertical mixing gradients, or periodic, daily or yearly cycles. Thus, H_2S may accumulate significantly in oceanic air when it passes over a local salt marsh area (8). The concentration changes may even reflect the local tidal or diurnal cycles which mainly release H_2S during low tide or at night. Measurements based on emission flux chambers show that the coastal zone may locally release H_2S at rates which are 10,000-fold above the global mean for sulfur gases. A large survey of such measurements in soils and wetlands indicates, however, also a more diffuse source (1). Coastal wetlands were estimated to contribute 40% to the total natural emission from the terrestrial environments. As the coastal zone comprises only 7% of the geographical area, it consequently has a 10-fold higher source strength than the inland soils. Some inland soils are found to be sinks for H_2S (9). This is not included in the terrestrial sulfur emission of 64 Tg S yr^{-1} estimated by Adams et al. (1). Such soils may still be sources of DMS, OCS, and CS_2 which are not as strongly absorbed as H_2S and $(CH_3)SH$. The net balance is, however, also influenced by the measuring technique. The latter authors used a sulfur-free sweep gas which did not allow any soil uptake of H_2S, while Jaeschke et al. (9) used closed chambers.

The composition of biogenic sulfur gases released to the atmosphere is dominated by DMS and H_2S, while OCS, CS_2, and $(CH_3)SH$ contribute less. A more precise, quantitative evaluation seems premature. The gases react with OH radicals in the atmosphere and are oxidized to SO_2 as well as to methyl sulfonic acid (CH_3SO_3H). Dimethyl sulfonic acid is an important product of DMS oxidation. OCS is the most abundant sulfur gas (about 500 ppbv) because it has the longest residence time. It is an important source of SO_2 in the upper troposphere and of sulfate aerosols in the stratosphere. An additional sulfur source from an unidentified sulfur gas was recently suggested by Rodhe and Isaksen (18). Among the sulfur gases collected in flux chambers were also unidentified species, although in minor concentrations (1).

The industrial SO_2 emission is the only significant, man-made source of atmospheric sulfur. This emission may also have a secondary effect on the natural sulfur flux because the excess sulfate which is introduced into soils and lakes via acid precipitation stimulates bacterial sulfate reduction. Artificial irrigation with sulfate-rich ground water or the flooding of paddy fields will lead to a similar effect, as will the eutrophication of coastal waters. In spite of a rapidly increasing number of measurements of sulfur pools and fluxes in the environment, the global, biogenic sulfur emission is still not well-known. Recent global models have suggested rather low fluxes of about 30 Tg S yr^{-1} (6), but the extensive results from flux chambers now suggest a diffuse terrestrial source which would bring the figure closer to the anthropogenic input of 80 Tg S yr^{-1}. The anthropogenic source will be dominating in the heavily industrialized areas of the northern hemisphere, while biogenic sources should be dominating in other areas, especially in the tropics.

HYDROGEN

Since hydrogen, H_2, plays a significant role both in the anaerobic biosphere and in atmospheric chemistry, it is briefly discussed here. Free hydrogen is an important electron carrier during the decomposition of organic matter in anoxic environments. As a precursor of biological methane formation as well as an important substrate for sulfate-reducing bacteria, the H_2 turnover in oxygen-deficient soils, sediments, and waters may account for 10-30% of the total, anaerobic electron flow. Very little of the biogenic H_2, however, escapes into the atmosphere due to its rapid consumption by both anaerobic and aerobic microorganisms.

Only few measurements are available on H_2 release from paddy fields, forest soils, etc. The global emission rates of a few Tg H_2 yr^{-1} from these biogenic sources seem to be small relative to other processes of atmospheric hydrogen production (20). The ocean is on the average 300% supersaturated with H_2 and contributes 2 Tg H_2 yr^{-1}. The main natural source of H_2 in the troposphere is the photochemical oxidation of methane via formaldehyde: $CH_2O + h\nu \rightarrow H_2 + CO$. This process yields 13 Tg H_2 yr^{-1}. In the present atmosphere there are contributions of similar magnitude from fossil fuel combustion, 21 Tg H_2 yr^{-1}, and from biomass burning, 15 Tg H_2 yr^{-1}.

Hydrogen has an atmospheric lifetime of three years and is very uniformly distributed in the troposphere. The average concentration is 560 ppbv. The main sink for H_2 is oxidation at the earth's surface, possibly by specialized hydrogen bacteria. One fourth of the hydrogen is oxidized within the troposphere by reaction with hydroxyl radicals: $H_2 + OH \rightarrow H_2O + H$ (20).

Acknowledgement. I thank R. Hallberg for his help in providing some recent literature.

REFERENCES

(1) Adams, D.F.; Farwell, S.O.; Robinson, E.; Pack, M.R.; and Bamesberger, W.L. 1981. Biogenic sulfur source strengths. Env. Sci. Tech. 15: 1493-1498.

(2) Almgren, T., and Hagström, I. 1974. The oxidation rate of sulphide in sea water. Water Res. 8: 395-400.

(3) Bremner, J.M., and Steele, C.G. 1978. Role of microorganisms in the atmospheric sulfur cycle. In Advances in Microbial Ecology, ed. M. Alexander, vol. 2, pp. 155-201. New York: Plenum.

(4) Crutzen, P.J. 1982. Atmospheric interactions - homogeneous gas reactions of C, N, and S containing compounds. In Interactions of the Biogeochemical Cycles, eds. B. Bolin and R. Cook. Stockholm: SCOPE, in press.

(5) Ehhalt, D.H., and Schmidt, U. 1978. Sources and sinks of atmospheric methane. Pageoph. 116: 452-464.

(6) Granat, L.; Rodhe, H.; and Hallberg, R.O. 1976. The global sulphur cycle. In Nitrogen, Phosphorus, and Sulphur Global Cycles. SCOPE Report 7, Ecol. Bull. (Stockholm) 22: 89-134.

(7) Hahn, J., and Junge, C. 1977. Atmospheric nitrous oxide: A critical review. Z. Naturf. 32: 190-214.

(8) Hitchcock, D.R.; Spiller, L.L.; and Wilson, W.E. 1978. Biogenic sulfur compounds in coastal atmospheres of North Carolina. Research Triangle Park, NC: EPA.

(9) Jaeschke, W.; Georgii, H.W.; Claude, H.; and Malewski, H. 1978. Contributions of H_2S to the atmospheric sulfur cycle. Pageoph. 116: 465-475.

(10) Jørgensen, B.B. 1982. Mineralization of organic matter in the sea bed - the role of sulfate reduction. Nature 296: 643-645.

(11) Jørgensen, B.B. 1982. Processes at the sediment-water interface. In Interactions of the Biogeochemical Cycles, eds. B. Bolin and R. Cook. Stockholm: SCOPE, in press.

(12) King, G.M., and Wiebe, W.J. 1978. Methane release from soils of a Georgia salt marsh. Geochim. Cosmochim. Acta 42: 343-348.

(13) Knowles, R.; Lean, D.R.S.; and Chan, Y.K. 1981. Nitrous oxide concentrations in lakes: Variation with depth and time. Limnol. Oceanogr. 26: 855-866.

(14) Liss, P.S. 1982. The exchange of biogeochemically important gases across the air-sea interface. In Interactions of the Biogeochemical Cycles, eds. B. Bolin and R. Cook. Stockholm: SCOPE, in press.

(15) Lovelock, J.E.; Maggs, R.J.; and Rasmussen, R.A. 1972. Atmospheric dimethyl sulfide and the natural sulfur cycle. Nature 237: 452-453.

(16) Molongoski, J.J., and Klug, M.J. 1980. Anaerobic metabolism of particulate organic matter in the sediments of a hypereutrophic lake. Freshw. Biol. 10: 507-518.

(17) Rasmussen, R.A., and Khalil, M.A.K. 1981. Increase in the concentration of atmospheric methane. Atmos. Env. 15: 883-886.

(18) Rodhe, H., and Isaksen, I. 1980. Global distribution of sulfur compounds in the troposphere estimated in a height/latitude transport model. J. Geophys. Res. 85: 7401-7409.

(19) Rudd, J.W.M., and Hamilton, R.D. 1978. Methane cycling in a eutrophic shield lake and its effects on whole lake metabolism. Limnol. Oceanogr. 23: 337-348.

(20) Schmidt, U.; Kulessa, G.; and Röth, E.P. 1980. The atmospheric H_2 cycle. In Proceedings of the NATO Advanced Study Institute on Atmospheric Ozone: Its Variation and Human Influences, ed. A.C. Aikin, Algarve, Portugal, October 1-13, 1979.

(21) Seiler, W., and Conrad, R. 1981. Field measurements of natural and fertilizer-induced N_2O release rates from soils. APCA J. 31: 767-772.

(22) Söderlund, R., and Svensson, B.H. 1976. The global nitrogen cycle. In Nitrogen, Phosphorus, and Sulphur - Global Cycles. SCOPE Report 7. Ecol. Bull. (Stockholm) 22: 23-73.

(23) Sørensen, J. 1978. Occurrence of nitric and nitrous oxides in a coastal marine sediment. Appl. Env. Microbiol. 36: 809-813.

(24) Vogels, G.D. 1979. The global cycle of methane. Antonie van Leeuwenhoek 45: 347-352.

(25) Zinder, S.H., and Brock, T.D. 1978. Methane, carbon dioxide, and hydrogen sulfide production from the terminal methiol group of methionine by anaerobic lake sediments. Appl. Env. Microbiol. 35: 344-352.

Atmospheric Chemistry, ed. E.D. Goldberg, pp. 231-249. Dahlem Konferenzen 1982.
Berlin, Heidelberg, New York: Springer-Verlag.

The Production and Fate of Volatile Molecular Species in the Environment: Metals and Metalloids

F. E. Brinckman, G. J. Olson, and W. P. Iverson
Chemical and Biodegradation Processes Group
National Bureau of Standards, Washington, DC 20234, USA

Abstract. Forms of volatile environmental metal(loid)s cannot be predicted from thermodynamic considerations of redox conditions. In their transport to and from the atmosphere, they may be degraded, sorbed, or regenerated by both chemical and biological events at rates largely unknown, though measurable. Trapping and escape of these elements at ubiquitous aquatic surface microlayers can now be reconciled with new progress in correlating molecular geometries with air-water transport and the biogeochemistry of microenvironments.

INTRODUCTION

Increasingly, we perceive that biogeochemical cycling of a broad range of elements has implications for understanding their transport to and from the atmosphere. Such processes can be either by purely chemical reactions or by biological transformations that may or may not resemble abiotic events. In either case, or in combination, environmental factors must play a large role in our assessments, especially in conditions where oxidation and reduction of substrates and transport agents are involved. It cannot be presumed that under reducing situations gaseous species form which incorporate metal(loid)s only in lower oxidation states. Rather, high oxidation states and coordination numbers are also involved. For example, both $(CH_3)_4Sn$ and $(CH_3)_3As$ occur ubiquitously in the environment as oxidized and reduced forms of the metal(loid)s, respectively. Thus, reduction of metal(loid)s occurs in anaerobic environments (either abiotically or biotically) as well as in oxygenated environments. A basic point is also raised concerning the relationship between macro- and microenvironmental influences (Fig. 1) on overall redox processes. The photochemical reduction of atmospheric N_2 to NH_3 catalyzed on the TiO_2 surfaces of desert sands, annually producing $\sim 10^{10}$ kg of NH_3 worldwide (17) suggests that large-scale anaerobic zones are not principal factors in assessing abiotic reductions in nature. Indeed, such global reducing processes, heretofore relegated to a primordial or prebiotic reducing atmosphere, may play a much larger role than expected in our modern postbiotic oxygenated atmosphere.

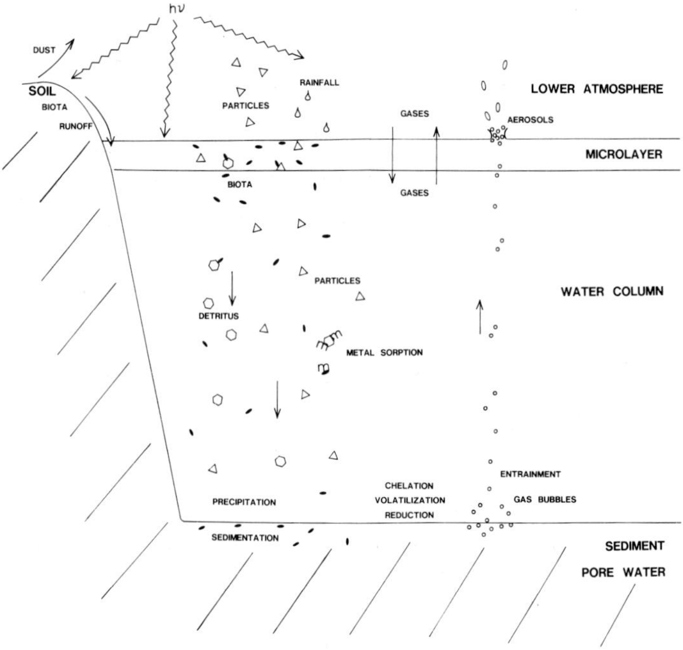

FIG. 1 - Macro- and microenvironmental compartments involved in the transport of metal(loid) species.

Similarly, caution must be used in applying thermodynamic principles to assessments of the forms and abundances of reduced or oxidized metal(loid) species in major or minor environmental compartments. Evaluation of kinetic aspects of transport or storage forms of the elements is necessary. Thus, bacteria and phytoplankton in the seas are responsible for global reduction of arsenate to arsenite, causing a much higher ratio of As(III) to As(V) in seawater than is predicted (3) on chemical grounds. Moreover, recent kinetic studies show that arsenite has a very long lifetime in aerobic waters (33).

Metal(loid)s in the atmosphere may result from direct anthropogenic inputs or originate from natural sources such as geothermal activity, dusts, aquatic aerosols (wave action, gas bubbles) (23), or biological formation of volatile metal(loid) hydrides or methylmetal(loid)s (35). Inputs of metal(loid)s to the hydrosphere from the atmosphere include aerosols, particles, rainfall, and gases. For some metals, such as mercury (12), rainfall is the major pathway.

It is the intent of this discussion to examine microenvironmental (cellular and molecular level) transformations resulting in volatile or mobile metal(loid) species, their synthesis, degradation, and transport.

PRODUCTION OF VOLATILE FORMS OF METAL(LOID)S

Microorganisms, whether aerobic, anaerobic, or facultative, are important factors in the production of some volatile metal(loid)-containing compounds in nature. Unfortunately, we have insufficient information regarding both the range and quantitative contribution of microbial metal(loid) transformations in the global cycles of these elements (12,35). Both true organometallic (metal-carbon bonded) and metallo-organic (metal-heteroatom bonded) compounds may be involved in atmospheric transport. The latter are generally more susceptible to hydrolysis and less hydrophobic or volatile but may explain reports of heavy metal volatilization by bacteria, for example, in the case of cadmium (19,29,35). Here, long-lived CH_3-Cd species are unlikely, owing to rapid protolysis by water. Rather, Cd-oxo-chelates may form from available organic keto-acids, or similar ligands, produced as metabolites to give complexes with significant vapor pressures. More information is available at present concerning true organometal(loid)s; consequently, we will focus our discussion on these.

A number of such volatile biogenic molecules are listed in Table 1 along with possible transient, unstable species.

TABLE 1 - Volatile biogenic methyl/hydride species of elements and possible transient molecules involved in transport.

Element	Methyl Form	Methyl/Hydride	Hydride
Antimony	$(CH_3)_3Sb$*	$(CH_3)_2SbH$*	SbH_3*
Arsenic	$(CH_3)_3As$	$(CH_3)_2AsH$ CH_3AsH_2	AsH_3
Halogen	CH_3I CH_3Br CH_3Cl		HI HBr HCl
Lead	$(CH_3)_4Pb$	$(CH_3)_3PbH$*	PbH_4*
Mercury	$(CH_3)_2Hg$ $(CH_3)_2Hg_2$*	CH_3HgH*	HgH_2*
Selenium	$(CH_3)_2Se$ $(CH_3)_2Se_2$ $(CH_3)_2SeO_2$	CH_3SeH* CH_3Se_2H*	H_2Se* H_2Se*
Tellurium	$(CH_3)_2Te$ $(CH_3)_2Te_2$	CH_3TeH* CH_3Te_2H*	H_2Te* H_2Te_2*
Thallium	$(CH_3)_3Tl$*	$(CH_3)_2TlH$*	
Tin	$(CH_3)_4Sn$	$(CH_3)_3SnH$ $(CH_3)_2SnH_2$ CH_3SnH_3*	SnH_4*

*Very unstable and/or not yet reported; analogous Si and Ge compounds merit consideration but are not listed.

These forms result from bioalkylation (biomethylation), in which methyl carbanions, carbonium ions, or radicals from biogenic methyl donors (e.g., methylcobalamin) are transferred to the metal(loid) (28), or hydridization where the metal(loid) accepts hydrogen atoms from a reductase. In some cases, both methylation and hydridization occur on the same element (20), though highly unstable products may result and decompose into involatile or new volatile products:

$$Sn(IV)_{aq} \xrightarrow[CH_3^-]{H^- \quad \textit{Pseudomonas}\ \text{spp.}} (CH_3)_nSnH_{4-n}\uparrow .$$

Many microorganisms enzymatically reduce Hg(II), possible via Hg-H bonds, to volatile Hg^o (19). Such volatile forms of metal(loid)s diffuse away from microbial cells, hence volatilization is viewed as a detoxifying mechanism. Microbial methylation may also occur indirectly in the absence of active cells by reaction of metabolic products with metal(loid)s (Table 2), for example, methyl iodide, produced by certain marine algae, methylates Pb(II) and Pb^o (21), producing methyllead salts and $(CH_3)_4Pb$.

TABLE 2 - Formation of volatile metal species by representative microbial metabolites.

Metabolite	Species/Conditions	Product
Acetate, methanol, ethanol	$Hg(II)_{aq}$/sunlight	$CH_3Hg^+_{aq}$
Propionic acid		$CH_3CH_2Hg^+_{aq}$
Methyl iodide	Pb^o, Pb(II)/water	$(CH_3)_nPb^{(4-n)+}_{aq}$
	Sn(II)/water	$(CH_3)_nSn^{(4-n)+}_{aq}$
Sulfide	CH_3Hg^+/water	$(CH_3)_2Hg_g$
	$(CH_3)_3Pb^+$/water	$(CH_3)_4Pb_q$
	$(CH_3)_3Sn^+$/water	$(CH_3)_4Sn_g$
Amino acids	$Hg(II)_{aq}$/ultraviolet	$CH_3Hg^+_{aq}$

In the presence of common microbial metabolic end products (acetate, methanol, ethanol) and ultraviolet light or sunlight, Hg(II) is methylated to CH_3Hg^+ (1). Biogenic H_2S, which forms insoluble metal sulfides, also causes redistribution reactions of involatile methylmetals through a covalent sulfide complex to highly volatile permethyl forms. For example, with aquated methylmercury cation (12,29),

$$2CH_3Hg^+ + H_2S \rightarrow [(CH_3Hg)_2S] + 2H^+ \rightarrow (CH_3)_2Hg\uparrow + HgS.$$

Trimethyllead and trimethyltin salts in water undergo analogous reactions with H_2S to form gaseous $(CH_3)_4Pb$ and $(CH_3)_4Sn$ (12,35). Finally, we cannot exclude the possibility that other known volatile, covalent forms of metal(loid)s may be produced by combinations of chemical and biotic events typified by Table 2. For example, global distributions of ethylene, NO, and CO suggest efforts to detect such well-known species as gaseous $Ni(CO)_4$ may be rewarded, since low-temperature formation of this toxic substance has long been a concern where traces of Ni and CO occur (7).

It should be stressed that microbial metal(loid) volatilizing processes are not limited to anaerobic environments. For many of the elements listed in Table 1, either aerobic or anaerobic conditions may lead to the production of volatile species. Additionally, in natural systems (Fig. 1), anaerobic microzones may exist in predominantly aerobic environments, so that a consideration of volatile metal(loid) species production should not be limited to an examination of anoxic basins, sediments, or soils.

LIFETIMES OF METAL(LOID) SPECIES IN WATER

Homogeneous Chemical Reactions

Once formed, metal(loid)-containing molecules may undergo slow decay, although derivations of aqueous solution properties from thermodynamic functions presume indefinite lifetimes for organometal(loid)s. While it is true that most of the coordinatively-saturated species we consider resist protolysis in pure water, solution conditions common in environmental waters may produce rapid and possible irreversible changes. For example, with As (9):

$$(CH_3)_3As + \frac{1}{2}\ O_2 \rightarrow (CH_3)_3AsO \quad \text{(oxidation)}$$
$$(CH_3)_3As + CH_3I \rightarrow (CH_3)_4As^+ + I^- \quad \text{(quarternization)}$$
$$(CH_3)_3As + Hg^{2+} \rightarrow (CH_3)_3As^{2+} + Hg^o \quad \text{(oxidation)}$$
$$(CH_3)_3As + \text{metal} \rightarrow (CH_3)_3As{\rightarrow}M \quad \text{(complexation)}.$$

Mainly, second and pseudo first-order kinetics are involved in such solution chemistry, though some species spontaneously decompose by first-order rates (35),

$$(CH_3)_3Sn^+ + TlCl_3 \rightarrow (CH_3)_2Sn^{2+} + [CH_3TlCl_2] \rightarrow CH_3Cl\uparrow + TlCl.$$

Such reductive elimination reactions are prevalent for metals in protic media and may constitute another source for halocarbons in the aquatic surface microlayer (Fig. 1) where high metal and organic concentrations arise. There, exposed to sunlight, competitive reactions may also occur:

$$Hg^{2+} + CH_3CO_2^- \xrightarrow{h\upsilon} CH_3Hg^+ + CO_2\uparrow$$
$$CH_3HgCl \xrightarrow{h\upsilon} Hg^o + CH_3Cl\uparrow + C_2H_6\uparrow.$$

Volatilization of "active" methyl groups from such a "methyl pool" (Table 2) may strongly shift these reactions to reduced forms of metals, even though in aerobic environments. We cannot yet say where the balance for such competing processes lies in air-water microzones.

In bulk seawater, a tentative estimate of $(CH_3)_3Sn^+_{aq}$ species' residence was calculated based on reported concentrations and a laboratory mechanism limited to demethylation by $HgCl_n^{(2-n)+}{}_{aq}$; this was 6×10^5 y, roughly comparable to residence times for total Sn and Hg in the oceans (8). The recent discovery (20) of methylstannanes in estuarine waters at $>10^{-10}$M, where chemistries of anthropogenic inputs are dominant, suggests that alternative decomposition pathways chiefly involving oxidation of polar Sn-H bonds may occur. Such reaction, however, may be slow in view of reported concentrations of organic oxidants in natural waters and rates of ozonization for alkyltins (2,24), resulting in half-lives of <0.5 d. Here again, the transport rate of such molecules in surface waters, effects of particulate adsorption, or sparging by microbubbles, and partition through the surface microlayer are unknown.

Biophysical Phenomena

Metal(loid) species in the aquatic environment are readily adsorbed on particles, sediment, detritus, and microbiota (12,22). Suspended, precipitated, or sedimented metal(loid)s may be mobilized by reduction, chelation, volatilization, or physical entrainment in gas (usually CH_4) microbubbles (Fig. 1). Of special note is the cellular component which is potentially involved in metal(loid) bioaccumulation (12). Microorganisms growing in a metal-stressed environment, for example, are grazed upon by higher organisms resulting in substantial food chain metal bioconcentration (5).

In some cases organometals are biodegraded to the inorganic metal and hydrocarbon, as with methylmercury which is degraded to Hg^o and CH_4 by sediment bacteria, or dimethylarsinic acid which is degraded to arsenite by marine microorganisms (3,32).

Microorganisms or microbial metabolic products may also sequester metal(loid)s, preventing potential volatilization processes. Cells may take up metal(loid)s, either for metabolic processes or as a mechanism of detoxification (29). Some metals are required as enzyme cofactors (Mg, Co, Fe, Mo, Zn, Cu) and are significantly concentrated from solution by cellular uptake mechanisms. Toxic metals may be bound and immobilized in the cell envelope or extracellularly by secreted polymers (19). Organometals are readily soluble in membrane lipids and may concentrate at the surfaces of cells (31). Metals are precipitated by biogenic hydrogen sulfide, but even metals in the most insoluble sulfides (Hg) may be slowly methylated (19). Microbial activities likewise can physically mobilize metal(loid)s by producing various gases under anoxic conditions.

Bubbles of biogenic gases, especially methane, are considered to be potentially important in producing metal-containing aerosols, but little is known of the metal content in aerosols (23).

PREDICTION OF MOLECULAR AIR-WATER TRANSPORT

Global interactions affecting the lower atmosphere chemistry of metal(loid)s depend upon bulk solution and desorption across the air-water interface as modified by microenvironmental events involving gases, aerosols, and particulates (23). The most promising array of chemical pathways for modeling abiotic and biogenic transport can be thought of in terms of air-water solvation and degassing. Thus, properties of molecular solubility, vapor pressure, or partition coefficients coupled with basic ideas about metal(loid) species' geometry and substituents offer tools for predicting behavior of many air-water transport systems. Requisite physicochemical data are scarce, hence we must evaluate these prospects in terms of individual molecular features.

Molecular Coordinative Saturation

A remarkable range of coordination numbers (CN) and molecular geometries is anticipated for many compounds (Table 1). The situation becomes more complex in cases where aquation results in major, even irreversible, alteration of the CN, stereochemistry, or oxidation state. Figure 2 compares documented changes by approximate structural class for some representative environmental molecules in gaseous (atmospheric) and aquated (hydrospheric) situations.

Generally, the bond polarity for various substituents, e.g., its ionizability or covalency, along with the overall stereorigidity of a molecule dictates the degree to which solvation occurs (34). In a compact, "spherical" molecule, such as $(CH_3)_4Pb$, the central metal is at maximum oxidation state and coordinatively saturated with four symmetric strong covalent metal-carbon σ bonds characteristic for many post-transition metal elements (11). These species resist protolysis by water.

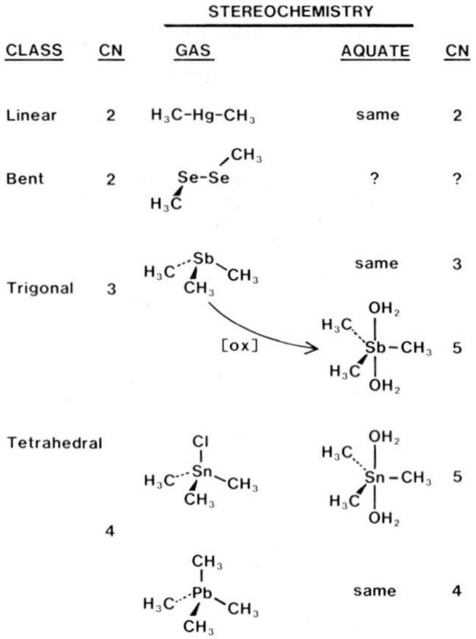

FIG. 2 - Molecular alterations between air and water for some known environmental methylmetal(loid)s.

Such molecules, including $(CH_3)_2Hg$, CH_3I, $(CH_3)_3As$, and $(CH_3)_4Sn$, are typically highly volatile in view of their molecular weights and show low water solubility.

Substitution of a methyl or similar alkyl group on these elements by a more polar ligand both reduces the symmetry of the molecule and introduces considerable bond ionicity. The case of $(CH_3)_3SnCl$ illustrates an important intermediate situation where the polar but covalent Sn-Cl bond still confers high volatility in air, though on exposure to water solvation and ionization occurs,

$$(CH_3)_3SnCl_{(g)} + H_2O_{(l)} = (CH_3)_3Sn^{+}{}_{aq} + Cl^{-}.$$

Cleavage of CH_3-Sn bonds does not occur, though extensive electronic and spatial reorganization takes place to yield the involatile trigonal bipyramidal cation depicted in Fig. 2 (36). Substitution of Cl by another polar ligand, hydride, commonly performed in the speciation of these environmental molecules (6,18,20), illustrates a further point,

$(CH_3)_3Sn^+_{aq} + H^-_{aq} \rightarrow (CH_3)_3SnH_{(g)}\uparrow$.

Here, the resultant methylstannane, bearing a polar covalent Sn-H bond, forms as a distorted tetrahedral molecule of great volatility and hydrophobicity which rapidly degasses from aqueous solution. These factors, coupled with the completeness of the reaction, underlie its analytical utility and imply an aquatic transport path for an otherwise reactive molecule. These considerations support the likelihood of additional methylmetal(loid) hydrides similar to tin (20) and arsenic (9) cases in environmental media (Table 1).

Implications of Molecular Geometry for Air-Water Transport

Spurred by Hansch's pioneering work (13), rapid progress in the past ten years has brought the development of quantitative means to correlate molecular architecture with solution properties relevant to environmental transport and biological activity. Earlier work united empirical additive-constitutive relationships with solubility, volatility, and toxicity, among other factors. Newest results go much further in successfully correlating detailed molecular topologies with hydrophobic properties of many classes of organic molecules (14). Logical extension of these predictive techniques to covalent organometal(loid)s is now appropriate and desirable.

Currently, emphasis is placed upon calculating "total surface area" (TSA), derived from bond distances, angles, and Van der Waal radii fitted within conformational or stereochemical possibilities expected for a hydrophobic molecule situated in a polar medium (38). Another approach treats "bond connectivities," that is, bases calculations on central atom stereochemistry (10). Either technique offers an elegant prospect for evaluating solution properties of simple neutral organometal(loid)s, and possibly even their more complex aquated ionic products (Fig. 2). Firmly grounded in solution thermodynamics (14,34), these calculations rely upon experimental data independent of solution measurements and serve to accurately predict equilibrium partitioning between disparate phases, viz.,

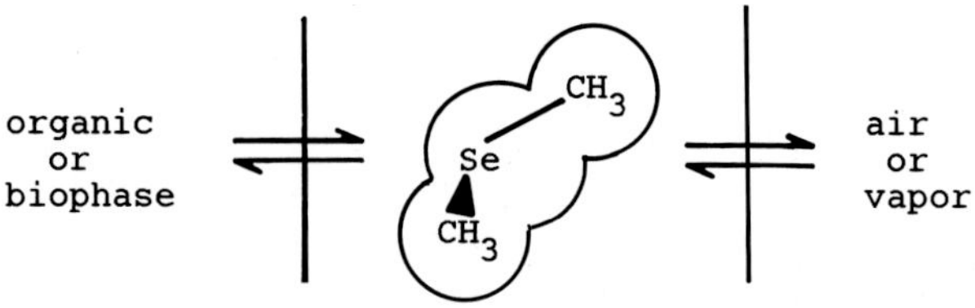

Not only do reported methods permit quantitative structure-activity relationships (QSAR) for predicting homogeneous solution behavior, but free-energy terms (activities) for estimating partition coefficients of solvent-surface and air-solvent transport are possible. For example, the expanding field of high performance liquid chromatography (HPLC) employs hydrophobic reverse bonded-phase microparticulate column substrates that allow direct equilibrium measurements for modeling lipid-water partition coefficients of sparingly soluble molecules (37). Other QSAR estimates of solubility (S_i) and vapor pressure (P_i^o) lead directly to values for mass transfer rate coefficients of a covalent molecule (i) via appropriate Henry's Law constants, H_i:

$$H_i = \frac{P_i^o M_i}{RT} \left(\frac{1}{S_i}\right), \qquad (1)$$

where M_i = molecular weight and RT is the usual gas constant term. Experimentally, a convenient marker or tracer compound, usually O_2, can be used to determine aeration rates in the laboratory or field relative to the molecule i. Under those conditions, where mass transfer is controlled by liquid film resistance ($H_i > 0.1$), the thin-film model predicts (26) that the overall mass transfer rate coefficient, β_i, is proportional to the molecular transfer coefficient of the volatile molecule k_i and O_2 (k_{O_2}). It is independent of species concentrations, temperature, and water turbulence, and is also proportional to solution molecular diffusion coefficients (D) (27):

$$\beta_i = \frac{k_i}{k_{O_2}} = \frac{D_i}{D_{O_2}} = \left(\frac{V_{c,O_2}}{V_{c,i}}\right)^{\delta}. \qquad (2)$$

Diffusion coefficients can also be roughly estimated from critical volumes V_c, thereby yielding an empirical constant δ (about 0.3 to 0.6). Reaeration coefficients have been determined for various streams and rivers in the United States; combined with laboratory-generated β values, volatilization rates for a number of organic pollutants were determined. Such measurements should be possible for organometal(loid)s, but it should be noted that competitive processes such as sorption on particulates or biological uptake deplete concentrations of such solutes in true solution (Fig. 1) and reduce the volatilization rate.

With the foregoing in mind, we tabulated from literature two correlations of molecular geometries of environmental organometal(loid)s with available solubility data and predicted that thin-film transfer rates typically needed information for estimating transport phenomena (Table 3).

Thus, prominent environmental methylmetal(loid)s correlate well with other methyl elements over a broad range of water solubility, based on available molecular topologies (Fig. 3). The expected relationship between molecular size (TSA) and V_c (26) allows approximate estimates of β_i from molecular dimensions

TABLE 3 - Correlation of organometal(loid) geometry with aqueous solubility and air-water transfer rate coefficients.

Compound	TSA, $Å^2$ [a]	Solubility[b,c] obsd.	calcd.	V_c, cm^3mol^{-1}	β_i [d]
CH_3Br	73.1	0.81	0.81	149	0.79
CH_3I	82.1	1.000	1.149	176	0.75
$(CH_3)_2Hg$	104.8	1.884	2.018	242	0.67
$(CH_3)_4C$	128.1	2.9	2.611	288	0.64
$(CH_3)_3Sb$	120.3	3.126	2.908	311	0.62
$(CH_3)_4Sn$	153.1	3.637	3.865	385	0.58

[a]From (38); [b]-log molal; [c]r = 0.975; [d]relative to O_2; δ = 1/3.

to be compared with experimental data, relative to O_2 and common pollutants such as benzene ($\beta = 0.53$), $CHCl_3$ (0.66), and CO_2 (0.89). As with organic nonelectrolytes, increasing salinity (or ionic strength) in the aqueous solution results in "salting out" or decreased solubility of organometal(loid)s as predicted by the Setchenov relationship for the salinity dependence of partition coefficients (34).

LIFETIMES OF METAL(LOID) SPECIES IN THE ATMOSPHERE

For certain metal(loid)s, especially Hg, As, and Se, biological processes may be significant in global element cycling, since large fluxes of these metal(loid)s from the sea surface are required for mass balancing (23). Andreae (4), however, calculates that mass balances for arsenic do not require significant biological volatilization processes in the oceans, and he did not detect organoarsenicals in rainwater, indicating that microbial As volatilization is not significant. The relative lifetimes of such transport molecules in water and air are important considerations. Gas phase oxidation of $(CH_3)_3As$ is much slower in air ($k_2 \sim 10^{-6}M^{-1}s^{-1}$) than in solution ($< 10^2M^{-1}s^{-1}$). In contrast, a very fast reaction occurs between gaseous $(CH_3)_3Sb$ and O_2 ($\sim 10^3M^{-1}s^{-1}$) (9). In field measurements, the arsine is detected in air whereas the stibine is not, suggesting a rough predictor for environmental transport from aquatic media.

Molecular Escape and Trapping at the Surface Microlayer

Volatile metal(loid) species must pass the aquatic surface microlayer enroute to the atmosphere (Fig. 1). The microlayer, though widespread, is an ill-defined and narrow (20 nm to 200 μm) zone containing high concentrations of microorganisms, organic compounds, hydrophobic molecules, and metal(loid)s compared to the underlying waters (25). Organometals may concentrate three orders of magnitude in the surface microlayer (Chau et al., personal communication). The high levels of reactants in the microlayer may promote concentration-dependent reactions, yet the surface film character and large surface area ensure that solar photolytic reactions occur. Thus, the microlayer is a dynamic system, both a sink and a source of volatile metal(loid)s. The production and escape of known atmospheric metal species from the microlayer is depicted in Fig. 3.

Partition between the air-water interface occurs by both physical (ejection) and chemical processes, and surviving molecules, such as the three depicted, are liberated into the atmosphere. Harrison and Laxen (15,16) detected volatilization of $(CH_3)_4Pb$ and subsequent atmospheric transport from biogenic sources in tidal flats; they estimate that OH serves as a primary sink for organolead gases with solar modulated breakdown of > 29 percent h^{-1}. Tetramethyltin is probably more resistant to photolysis and attack by OH, while methylstannanes would be more susceptible, though similar decomposition pathways are reasonable. Evidence for extensive but sporadic washout of

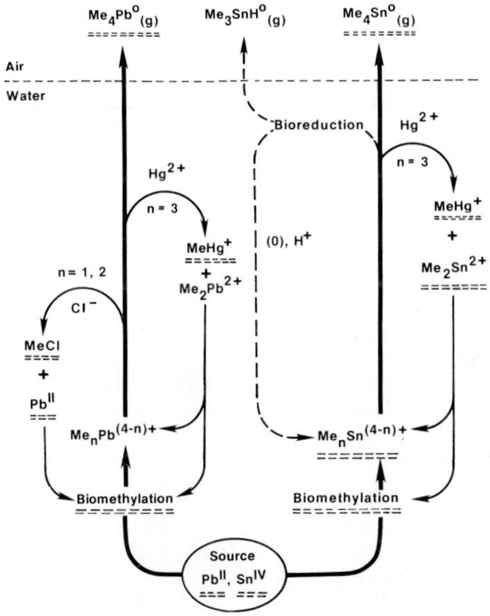

FIG. 3 - Known (= = =) species and possible pathways for tin and lead flow from particulate sources to the atmosphere.

$(CH_3)_nSn^{(4-n)+}$ species during rainfalls has been presented (6), and suggests a very profitable line of future research.

CONCLUSIONS

Adequate information about the production, interchange, and fate of volatile metal(loid) molecules is lacking, making global flux estimates difficult (23). The full range of such species supporting environmental transport of elements is unknown, and current chemical experience from the laboratory has not until recently provided guidance or directions for new studies. In cases where we know of certain metal(loid) volatilization reactions (particularly in aqueous media), the environmental significance is diminished because competing biotic and abiotic reactions in the natural environment are unknown. Our current position can be advanced, however, by determining molecular and thermodynamic properties of diagnostic or rate-determining transport species, establishing the relative contributions from abiotic and biotic processes, and developing appropriate ultratrace measurement and identification methods for the field.

REFERENCES

(1) Akagi, H.; Fujita, Y.; and Takabatake, E. 1975. Photochemical methylation of inorganic mercury in the presence of mercuric sulfide. Chem. Lett. (Japan): 171-176.

(2) Aleksandrov, Y.A., and Tarunin, B.I. 1974. Ozonolysis of tetraalkylstannanes. Yh. Obshch. Khim. 44: 1835.

(3) Andreae, M.O. 1979. Arsenic speciation in seawater and interstitial waters: The influence of biological-chemical interactions on the chemistry of a trace element. Limnol. Oceanogr. 24: 440-452.

(4) Andreae, M.O. 1980. Arsenic in rain and the atmospheric mass balance of arsenic. J. Geophys. Res. 85: 4512-4518.

(5) Berk, S.G., and Colwell, R.R. 1981. Transfer of mercury through a marine microbial food web. J. Exp. Mar. Biol. Ecol. 52: 157-172.

(6) Braman, R.S., and Tompkins, M.A. 1979. Separation and determination of nanogram amounts of inorganic tin and methyltin compounds in the environment. Anal. Chem. 51: 12-19.

(7) Brief, R.S.; Blanchard, J.W.; Scala, R.A.; and Blackev, J.H. 1971. Metal carbonyls in the petroleum industry. Arch. Environ. Health 23: 373-384.

(8) Brinckman, F.E. 1981. Environmental organotin chemistry today: experiences in the field and laboratory. J. Organometal. Chem. Library 12: 343-384.

(9) Brinckman, F.E.; Parris, G.E.; Blair, W.R.; Jewett, K.L.; Iverson, W.P.; and Bellama, J.M. 1977. Questions concerning environmental mobility of arsenic: needs for a chemical data base and means for speciation of trace organoarsenicals. Environ. Health Perspect. 19: 11-24.

(10) Cammarata, A. 1979. Molecular topology and aqueous solubility of aliphatic alcohols. J. Pharm. Chem. 68: 839-842.

(11) Cotton, F.A., and Wilkinson, G. 1980. Advanced Inorganic Chemistry, 4th ed. New York: Wiley Interscience.

(12) Craig, P.J. 1980. Metal cycles and biological methylation. In The Handbook of Environmental Chemistry, Part A, ed. A.O. Hutzinger, vol. 1, pp. 169-227. New York: Springer-Verlag.

(13) Hansch, C. 1969. A quantitative approach to biochemical structure-activity relationships. Acc. Chem. Res. 2: 232-239.

(14) Hansch, C.; Quinlan, J.E.; and Lawrence, G.L. 1968. The linear free-energy relationship between partition coefficients and the aqueous solubility of organic liquids. J. Org. Chem. 33: 347-350.

(15) Harrison, R.M., and Laxen, D.P.H. 1978. Natural source of tetraalkyllead in air. Nature 275: 738-739.

(16) Harrison, R.M., and Laxen, D.P.H. 1978. Sink processes for tetraalkyllead compounds in the atmosphere. Env. Sci. Technol. 12: 1384-1392.

(17) Henderson-Sellers, A., and Schwartz, A.W. 1980. Chemical evolution and ammonia in the early Earth's atmosphere. Nature 287: 526-528.

(18) Hodge, V.F.; Seidel, S.L.; and Goldberg, E.D. 1979. Determination of tin(IV) and organotin compounds in natural waters, coastal sediments, and macro alqae by atomic absorption spectrometry. Anal. Chem. 51: 1256-1259.

(19) Iverson, W.P., and Brinckman, F.E. 1978. Microbial metabolism of heavy metals. In Water Pollution Microbiology, ed. R. Mitchell, vol. 2, pp. 201-232. New York: John Wiley & Sons.

(20) Jackson, J.A.; Blair, W.R.; Brinckman, F.E.; and Iverson, W.P. 1982. Gas-chromatographic speciation of methylstannanes in the Chesapeake Bay using purge and trap sampling with a tin-selective detector. Env. Sci. Technol. 16: 110-119.

(21) Jarvie, A.W.P., and Whitmore, A.P. 1981. Methylation of elemental lead and lead(II) salts in aqueous solution. Env. Technol. Lett. 2: 197-204.

(22) Kelly, D.P.; Norris, P.R.; and Brierley, C.L. 1979. Microbiological methods for the extraction and recovery of metals. In Microbial Technology: Current State, Future Prospects, eds. A.T. Bull, D.C. Ellwood, and C. Ratledge, pp. 263-308. Cambridge University Press.

(23) Lantzy, R.J., and Mackenzie, F.T. 1979. Atmosphere trace metals: global cycles and assessment of man's impact. Geochim. Cosmochim. Acta 43: 511-525.

(24) Larson, R.A.; Symkowski, K.; and Hunt, L.L. 1981. Occurrence and determination of organic oxidants in rivers and waste waters. Chemosphere 10: 1335-1338.

(25) Lion, L.W., and Leckie, J.O. 1981. The biogeochemistry of the air-sea interface. Ann. Rev. Earth Plan. Sci. 9: 449-486.

(26) Matter-Müller, C.; Gujer, W.; and Giger, W. 1981. Transfer of volatile substances from water to the atmosphere. Water Res. 15: 1271-1279.

(27) Rathburn, R.E., and Tai, D.Y. 1981. Technique for determining the volatilization coefficients of priority pollutants in streams. Water Res. 15: 243-250.

(28) Ridley, W.P.; Dizikes, L.J.; and Wood, J.M. 1977. Biomethylation of toxic elements in the environment. Science 197: 329-332.

(29) Robinson, J.W., and Kiesel, E.L. 1981. Methylation of cadmium with vitamin B_{12}: a possible method of detoxification. J. Env. Sci. Health A16: 341-352.

(30) Rowland, I.R.; Davies, M.J.; and Grasso, P. 1977. Volatilization of methylmercuric chloride by hydrogen sulphide. Nature 265: 718-719.

(31) Silverberg, B.A.; Wong, P.T.S.; and Chau, Y.K. 1976. Ultrastructural examination of Aeromonas cultured in the presence of organic lead. Appl. Env. Microbiol. 32: 723-725.

(32) Spangler, W.J.; Spigarelli, J.L.; Rose, J.M.; and Miller, H.M. 1973. Methylmercury: bacterial degradation in lake sediments. Science 180: 192-193.

(33) Tallman, D.E., and Shaikh, A.U. 1980. Redox stability of inorganic arsenic(III) and arsenic(V) in aqueous solution. Anal. Chem. 52: 196-199.

(34) Tanford, C. 1980. The Hydrophobic Effect: Formation of Micelles and Biological Membranes, 2nd ed. New York: Wiley Interscience.

(35) Thayer, J.S., and Brinckman, F.E. 1982. The biological methylation of metals and metalloids. Adv. Organometal. Chem. 20: 313-356.

(36) Tobias, R.S. 1966. σ-bonded organometallic cations in aqueous solutions and crystals. Organometal. Chem. Rev. 1: 93-129.

(37) Unger, S.H.; Cook, J.R.; and Hollenberg, J.S. 1978. Simple procedure for determining octanol-aqueous partition, distribution, and ionization coefficients by reverse-phase high-pressure liquid chromatography. J. Pharm. Sci. 67: 1364-1367.

(38) Valvani, S.C.; Yalkowsky, S.H.; and Amidon, G.L. 1976. Solubility of nonelectrolytes in polar solvents. VI. Refinements in molecular surface area computations. J. Phys. Chem. 80: 829-835.

Group on
Biogenic Contributions to Atmospheric Chemistry

Standing, left to right:
Neil Anderson, Fabrizio Bruner, Heinz Bingemer, Ulrich Schmidt, Fred Brinckman, Bo Jørgensen, and Dieter Ehhalt.

Seated, left to right:
Peter Liss, Jim Lovelock, Wolfgang Balzer, Rolf Hallberg, and Andy Andreae.

Atmospheric Chemistry, ed. E.D. Goldberg, pp. 251-272. Dahlem Konferenzen 1982.
Berlin, Heidelberg, New York: Springer-Verlag.

Biogenic Contributions to Atmospheric Chemistry Group Report

M. O. Andreae, Rapporteur
N. R. Andersen, W. Balzer, H. G. Bingemer, F. E. Brinckman,
F. Bruner, D. H. Ehhalt, R. O. Hallberg, B. B. Jørgensen,
P. S. Liss, J. E. Lovelock, U. Schmidt

INTRODUCTION

With a subject matter spanning the range from microbial metabolism and physiology to the physical and chemical dynamics of wet and dry deposition on plant surfaces, our selection of topics had to be an eclectic one, constrained as much by the range of individual expertise assembled as by the time available for discussion. A widely felt concern about the influence of methodological considerations on the validity of the results led us to consider in some detail the requirements of experimental design while not closing our eyes to the gifts of serendipity. When examining the scales of the biological processes interacting with the world atmosphere, we found it difficult to maintain traditional dualistic views: the concepts of oxic vs. anoxic environments, of microbiological vs. macrobiological processes turned into continua of physicochemical and biological scales with the respective end members having only a conceptual significance at best.

The title of this group report appears to put a large emphasis on biogenic sources; in fact this was reflected by the amount of our discussion spent on this aspect of the element cycles. This emphasis was forced upon us by the fact that the sources, as well as sometimes the ultimate sinks, are in most cases the least uncertain part of the cycles of biologically influenced material cycles. We discovered the most serious gaps in our knowledge to be in the fate of biogenic substances once they have left their immediate source environment. This refers, in particular, to the processes that modulate the emissions of biologically produced substances into the various compartments of the world atmosphere (continental/marine troposphere, boundary layer(s), free troposphere, stratosphere, etc.), as well as to the processes which interconvert different chemical species of the elements and interconnect the different element cycles. A great need was felt to deepen our understanding of the biosphere as a sink for atmospheric constituents. This includes both an interest in the impact of chemicals of natural and man-made origin on the biosphere (including, of course, man and man-made environments) and in the role of biota in the removal of atmospheric constituents.

A number of important compounds, influenced at least on long time scales by biological activities, had to be omitted from our discussions for various reasons: the main atmospheric constituents N_2 and O_2, as well as CO_2, and the direct releases of biogenic particulate matter, e.g., organic material in sea-spray and the waxy particles flaking off plant surfaces. This group report contains a bird's-eye view of the discussion of our group and the contributions made by members of other working groups during these discussions (details can be found in the background papers and references). We separated our debate into a section on general concepts and one on case studies; the separation will be retained in this report.

GENERAL CONCEPTS

Element Cycles

For some time the concept of element cycles has been one of the organizing principles of marine and atmospheric science. It structures processes in the environment into compartments, sources, pathways, and sinks, in a sense comparable to the chemical engineer's containers, reactors, pipes, and valves. This provides a valuable tool in organizing the complex interactions within and between the realms of the lithosphere, hydrosphere, biosphere, and atmosphere. Quantification of the reservoirs and fluxes within the cycles allows the introduction of time scales and mass balances and supplies us with educated guesses about which components may be dominant and which trivial. When mass balances show significant discrepancies between sources and sinks, they may point out important processes which have not yet been considered.

However, when making the step from considering the abstract concept of element cycles to attempting to put together a cycle for a specific element, we encounter serious difficulties: the compartments have to be defined and boundaries drawn, processes and fluxes have to be identified, and estimates for the sizes of compartments and fluxes have to be made. Often it is not clear at all where to put the boundaries: when considering the release of N_2O by soil biota, should we put the boundary within the soil at the top of the productive zone, or at the soil surface, or at the top of the leaf canopy, or maybe at the top of the surface boundary layer, or at the top of the planetary boundary layer? The answer to this question requires consideration of the objective of constructing our model and cannot be given in general. Putting together the relevant fluxes between the compartments is a similarly vexing task, especially when quantitative estimates are sought: it implies that the relevant processes are known, which they often are not, and that accurate and representative data are available, which usually is not the case for the cycles of interest. The results of such endeavors thus have to be viewed with much caution, as there is often a tendency to mold the uncertainty of the data in order to make a point. With these caveats in mind, however, we feel that the concept of elemental cycles has great power as an organizing principle and investigative tool, both for the study of global and local cycles.

Transport Processes

Transport from one compartment to another normally requires crossing some interface between the compartments, e.g., the air/sea interface, the boundary between anoxic and oxic environments, atmospheric inversion layers, etc. The characteristics of this interface, and the physical, thermodynamical, and biological processes associated with it, limit the transfer across compartment boundaries. In this context, the specific chemical and biochemical characteristics of the transfer species have to be considered as well as those of the interface. The massive bacterial production of H_2S in anoxic sediments, even in tidal flats exposed to the atmosphere, does not result in a comparable emission of H_2S to the atmosphere: a bacterial layer, often only a fraction of a millimeter thin, lives on the oxidation of this gas, preventing its emission to the atmosphere. On the other hand, volatile substances produced in much smaller amounts in the same anoxic environment, e.g., CS_2, COS, dimethylsulfide (DMS), etc., may diffuse through this bacterial barrier without much loss. In fact, their low concentrations in the surface ocean may contribute to their successful escape, as no bacterial population may find it worthwhile to specialize in their consumption. Similar observations have been made for other gas species: in fertilized soils, a significant microbial production of N_2O has been observed just a few centimeters below the soil surface; essentially all of this N_2O is scavenged before it can escape from the soil into the atmosphere. From these observations it is very clear that extreme caution has to be used not to disturb or modify such boundaries when making experiments designed to measure intercompartment fluxes, e.g., with flux chambers.

The permeability of the compartment boundaries may be strongly influenced by environmental variations. For example, the production of oxygen in algal mats prevents the escape of sulfur gases during the day, while at night anoxia pervades the algal layer and the sulfur gases are free to diffuse into the overlying water. In such situations it will be very important to make measurements on a diurnal basis: the most important processes may be missed by the observer who works only "from nine to five."

A considerable amount of uncertainty still exists in the quantitative application of transfer models for gases across the air/sea interface, introducing an uncertainty of as much as a factor of two into estimates of global fluxes of gases into or out of the ocean. In view of the increasing degree of accuracy to which we can estimate worldwide averages of surface ocean concentrations of some volatile substances, e.g., DMS, improvements in the accuracy and precision of transfer rate estimates, especially on a local scale, are highly desirable. Cooperation with physical and chemical oceanographers studying air/sea exchange processes should be encouraged. The influence of possible near-surface gradients in the ocean, including those in the sea-surface microlayer, and their influence on gas exchange rates should be investigated. In addition, transport processes across the soil/air interface and across the surfaces of plants are poorly understood and deserve intensive study.

Once in the atmosphere, transport and fate of the compounds produced by the biosphere depend on atmospheric mixing dynamics and chemical, especially photochemical, reactivity. Only the most stable compounds are able to be subject to long-range transport and to reach the stratosphere. Unusually low photochemical activity, e.g., under large cloud regions or in the polar night, may allow the long-range transport of otherwise rather unstable substances, e.g., HCHO and peroxyacetylnitrate (PAN).

The Redox Dimension: Oxic Vs. Anoxic Environments

It would appear natural to separate the world into oxic and anoxic environments: this separation reflects the history of planetary evolution and seems to provide an organizing principle for the type of biochemistry expected to produce reduced or oxidized volatile species. It must be recognized, however, that biologically active oxic environments, especially those that contain photosynthetic organisms, have abundant reducing capability in the form of enzymatic systems, e.g., NADP/NADPH, which are able to perform any reduction observed in anoxic environments. It is in fact the oxic environments which supply the reducing power to the anoxic ones in the form of reduced carbon. This carbon can be returned rather effectively by anoxic biota to the oxic zone, as in the case of methanogens in lake sediments which return 95% of the carbon supply in the form of methane. Most of this methane is oxidized in the water column, however, and does not reach the atmosphere. This system provides another example of the importance of transfer processes for the modulation between biological production and the emission to the atmosphere, especially in view of the fact that the presence of aquatic macrophytes can provide an important bypass for the escape of methane from lake bottoms to the atmosphere.

Biogenic volatile compounds, most of which are reduced with respect to the oxidation state thermodynamically stable in the presence of oxygen, can thus be produced both in anoxic and in oxic environments. In many instances, as almost certainly in the case of sulfur, the release from the oxic systems in the surface ocean and terrestrial vegetation and soils will contribute much more to the global sulfur cycle than the emissions from anoxic muds and soils. The situation is further complicated by the presence of microanaerobic environments which allow anoxic biochemistry to proceed in an otherwise oxic environment. The importance of such processes for atmospheric chemistry remains largely unknown, however. The transitional zone between the region containing free oxygen and the highly reducing environments characterized by H_2S and methane production, often called the suboxic zone, can be important for the production of a number of gaseous species, especially oxides of nitrogen (N_2O, NO).

The absence of a clear separation between oxic and anoxic environments in time and space and the importance of transitions between these environments both in space (suboxic regime) and in time (diurnal changes) suggest that further studies on the relationship between oxic and anoxic systems and its influence on biogenic emissions to the atmosphere should be conducted.

Time and Space Scales

The environments of biological production and scavenging of atmospheric trace species tend to be patchy at all discernible temporal and spatial scales. This applies less to the rather long-lived species, e.g., methane, but even for these compounds significant small scale variation occurs. This makes it important to conduct measurement programs which will eventually cover all time and space scales which are important for a given species. Usually, measurements at different time scales will reveal different types of information: fluctuations at the time scale of atmospheric turbulence (seconds to minutes) can detect gas fluxes when used in eddy correlation techniques. Diurnal variations may reflect changes in source strengths, transport modulation, or rates of decomposition (sink strengths). Seasonal changes relate to the changes of these parameters on an extended scale, whereas secular variations may yield information about man-made or natural changes in the global environment. It has to be emphasized that to neglect consideration of any of these scales may lead to strongly biased results, especially when the periodicity of the measurement schedule happens to be related to the periodicity of the fluctuations of the measured parameter. The simplest case for such a problem is the performance of measurements only during the day or only in summer, etc.

An interesting potential for the investigation of changes on longer temporal scales is presented by ice cores taken from the polar regions. Due to homogenization processes during the ice formation process, the resolution appears to be on the order of time spans in excess of a decade. This core material is extremely expensive to collect and provides a unique and valuable resource. It is at this time not clear what the influence of the ice formation processes is on the information contained in the sample about the gases present in the atmosphere under which it was deposited. For this reason, studies involving comparisons of current atmospheric gas concentrations with their record in recent ice and firn should be undertaken before valuable core material is sacrificed with uncertain interpretability.

Spatial scales ranging from the micron range to global dimensions must be taken into account. This will require close collaboration between microbiologists, ecologists, geographers, oceanographers, meteorologists, and atmospheric chemists. Close relationships between ecological zones, including agricultural environments, and the rates and types of biogenic emissions have to be expected. The range of atmospheric transport, the propagation of local effects, and the degree of lateral and vertical mixing depend on the environmental lifetimes of the species concerned. But even if the substances emitted directly are very short-lived in the atmosphere, they may give rise to long-lived compounds which can propagate globally. An example for this can be found in the emission of the short-lived hydrocarbon isoprene (lifetime on the order of days), which can be photooxidized to the relatively long-lived species carbon monoxide (lifetime on the order of a few months).

An understanding of the principles controlling spatial variability is an important ingredient for the design of representative sampling strategies: while this understanding is likely to be absent during initial investigations of an individual species, attempts should be made to obtain an understanding of the relationship of ecological and geographical parameters to the emissions of biologically produced species as a guiding principle to further sampling.

Among the types of world ecozones currently under investigation, the oceans have received considerable attention, in part because of their apparent homogeneity and because of their importance in terms of surface area. In the future we will require more intensive studies into the role of land biota on the chemistry of the atmosphere, both in natural and agricultural ecozones. This work will have considerable importance for our understanding of the carbon, nitrogen, and sulfur cycles. Interactions of trace elements, including metals and metalloids, in the atmosphere with terrestrial biota should also be emphasized. This work will require the compilation of information on biogeography, soil and vegetation characteristics, etc., in the form of atlases or computer data files in such a way that they can be of use to atmospheric scientists.

Site selection for the collection of samples and data for air chemical investigations remains a very important problem. This applies, in particular, to continental sites, where site-specific problems and influences tend to be more important than at marine sites. For investigations on biological production and release of trace atmospheric species, it is important to be able to separate biological and anthropogenic sources. This applies most obviously to substances which are dominant pollutants, e.g., SO_2 and organoleads. It is strongly recommended that such studies be conducted at remote sites with continuous monitoring for potential pollution influences, as well as for the influence of contamination created by the experimenter.

Certain sites, at which continuous programs of monitoring and scientific investigations are conducted, are especially useful in this context, e.g., the Australian Baseline Station at Cape Grim, Tasmania, and the NOAA station at Samoa. The information accumulated at these sites and the experience of other investigators with site-specific problems provide an invaluable resource for future investigations.

Methodological Considerations

The compounds produced by biological processes are present usually only at minute concentrations, often in the part per trillion range. This makes their determination difficult, especially when species selectivity is required. In addition, contamination of the samples during or after sampling is a most serious problem, as well as decomposition or loss of the analyte during storage or analytical manipulations. It is therefore of utmost importance to maintain strictest quality control during the sampling and analysis process. Blank determinations, procedural controls, sterile controls, calibration by standard addition, and use of two independent methods for the determination of a substance are some of the methods which should be employed to verify the results and to avoid measurement artefacts. The use of standard reference materials is to be encouraged. If at all possible, it is advisable to perform most of the analyses on site to avoid storage artefacts and to make possible an immediate feedback between measurement results and sampling strategies.

While the measurement of individual chemical species should be emphasized, it is also advisable to perform measurements of total element concentrations. This makes it possible to identify serious analytical errors. It may also alert us to the presence of important species which may have been missed because they had not been looked for previously.

In the context of methodologies, it is important to emphasize the value of collecting supporting data which will be required for the interpretation of the chemical data. Determinations of UV and visible light flux, oxidant gases, meteorological data, etc., should be available. Measurements at continuously maintained monitoring stations, e.g., Cape Grim, have the benefit that such data are routinely available at these sites. For many investigations into vertical fluxes of atmospheric substances, the availability of a vertical component anemometer would be highly beneficial.

Recommendations

Before we discuss individual element cycles, some general recommendations can be made:

1) Analytical quality control and the acquisition of supporting data must be given a very high priority during all investigations of biogenic trace substances.
2) The metabolic pathways leading to the formation of volatile substances which may enter the atmosphere deserve further study.
3) The chemical and photochemical controls of the cycles of biogenic gases in surface waters and in the atmosphere directly above the water surface should be investigated.
4) Studies of the role of continental, including agricultural, ecosystems should be intensified.
5) The importance of anthropogenic disturbances on the rates of biogenic emissions should be studied.
6) More information on the role of sinks of atmospheric substances is needed. This includes both studies on sinks for biogenic substances and on the biosphere as a sink for atmospheric components.

ELEMENTS AND PROCESSES: CASE STUDIES

For specific discussion, the cycles of the biologically important elements (sulfur, nitrogen, and carbon) and of some selected metals and metalloids have been chosen. Because of its significant involvement with the biosphere, a short discussion of the hydrogen cycle has been added. Biomass burning has a special relevance to all nutrient element atmospheric cycles and will be discussed in a separate paragraph.

Sulfur

Biogenic release of sulfur gas is a commonplace event and can be appreciated by anybody who has ever eaten an oyster (DMS) or opened a rotten egg (H_2S). Its role for the global sulfur cycle has been the object of considerable debate since it has first been deduced from discrepancies in the global atmospheric sulfur budget. Estimates of the biogenic contribution to the atmospheric sulfur budget have been gyrating between 200 and 6 Tg/yr. Recent evidence on the abundance of dissolved volatile substances allows us to place some limits on this contribution. Let us review the actors and their roles:

H_2S (hydrogen sulfide): This gas is abundantly produced by bacteria in anoxic environments. Because of its desirability as a substrate by other bacteria, little of it escapes into the atmosphere. Further rapid oxidation in the water column also helps to prevent this gas from reaching the atmosphere. Some release occurs from terrestrial biota; this source requires better quantification. Source strength estimates range from a few Tg/yr to some tens of Tg/yr. Its lifetime in the atmosphere is relatively short, on the order of two days.

COS (carbonyl sulfide): This gas is thought to have a lifetime on the order of a year and may thus make a considerable contribution to stratospheric sulfur chemistry. Its potential sources are the burning of biomass, the photooxidation of CS_2, and the release from the oceans. The surface ocean has been observed by three independent groups to be supersaturated with COS; current estimates of the releases from this source are on the order of 1 Tg/yr. While COS has been observed as a result of bacterial decomposition of organic matter under oxic conditions, its mode of formation in the surface ocean has not been established. Currently there are no known tropospheric sinks. Investigations should continue to improve the quantification of known sources and sinks, and the possibility of additional tropospheric sinks should be explored.

CS_2 (carbon disulfide): The lifetime with respect to OH oxidation is about 2 weeks. Its sources are largely unknown; releases from the land biota, from estuarine systems, and to a small extent from the open ocean may play a role. There may be a significant anthropogenic component. Studies on its concentration in seawater suggest a marine source of only ca. 0.15 Tg S/yr. The importance of the emissions from terrestrial biota and from estuaries needs to be assessed.

CH_3SH (methyl mercaptan, methanethiol): This is a highly reactive gas which should have a lifetime of only a few hours. It is often present in surface seawater, especially from coastal regions. No good quantitative data are yet available; estimates of seawater concentrations are in the range of up to 10 ng/L (about 10 times less than DMS). It may thus have a significant marine source. Accurate data on its distribution in the oceans need to be acquired, and the source strength from estuarine and continental sources will have to be assessed.

CH_3SSCH_3 (dimethyldisulfide): This is another highly reactive gas whose OH lifetime is a few hours at best. Concentrations in all known source environments are much lower than those of DMS and of CH_3SH (from which it is probably formed). This and its lower volatility make it a trivial source of sulfur to the atmosphere.

$(CH_3)_2S$ (dimethylsulfide): This is the most abundant volatile sulfur compound detectable in seawater, with an OH lifetime estimated to be two days. On the basis of 300 samples of surface seawater over a large region of the world ocean and current gas exchange models, the source of this substance to the world atmosphere from the oceans is estimated to be 30 to 50 Tg S per year. A release of similar magnitude may occur from continental biota. The concentrations of DMS in the marine atmosphere suggest a sink mechanism which is much more rapid than OH oxidation and which is non-photochemical. Investigations into the sink mechanisms (oxidation) and the continental sources of DMS should be given high priority. Determinations of methanesulfonic acid, a potential major oxidation product of DMS, may shed light on the sink mechanisms for this substance.

Current estimates of the deposition rate of non-seaspray derived sulfate aerosol over remote regions, where anthropogenic SO_2 is not expected to be a significant source, are on the order of 50% of the input estimate of DMS in these areas. The presence of relatively constant concentrations of SO_2 over remote marine regions also supports the presence of a significant seawater source of volatile sulfur.

The sulfur cycle in the remote marine boundary layer thus presents a roughly consistent picture with an approximate match of input and output estimates for biogenic sulfur. Large uncertainties persist in the fluxes of the minor species and for COS which represents a minor tropospheric flux but plays a significant stratospheric role. The most problematic area is the system of atmospheric reactions which interconnects the atmospheric species and which links the sulfur cycle to the other element cycles.

We should, however, keep our eyes open for additional sulfur species which have not yet been measured in the environment, e.g., S_2O. The possibility of sulfate and methanesulfonic acid measurements in polar glaciers to place limits on the importance of biogenic sulfate production should be investigated. More information is required on the role of deposition of SO_2 and of acidic sulfate particles on plant surfaces and in the human lung.

Nitrogen

N_2O, NH_3, and NO are trace gases which influence the chemistry of the atmosphere in important ways. N_2O is the major source of NO_x in the stratosphere and thereby exerts a controlling influence on stratospheric ozone. NO regulates OH concentration in the troposphere as well as O_3 production in the troposphere. And NH_3, which is returned to the earth's surface as NH_4^+ in rain, is important in neutralizing aerosols and precipitation. All of these gases have strong biogenic sources, largely resulting from the activities of soil microorganisms.

N_2O is almost exclusively produced biologically. Only about 2 to 5 Tg N/yr of N_2O are emitted from the oceans. This source is largely confined to a few zones of oceanic upwelling, while most of the ocean surface is undersaturated slightly with respect to atmospheric concentrations. The total source strength of the biosphere for N_2O is considered to be about 20 Tg N/yr, with a considerable uncertainty (about a factor of three) attached to this number. Most of the release is thought to occur on land, and the relationships between ecozones and N_2O release are not adequately understood to permit a reliable integration. Nitrogen fertilization of agricultural soils may cause a significant increase in N_2O release. Investigations have shown this to be a highly complex issue with significant temporal variability and dependence on numerous factors, e.g., type of fertilizer, crop type, soil type, and the net effect of fertilizer use on global N_2O emissions cannot yet be assessed. The emission of N_2O from land areas and its controlling parameters require substantial further study.

NH_3 is almost exclusively from biological origin. Most of it is released from land surfaces, where the major source seems to be provided by conversion of animal excreta. The net effect of the oceans on the atmospheric NH_3 budget is not known, since the necessary field measurements have not yet been made.

The largest uncertainties of the role of the biosphere in the atmospheric nitrogen budget exist for NO. It appears, however, that the microbial emission, though potentially important, is only one of a number of major sources. Substantial releases of NO from soils have been observed, averaging 0.1 ±0.05 g $N/m^2/yr$. In addition, there is the possibility of a small oceanic release of NO, although at present it is not clear if the ocean will act as a sink or a source of NO. Nor is it certain if the oceanic NO release is controlled by biological factors. At present it is thought that the NO available for potential release is formed near the surface through photolysis of NO_2^-.

There are a few other nitrogen compounds which have been observed recently in the atmosphere in significant concentrations (e.g., HCN, CH_3CN, and CH_3ONO) or are known to be released to the atmosphere (e.g., di- and trimethylamines). But neither the chemical role of these substances in the atmosphere nor the extent of biological production has been established. On the whole, we may state that although it is recognized that the biosphere plays a very important role in the emission of nitrogen compounds, it is difficult to quantify this role. The reason lies in the sporadic nature of the strong sources which show large temporal and spatial variations, and in the fact that ecosystems of large extent, such as the oceans, may act both as a weak source and as a weak sink. Future research on the role of the biosphere in the atmospheric nitrogen cycle should therefore be focused on studies in continental regions. In particular, measurements of emission rates of NH_3 and NO from just about any ecosystem are required. The atmospheric concentrations of NO_x need to be monitored over extended time scales.

Carbon

Almost by definition, the biosphere dominates the atmospheric carbon cycle. A tremendous variety of substances is produced and taken up, and our discussions could barely touch the surface of this subject. Of course, there is CO_2 (which we decided to leave aside so as to avoid starting another Dahlem Workshop right there) and CH_4, both of which have massive fluxes through the atmosphere. The hydrocarbons, especially those with carbon numbers from two to four, isoprene, and the terpenes are an important class of biogenic substances released to the atmosphere, in particular from land biota. Carbon monoxide plays a central role in atmospheric chemistry. In the following paragraphs a few views and needs are expressed for some of the important species and substance classes.

CH_4 is essentially all biogenic; it is produced by methanogenic bacteria in sediments and by bacterial symbionts inside animals, especially ungulates and possibly termites. A recent estimate suggests that as much as a third of terrestrial primary production is recycled through termites. There remains, however, a prominent need for better quantification of methane production by biota.

While ethane results predominantly from fossil sources, most of the ethylene is considered to be produced by biota. The fluxes of this gas are probably too small to have an effect on the atmospheric cycle. Some ethylene as well as propylene is observed over the open oceans, suggesting the existence of a detectable oceanic source. The C_5 compound isoprene is one of the more important substances released by land plants. The related compound class of the terpenes is seen prominently in experiments on the release of substances from plants. Its abundance in terrestrial atmospheres is comparatively low, suggesting rapid conversion. Measurements over the oceans show significant amounts of hydrocarbons extending up to very high carbon numbers. An oceanic biological source for these substances appears established.

Hydrocarbons with some oxygen-containing functional groups are widely observed in the atmosphere. It is not clear to what extent these compounds could be directly released from biota and to what degree they result from atmospheric oxidation of biogenic exudates. Many of the substances discussed in this section form the precursor compounds both of gases, such as CO, and of carbon-containing atmospheric aerosols.

There is a great need for further research into the release rates and fates of atmospheric hydrocarbons of biogenic origin. Special emphasis should be placed on the lightweight hydrocarbons, isoprene, the terpenes, and their reaction products, but higher molecular-weight hydrocarbons should be measured simultaneously to assess the potential of gasoline-related contamination. Measurements on longer time scales are especially important for methane and the light hydrocarbons. The processes of gas-to-particle conversion resulting in organic aerosols will require considerable attention.

Carbon monoxide is produced in roughly comparable amounts by methane oxidation, oxidation of non-methane hydrocarbons, biomass burning, and technological sources, with some additions from plants and the ocean surface. It is largely consumed by oxidation with OH, but microbial destruction at the soil surface plays an important role. To improve the source estimates, improved measurements of hydrocarbon abundances and oxidation efficiencies are needed. The removal by biological decomposition at soil surfaces is characterized by first-order kinetics. Current removal rates have to be based on CO concentrations over the oceans because of the lack of representative CO data over continental areas where the levels may be as much as twice as high. Such data are urgently needed to refine the CO budget.

Metals, Metalloids, and Miscellaneous Elements

A rather intriguing field of biogeochemistry relates to the biological volatilization of metals and metalloids. For many elements (arsenic, tin, mercury, selenium, lead, antimony, etc.), biological methylation has been observed at least in the laboratory. The formation of volatile methylated compounds of most of these elements has also been detected in highly polluted environments: this is the case for arsenic, tin, mercury, selenium, and maybe lead. For the latter element, any study of biomethylation in the environment bears a strong burden of proof with regard to the absence of contamination from the pervasive presence of gasoline additives. For another sequence of elements (arsenic, antimony, tin, and germanium), organic species, in particular ionic methyl compounds of low volatility, have even been seen in uncontaminated environments. So far, the only strong candidates for a significant atmospheric flux by biomethylation are the elements tin and mercury. For arsenic, the contribution of biomethylation to the atmospheric budget remains certainly below five percent. Some reports exist which suggest the biogenic formation of volatile forms of cadmium of as yet inadequately characterized constitution: their potential role in the environment remains to be investigated. No volatile forms of selenium have been found in surface ocean waters, even though nonvolatile organic compounds of this element are present in some natural waters.

Future research in this field has to place an emphasis on the direct detection of organic species in the atmosphere, rather than on their deduction from inferential information. Rainwater may be a convenient source of samples for this purpose. While the in vitro chemistry of most of the organometallics is reasonably well studied, there remains a large need for field data, especially from outside of the highly contaminated areas.

There is some interesting evidence for the possibility of the biogenic volatilization of phosphorus. Studies involving the identification of the transport species are required here.

Fluorine shows some surprising biochemical activity. Fluoroacetate is an important metabolite in metazoans and some plants. The biological production of vinyl fluoride has been reported. The potential of these compounds for the atmospheric chemistry of fluorine remains to be explored.

Hydrogen

Molecular hydrogen is one of the most abundant trace gases in the atmosphere, having a global average mixing ratio of about 0.6 ppmv. Its tropospheric cycle is mostly controlled by non-anthropogenic processes, the largest source being photochemical production as one of the major end products of hydrocarbon oxidation in the atmosphere (~50%). Additional natural sources (accounting for ~10%) have been identified (volcanic emissions, ocean surface). Anthropogenic sources constitute the remaining 40%; about one half of it is incomplete combustion of fossil fuels and the other half may be biomass burning. Sources having individual errors of about a factor of two sum up to a total of about 80 Tg H_2/yr. They are balanced by two dominant sink processes: reaction with OH radical (~25%) and microbial consumption at the soil surface (~75%). Assuming the cycle to be in a steady state, the tropospheric residence time is on the order of 2 - 3 years.

Biomass Burning

The combustion of biomass, much of it intentional, as part of agricultural practices, releases large amounts of carbon, nitrogen, sulfur, potassium, and possibly other elements into the atmosphere. These emissions are generally in the less industrialized regions of the globe, which makes them an even more prominent source of atmospheric contamination in these regions. Some substances and substance classes (carbon monoxide, NO_x, HCN, acetonitrile (CH_3CN), methyl nitrite (CH_3ONO), soot carbon, as well as a large variety of hydrocarbons, among them ethane, ethylene, propane, and propylene, and a vast array of other organic compounds) have been recognized as having potentially significant sources in biomass combustion. The identity of many of these compounds remains to be explored. The rates of production of the dominant compounds need to be quantified, their fate in the atmosphere needs to be investigated, and their role for the global atmosphere needs to be assessed.

Anthropogenic Pollutants

The workshop recognized the role of the atmosphere as an important vector for the transport of anthropogenic pollutants, e.g., DDT, PCB, and toxaphene, to terrestrial and oceanic ecosystems over large distances. A strong need is perceived to determine the importance of this pathway for global pollution and its impacts on the environment. The atmosphere offers a significant analytical opportunity for the investigation of the dispersal of substances which are otherwise difficult to observe.

EPILOGUE

The working group reviewed a wide spectrum of information of interaction of the biosphere and the atmosphere. We were impressed with the vast amount of exciting information which has resulted from recent research in this area, and also with the number of even more exciting avenues for further research which this work has opened up. The recommendations and suggestions for future work pertaining to the individual element cycles have been included in the individual paragraphs rather than assembled at the end. This report represents the results of a group effort, and much of the information given is still unpublished. Other information was presented in discussions without the precise identification of its sources. Therefore this report has no "references." A short list of "suggested readings" follows. The reader interested in more detail will find the pertinent references in these publications and in the background papers.

BIBLIOGRAPHY

(1) Andreae, M.O. 1982. The biological production of dimethylsulfide in the ocean and its role in the global atmospheric sulfur budget. Ecol. Bull. (Stockholm), in press.

(2) Balzer, W. 1981. Organic sulphur in the marine environment. In Marine Organic Chemistry, ed. E.K. Duursma and R. Dawson, pp. 395-414. Amsterdam: Elsevier.

(3) Bolin, B., and Cook, R.B., eds. 1982. SCOPE 24: The biological cycles of C, N, P, and S and their interactions. Ecol. Bull. (Stockholm), in press.

(4) Bremner, J.M., and Steele, C.G. 1978. Role of microorganisms in the atmospheric sulfur cycle. In Advances in Microbial Ecology, ed. M. Alexander, vol. 2, pp. 155-201. New York: Plenum Press.

(5) Brinckman, F.E., and Bellama, J.M., eds. 1978. Organometals and Organometalloids: Occurrence and Fate in the Environment, vol. 82. Washington, DC: American Chemical Society Symposium Series.

(6) Cicerone, R.J., and Shetter, J.D. 1981. Sources of atmospheric methane: measurements in rice paddies and a discussion. J. Geophys. Res. 86: 7203-7209.

(7) Crutzen, P.J.; Heidt, L.E.; Krasnec, J.P.; Pollock, W.H.; and Seiler, W. 1979. Biomass burning as a source of atmospheric gases CO, H_2, N_2O, NO, CH_3Cl and COS. Nature 283: 253-256.

(8) Cullis, C.F., and Hirschler, M.M. 1980. Atmospheric sulphur: natural and man-made sources. Atmos. Env. 14: 1263-1278.

(9) Dawson, G.A. 1977. Atmospheric ammonia from undisturbed land. J. Geophys. Res. 82: 3125-3133.

(10) Ehhalt, D.H., and Drummond, J.W. 1982. The tropospheric cycle of NO_x. In Proceedings of the NATO Advanced Study Institute on Chemistry of the Unpolluted and Polluted Troposphere, Corfu, 1981. Dordrecht: Reidel, in press.

(11) Hansen, M.H.; Ingvorsen, K.; and Jørgensen, B.B. 1978. Mechanisms of hydrogen sulfide release from coastal marine sediments to the atmosphere. Limnol. Ocean. 23: 68-76.

(12) Logan, J.A.; Prather, M.J.; Wofsy, S.C.; and McElroy, M.B. 1981. Tropospheric chemistry: a global perspective. J. Geophys. Res. 86: 7210-7254.

(13) National Academy of Sciences. 1978. The Tropospheric Transport of Pollutants and Other Substances to the Oceans. Washington, DC.

(14) Slinn, W.G.N.; Hasse, L.; Hicks, B.B.; Hogan, A.W.; Lal, D.; Liss, P.S.; Munnich, K.O.; Sehmel, G.A.; and Vittori, O. 1978. Some aspects of the transfer of atmospheric trace constituents past the air-sea interface. Atmos. Env. 12: 2055-2087.

Atmospheric Chemistry, ed. E.D. Goldberg, pp. 273-300. Dahlem Konferenzen 1982.
Berlin, Heidelberg, New York: Springer-Verlag.

Physics and Chemistry of Atmospheric Ions

F. Arnold
Max-Planck-Institut für Kernphysik
6900 Heidelberg, F. R. Germany

Abstract. Atmospheric gaseous ions not only control the electrical properties of the atmosphere but also play a role in trace gas and aerosol processes. They react selectively with certain neutral trace gases, and thus in situ ion composition measurements can be used as a powerful tool for trace gas detection. This analytical application of atmospheric ions has attracted great interest and already provided a wealth of new information on trace gases and aerosols. During recent years the exploration of the atmospheric ion composition has progressed downwards in altitude to the tropopause region. The present paper reviews recent progress in our understanding of stratospheric ions and attempts an assessment of tropospheric ion chemistry.

INTRODUCTION

Atmospheric gaseous ions are important as they not only control atmospheric electrical properties but also may play a role in aerosol formation (4,5,24,33) and trace gas processes. Apart from their role in atmospheric processes, gaseous ions are of great importance as they can be used as powerful probes for neutral trace gas detection (7).

The ultimate physical cause of the importance of atmospheric ions lies in their very efficient interaction with atmospheric atoms, molecules, ions, and electrons. This interaction, of course, is due to long-range charge-dipole and Coulomb attraction forces which give rise to large collision cross sections, increase chemical reactivity, and promote the formation of molecular clusters by electrostatic bonding.

Our current understanding of the chemical nature of atmospheric ions and their role in atmospheric aerosol and trace gas processes derives primarily from in situ ion composition measurements using rocket- and balloon-borne mass spectrometers as well as from laboratory studies of ion-molecule reactions and ion nucleation.

In situ ion composition measurements were first made in the upper atmosphere using rocket-borne instruments. Almost two decades ago it became technically feasible to extend positive ion composition measurements (28) into the mesosphere to heights as low as about 60 km. Mesospheric negative ion composition measurements (12,29) were started in 1970. Considerable progress in our understanding of mesospheric ions has been made, particularly during the last decade, and this field still remains very active and fruitful.

Stratospheric in situ ion composition measurements became technically feasible only during recent years mostly using balloon-borne instruments (3,6). In this relatively short period our understanding of stratospheric ions has made enormous progress. Evidence was obtained that stratospheric ions play a role in aerosol formation and possibly also in trace gas processes. Most important, however, it was found that a wealth of interesting information on neutral trace gases and aerosols can be obtained from in situ ion composition measurements.

Very recently (18), active chemical ionization mass spectrometry was introduced as a new analytical tool for stratospheric trace gas detection.

Also very recently, a group at the Max-Planck-Institut für Kernphysik succeeded in conducting the first in situ composition measurements of positive ions in the upper troposphere (Arnold et al., unpublished) yielding rough information on ion masses. In view of this success, it seems that detailed in situ composition measurements of tropospheric ions may become technically feasible in the near future.

The purpose of the present paper is to review current progress in our understanding of stratospheric ion chemistry and to attempt an assessment of the possible role tropospheric ions play in aerosol and trace gas processes.

Sources and Sinks of Stratospheric Ions

The only important source of ionization in the stratosphere, under most conditions, is galactic cosmic rays (36). These are mostly energetic protons having average kinetic energies of about 100 - 1000 MeV corresponding to an atmospheric penetration depth of about 10 - 15 km. Thus, the galactic cosmic ray ionization rate, Q, reaches a maximum of about 10 - 100 cm^{-3} s^{-1} around this altitude.

Since the galactic cosmic rays are partially shielded by the interplanetary and earth's magnetic fields, Q undergoes temporal and spatial variations. It decreases as solar activity increases, and it increases with increasing geomagnetic latitude.

Primary charged species formed by galactic cosmic ray ionization are N_2^+, O_2^+, O^+, N^+, and free electrons. The latter are rapidly attached to gas molecules, giving rise to simple negative ions, mostly O_2^-. Subsequent ion molecule reactions of primary positive and negative ions lead to complex positive and negative cluster ions. Ultimately these are removed by ion-ion recombination involving either a binary or a ternary mechanism (32).

Considerable experimental (14) and theoretical (31) progress has recently been made in determining stratospheric ion-ion recombination coefficients. In particular, it was found that binary ion-ion recombination can be enhanced as the gas pressure increases ("pressure enhanced binary ion-ion recombination").

Taking recent effective ion-ion recombination coefficients α and measured Q-values, the total ion concentration, n, can be calculated from the simple steady-state continuity equation $n = (Q/\alpha)^{1/2}$. Resulting n-values range between about 10^3 - 10^4 cm^{-3} and are in reasonable agreement with measured n-values recently obtained from improved Gerdien-condenser experiments (34) (Fig. 1).

Ion-recombination lifetimes $t_R = (\alpha n)^{-1}$ range between about 10^2 - 10^4 s (Fig. 2). When compared with ion lifetimes t_A against collision with aerosols (21) (also shown in Fig. 2), t_R is much smaller throughout the stratosphere.

Consequently, ion attachment to aerosols should not be an efficient loss process for stratospheric ions.

COLLISION PROCESSES OF STRATOSPHERIC IONS

Due to the very large abundance ratio of neutral molecules to ions (about 10^{13} - 10^{14} in the middle stratosphere), ion interactions are dominated by ion-molecule collisions. These have

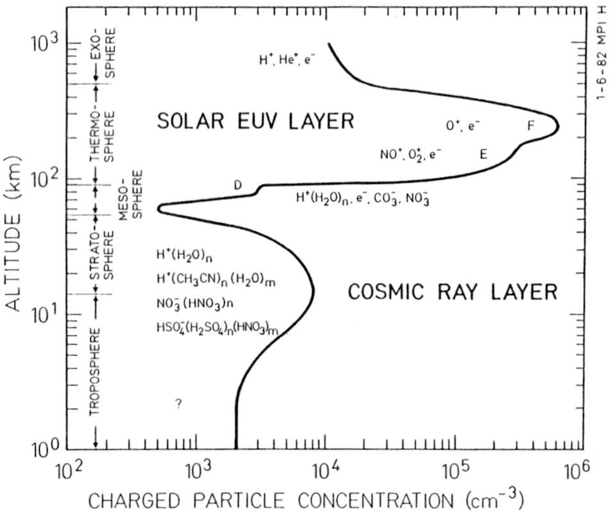

FIG. 1 - Schematic representation of atmospheric ionization layers.

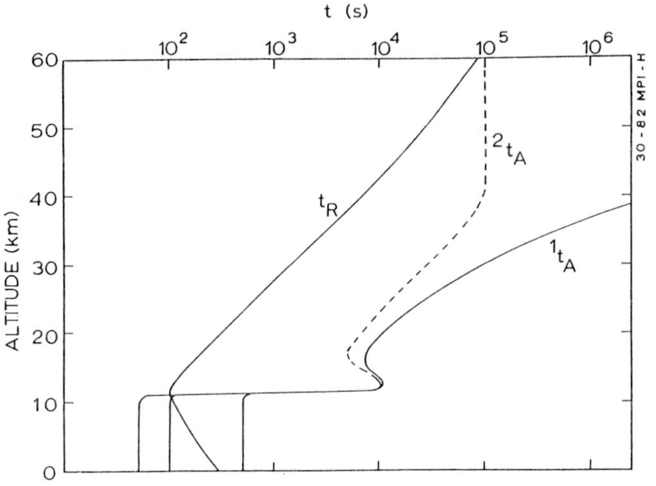

FIG. 2 - Typical ion-recombination lifetimes (t_R) compared with free-ion lifetimes against attachment to aerosols (^1t_A, ^2t_A "meteor smoke particles" added).

cross sections being about 100 times larger than those for molecule-molecule collisions due to relatively long-ranging charge-dipole attraction forces.

Another consequence of these forces is that reactive ion-molecule collisions ("ion-molecule reactions"), in contrast to neutral gas reactions, in many cases do not possess activation energy barriers and thus proceed at the collision rate (16).

Finally, charge-dipole attraction also gives rise to clustering of molecules to ions (16), leading to large cluster ions which ultimately may become condensation nuclei (5). Interestingly, even cluster ions which contain only a relatively small number of molecules, e.g., 5 - 10, develop properties similar to those of macroscopic liquid droplets (30).

Under stratospheric conditions, ion clustering proceeds via a ternary mechanism,

$$A^+ + B + M \rightarrow A^+B + M, \qquad (1)$$

where M is a collision partner. If there is no kinetic limitation, a thermodynamic equilibrium between processes such as Eq. 1 and thermal dissociation (reverse of Eq. 1) determines the size distribution of $A^+(B)_n$ cluster ions.

A new class of ion-molecule reactions, so-called ion-catalyzed reactions which were only recently (19) studied in the laboratory, may occur in the stratosphere. They involve the reaction of a molecule which is clustered to an ion with a gas-phase atom or molecule. Thus "ion-catalyzed reactions" involving larger cluster ions are somewhat similar to surface catalysis. The reactant molecule may be regarded as being adsorbed on the "surface" of the cluster ion.

When compared with its homogeneous gas-phase analog, an ion-catalyzed reaction may have a rate coefficient which is larger by orders of magnitude. The reactions (19)

$$O_3 + NO \rightarrow NO_2 + O_2 \tag{2}$$

and

$$Na^+O_3 + NO \rightarrow NO_2 + O_2 + Na^+, \tag{3}$$

e.g., have rate coefficients of $1.3 \cdot 10^{-14} - 2.1 \cdot 10^{-17}$ $cm^3 s^{-1}$ (depending on temperature) and $6.5 \cdot 10^{-11}$ $cm^3 s^{-1}$, respectively.

Besides interacting with molecules, stratospheric ions preferably interact with oppositely-charged ions due to very long-ranging Coulomb-attraction forces.

Besides giving rise to free ion removal, ion-ion recombination may also lead to the formation of stable ion pairs (15) which by ion attachment may grow to polyions (8). The polyions are still hypothetical as they have neither been detected in the laboratory nor in the atmosphere. If existent, they may play a role as precondensation nuclei (4).

Thus, ions may promote aerosol formation in two ways, by conventional ion nucleation involving ion clustering and by polyion nucleation.

ION CHEMISTRY

Positive Ions

The most abundant stratospheric positive ions are $H^+(H_2O)_n$ and $H^+(CH_3CN)_\ell(H_2O)_m$ cluster ions with the former dominating above about 35 km and the latter becoming most prominent below this altitude (Fig. 3).

Besides these major ions, various minor ion species were detected (Table 1), possibly also containing molecules such as CH_3OH. In addition, very massive positive ions (up to about 320 atomic mass units) were detected but could as yet not be identified.

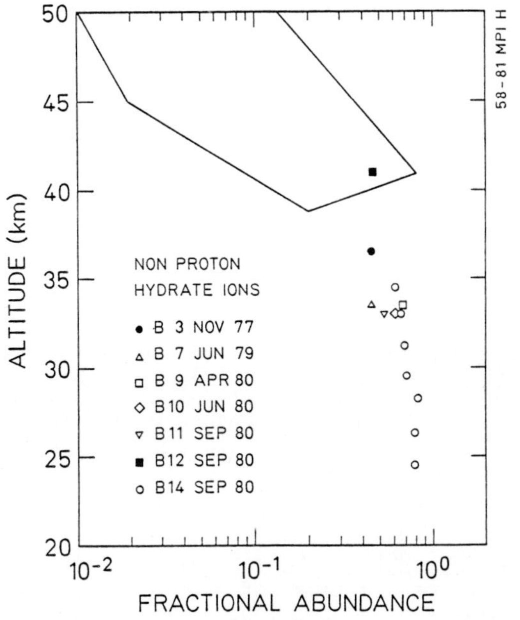

FIG. 3 - Fractional abundance of positive ions not belonging to the $H^+(H_2O)_n$ family (Schlager and Arnold, submitted for publication).

TABLE 1 - Stratospheric positive ion species detected by balloon-borne mass spectrometers (Arnold and Bührke, submitted for publication, and Viggiano, Schlager, and Arnold, submitted for publication).

Ion	Mass	Ion	Mass
H_3O^+	19	$H^+CH_3CN(H_2O)_2$	104±1
Na^+	23±1	$H^+(H_2O)_6$	109
H^+HCN	29±2	$H^+CH_3OH(H_2O)_2$	110±1
$H^+(H_2O)_2$	37	$H^+CH_3CN(H_2O)_4$	114±1
H^+CH_3CN	42		117±1
$H^+HCN\cdot H_2O$	45±1	$H^+(CH_3CN)_2(H_2O)_2$	119
$H^+CH_3OH\cdot H_2O$	49±1		121±1
$H^+(H_2O)_3$	55		125±1
$Na^+(H_2O)_2$	58±1		128±1
$H^+CH_3CN\cdot H_2O$	60		134±2
$H^+HCN(H_2O)_2$	63±1		139±2
$H^+CH_3OH(H_2O)_2$	67±1	$H^+H_2O(CH_3CN)_3$	142±1
$H^+(H_2O)_4$	73		151±2
$H^+CH_3CN(H_2O)_2$	78		158±3
$H^+HCN(H_2O)_3$	81±1		168±3
$H^+(CH_3CN)_2$	83		179±3
$H^+CH_3OH(H_2O)_3$	86±1		182±3
$H^+(H_2O)_5$	91		186±3
$H^+CH_3CN(H_2O)_3$	96		190±3
$H^+(CH_3CN)_2H_2O$	101		202±3

The chemical evolution of stratospheric positive ions may be viewed as proceeding in three stages. The first stage involves reactions of primary positive ions (N^+, O_2^+, O^+, N^+) with major gases (N_2, O_2) leading to O_4^+ and NO^+ (Fig. 4). The fractional rates of formation for these ions are about 90% and 10%. The time scale for stage one is only on the order of 10^{-5} s in the middle stratosphere (if not otherwise indicated, time scales given hereafter refer to an altitude of 35 km). Stage two involves reactions of major trace gases, mostly H_2O leading to $H^+(H_2O)_n$ cluster ions. It is essentially the large proton affinity of the water molecule and the strong bonding

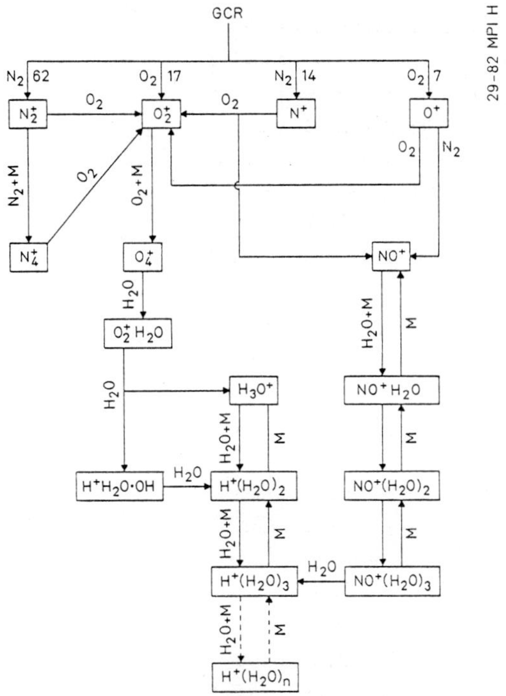

FIG. 4 - Stratospheric positive ion reaction scheme (stages one and two) (16).

of H_2O-molecules to the hydronium ion, H_3O^+, which drive the formation of $H^+(H_2O)_n$ ions. Due to the relatively large abundance of water vapor a quasi-equilibrium size distribution of $H^+(H_2O)_n$ is rapidly established. Usually, the distribution peaks around n = 4 or 5 in the middle stratosphere, depending somewhat on temperature. The time scale for stage two is only on the order of 10^{-3} s, which is much smaller than the ion-recombination lifetime, t_R (see Fig. 2).

Stage three (Fig. 5) involves reactions of very low abundance trace gases having time scales which are on the order of t_R or even larger. Consequently, the abundance ratio for product ions and precursor ions never becomes very much larger than one.

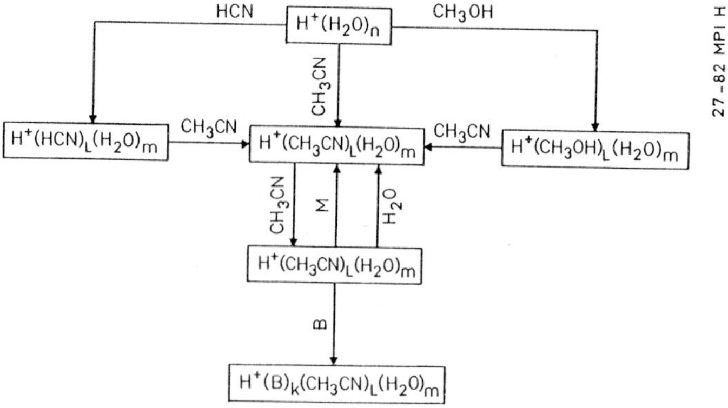

FIG. 5 - Stratospheric positive ion reaction scheme (stage three).

These trace gases all have proton affinities substantially larger than that of H_2O (170 kcal $mole^{-1}$) and thus can react with $H^+(H_2O)_n$ ions via

$$H^+(H_2O)_n + C \rightarrow H^+C\ (H_2O)_{n-1} + H_2O. \tag{4}$$

Equation 4 is not a simple proton-transfer reaction but rather involves displacement of an H_2O molecule contained in an $H^+(H_2O)_n$ cluster ion by a molecule C. Usually, C must have a substantially larger proton affinity than H_2O (170 kcal $mole^{-1}$) in order to make Eq. 4 exothermic ((8), and Ferguson, private communication). This is true because H_2O-molecules mostly bond more strongly to H_3O^+ than to H^+C.

Among the reactant molecules, C acetonitrile, CH_3CN, which has a proton affinity PA (CH_3CN) = 186 kcal $mole^{-1}$ seems to be the most important ((25), and Schlager and Arnold, submitted for publication). Reactions such as Eq. 4 involving CH_3CN lead to $H^+CH_3CN(H_2O)_m$ mixed cluster ions. Subsequent displacement of H_2O-ligands by CH_3CN via

$$H^+(CH_3CN)_\ell(H_2O)_m + CH_3CN \rightarrow H^+(CH_3CN)_{\ell+1}(H_2O)_{m-1} + H_2O \tag{5}$$

may occur. According to in situ observations and laboratory studies (Ferguson, private communication), however, it seems that the exothermicity of Eq. 5 decreases as ℓ increases.

Under stratospheric conditions ℓ is usually not larger than two to three. The total number of molecules attached to H^+CH_3CN is about the same as the number of H_2O-ligands attached to H_3O^+-cores.

It was also found from laboratory studies (Ferguson, private communication) that $H^+(CH_3CN)_3H_2O$ is a particularly stable cluster. This is probably due to a symmetric structure with an H_3O^+-core having a delocalized positive charge shared by its three hydrogen atoms to each of which a CH_3CN molecule is attached.

Mixed cluster ions of the type $H^+CH_3CN(H_2O)_n$ react also with other large proton affinity molecules such as CH_3OH via

$$H^+CH_3CN(H_2O)_n + CH_3OH \rightarrow H^+CH_3CN \cdot CH_3OH(H_2O)_{n-1} + H_2O, \qquad (6)$$

leading to even more complex heteromolecular cluster ions.

Generally, the competitive gas-phase solvation of the proton seems to favor H_2O in the outer "shell" of the cluster. This is probably due to the strong hydrogen bonding of H_2O, which is also responsible for the large heat of vaporization of bulk water.

If present in the stratosphere, NH_3, due to its very large proton affinity, should efficiently react with $H^+(CH_3CN)_\ell(H_2O)_m$ ions. Recent laboratory studies (20) indicate large rate coefficients for these processes.

Other molecules with a high proton affinity which have been discussed (22) as potential reactants for stratospheric positive cluster ions are metal compounds such as NaOH or NaCl. It is thought that these are formed in the mesosphere from meteor ablation material which is mixed downwards into the stratosphere.

According to laboratory studies (11), NaOH and NaCl both react with $H^+(H_2O)_n$ ions, leading to $Na^+(H_2O)_n$ cluster ions.

From the failure to detect the latter ion species in the middle stratosphere, it was concluded (10,23) that NaOH and NaCl are not present in gaseous form but probably are converted to some kind of aerosols before they reach the middle stratosphere.

An alternative possibility is that strongly polar compounds such as NaCl form dimers or higher polymers which by reaction with $H^+(H_2O)_n$ ions do not lead to simple $Na^+(H_2O)_n$ cluster ions but to more complex clusters containing two or more metal atoms.

Acetonitrile, the most important reactant molecule involved in stage three of the positive ion evolution, seems to originate from the troposphere (see section on Analytical and Diagnostic Applications of In Situ Ion Composition Measurements) as discussed by Henschen and Arnold (25). Recently, this view received strong support from laboratory studies suggesting that CH_3CN hardly reacts with OH (Moortgart, private communication) and is not photodissociated by UV-radiation reaching the middle stratosphere (Crutzen, private communication). Thus, if formed in the troposphere, CH_3CN should be capable of reaching the middle stratosphere. However, the possibility exists that CH_3CN, due to its relatively large solubility in water, may efficiently be rained or washed out in the troposphere (1).

The tropospheric source of CH_3CN as yet has not been identified with certainty. Biomass burning has been discussed (1) as a potential CH_3CN source.

Kinetic and thermodynamic data on ion reactions involved in stages one and two are to a large extent available (9). By contrast, only few data are available for stage three processes.

In summary, it seems that our understanding of the stratospheric positive ion chemistry is far from satisfactory because independent information on both underlying processes and reactant trace gases is largely lacking.

Negative Ions

The most abundant negative ion species observed in the

stratosphere can be grouped (26) in two main families, $NO_3^-(HNO_3)_n$ and $HSO_4^-(H_2SO_4)_\ell(HNO_3)_m$, with the latter being dominant above about 25-30 km (Fig. 6). Besides these major ions various minor negative ion species (2,13) were detected (Table 2). These are mostly cluster ions containing high electron affinity core molecules such as CN and CO_3 and high gas phase acidity ligand molecules such as HCl, HNO_2, HOCl, and HSO_3. Water molecules are also present as ligands.

Usually, the $NO_3^-(HNO_3)_n$ ions contain about two to three HNO_3-ligands. Major $HSO_4^-(H_2SO_4)_\ell(HNO_3)_m$ ions may contain up to three H_2SO_4 ligands.

Recently (17) it was found that larger $HSO_4^-(H_2SO_4)_n$ ions are markedly hydrated, suggesting an increase of the H_2O-bond energy for increasing n.

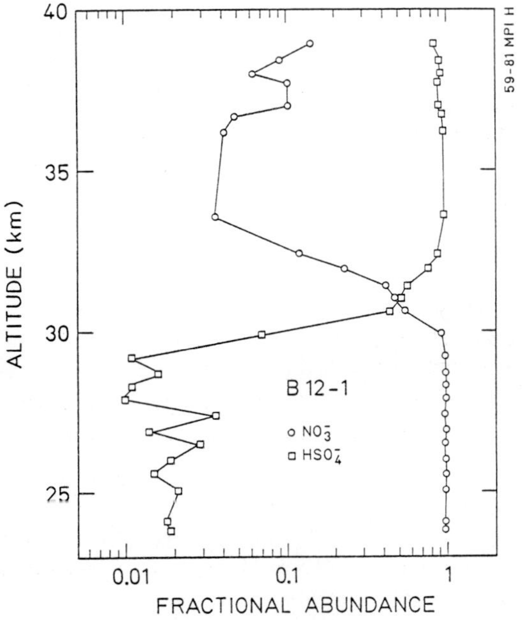

FIG. 6 - Fractional abundance of major negative cluster ion families containing NO_3^- and HSO_4^- cores (35).

TABLE 2 - Stratospheric negative ion species detected by balloon-borne mass spectrometers (2,35).

Mass	Ion	Mass	Ion
26±2	CN^-	223±1	$HSO_4^-(HNO_3)_2$
43±2	CN^-H_2O	241±1	$HSO_4^-(HNO_3)_2H_2O$
61±1	CO_3^-, NO_3^-	251±1	$NO_3^-(HNO_3)_3$
80±2	$CO_3^-H_2O$, $NO_3^-H_2O$	258±1	$HSO_4^-\cdot H_2SO_4\cdot HNO_3$
97±1	HSO_4^-	274±1	$HSO_4^-(HNO_3)_2HClO$
109±2	$NO_3^-HNO_2$		$HSO_4^-\cdot H_2SO_4\cdot HSO_3$
125	$NO_3^-HNO_3$	286±1	$HSO_4^-(HNO_3)_3$
133±1	$HSO_4^-(H_2O)_2$	293±1	$HSO_4^-(H_2SO_4)_2$
	HSO_4^-HCl	374±1	$HSO_4^-(H_2SO_4)_2HSO_3$
143±1	$NO_3^-HNO_3H_2O$	391±1	$HSO_4^-(H_2SO_4)_3$
148±2	HSO_4^-HOCl	409±1	$HSO_4^-(H_2SO_4)_3H_2O$
160	$HSO_4^-HNO_3$	427±1	$HSO_4^-(H_2SO_4)_3(H_2O)_2$
174±1	$NO_3^-HNO_3HNO_2$	454±1	$HSO_4^-(H_2SO_4)_3HNO_3$
188	$NO_3^-(HNO_3)_2$	472±1	$HSO_4^-(H_2SO_4)_3HSO_3$
195	$HSO_4^-H_2SO_4$	489±1	$HSO_4^-(H_2SO_4)_3HSO_3H_2O$
206±1	$NO_3^-(HNO_3)_2H_2O$		

The chemical evolution of stratospheric negative ions, like that of positive ions, may be viewed as proceeding in three stages.

Stage one (Fig. 7) involves reactions of the major gas, O_2, leading to O_4^- (time scale: 10^{-3} s). Stage two (Fig. 7) involves reactions of relatively abundant trace gases such as CO_2, O_3, and H_2O, leading mostly to CO_3^- ions and its hydrates (time scale $\leq 10^{-3}$ s). Since the latter time scale is usually much smaller than the ion recombination lifetime, t_R, CO_3^- and its hydrates are further converted by reactions with less abundant trace gases, mostly NO_x (NO, NO_2, HNO_3, N_2O_5). This leads to the formation of $NO_3^-(HNO_3)_n$ cluster ions (stage three, Fig. 8).

Since the HNO_3-vapor concentration is sufficiently large in the stratosphere, a quasi-equilibrium size distribution of $NO_3^-(HNO_3)_n$ is established, peaking around n = 2 or 3.

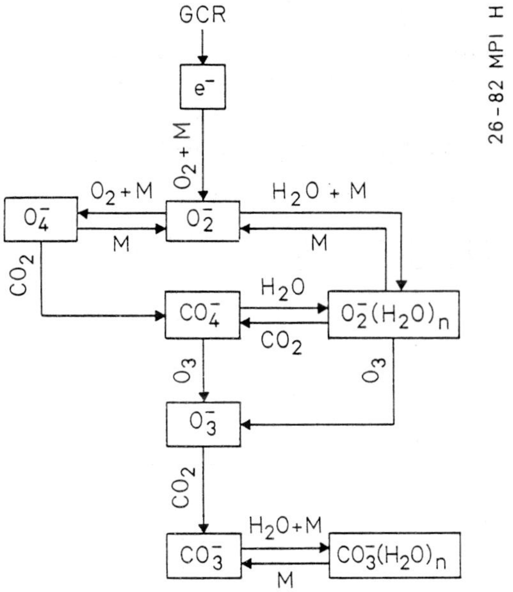

FIG. 7 - Stratospheric negative ion reaction scheme (stages one and two) (16).

The second part of stage three involves reactions with sulfur-bearing gases, mostly H_2SO_4 and HSO_3, leading to HSO_4^--cores. Subsequently, H_2SO_4 displaces HNO_3-ligands, leading to $HSO_4^-(H_2SO_4)_\ell(HNO_3)_m$ mixed cluster ions. As ℓ becomes larger than about two, these cluster ions are markedly hydrated. This probably reflects strong cooperative bonding effects between H_2SO_4- and H_2O-ligands. In this respect, the cluster ion already resembles a small solution droplet composed of an H_2SO_4/H_2O mixture which has a large heat of mixing.

The NO_x-reactant gases leading to $NO_3^-(HNO_3)_n$ ions are formed in the stratosphere by photochemical processes mostly from the precursor gas N_2O which is of tropospheric origin. The sulfur-bearing reactant molecules H_2SO_4 and HSO_3 are also formed in the stratosphere by photochemical and/or heterogeneous processes. Here, the most important precursor gas seems to be OCS which is also of tropospheric origin (27). During

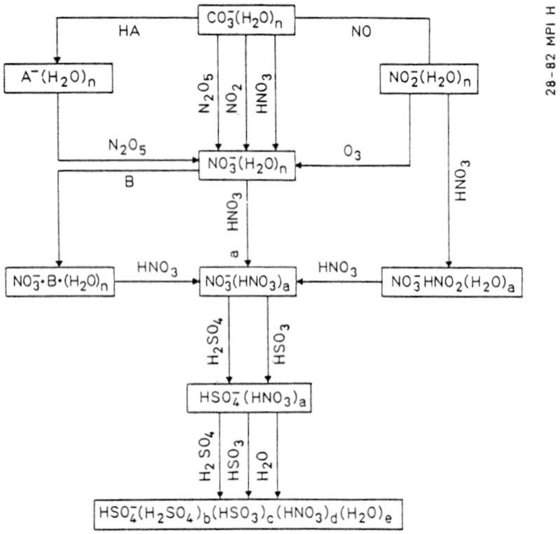

FIG. 8 - Stratospheric negative ion reaction scheme (stage three).

volcanically very active periods SO_2 may also become an important precursor gas.

Kinetic and thermodynamic data for negative ion reactions involved in stages one and two are to a large extent available from laboratory studies (9). Stage three data are mostly lacking.

Thus, our understanding of stratospheric ion chemistry, like that of positive ion chemistry, is far from satisfactory due to the lack of independent information on reactant trace gases and laboratory data.

Potential Role of Ions in Stratospheric Trace Gas and Aerosol Processes

Trace gases can be formed, destroyed, or removed from the gas phase by ion processes including ion-molecule reactions, ion-ion recombination, and ion nucleation.

Trace gas destruction is limited by the relatively small total ion concentration, n. Taking an upper limit of $k = 10^{-9}$ cm^3s^{-1}

as the rate coefficient for an ion-molecule collision, a lower limit of about one day is obtained for the lifetime of a molecule against destruction or removal by ion processes. For noncatalytic ion processes the lifetime can be much larger.

Trace gas formation by noncatalytic ion processes is limited by the total ionization rate, Q, being on the order of 10 - 100 $cm^{-3}s^{-1}$. For catalytic processes the upper limit to the rate of trace gas molecule formation is k n [B], or about 10^{-5} [B], where [B] is the concentration of the reactant molecule.

Ions may also promote aerosol formation via ion nucleation. Basically two processes, conventional ion nucleation, IN (5, 33), and polyion nucleation, PIN (4), have been discussed (see also section on COLLISION PROCESSES OF STRATOSPHERIC IONS). IN, although suffering from severe kinetic limitations, may represent a potential source for stratospheric condensation nuclei (5). IN rates may become particularly large around 30 - 35 km altitude during sudden winter coolings following major stratospheric warmings. Under these conditions, due to large [H_2SO_4] values, kinetic limitations of IN may be markedly reduced.

Thus, a seed layer of condensation nuclei may be formed around 30 - 35 km altitude. Even under conditions of a condensation-evaporation equilibrium, IN may be sufficiently efficient to maintain the stratospheric aerosol layer (5) (Fig. 9).

If ions were, in fact, involved in stratospheric aerosol formation, a physical link between solar activity and the stratospheric aerosol layer may exist (5).

Analytical and Diagnostic Applications of In Situ Ion Composition Measurements

Atmospheric ions react selectively with certain neutral trace gases leading to characteristic product ion species. By measuring reactant and product ions in situ, the number densities of the reactant trace gases can be inferred with great

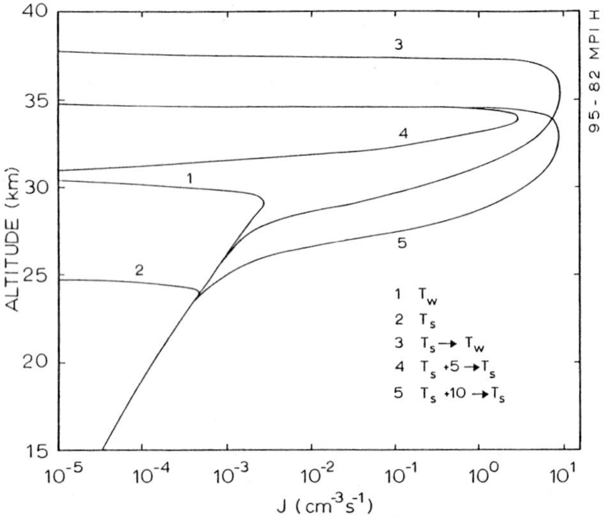

FIG. 9 - Ion-nucleation rates as calculated by Arnold (see (5)) for winter (1), summer (2), and stratospheric winter warming-cooling events (3,4,5).

sensitivity. This indirect method for atmospheric trace gas detection was termed PACIMS (passive chemical ionization mass spectrometry) (6).

The extremely large sensitivity of PACIMS is due to both the relatively large ion-recombination lifetime, t_R, and the large rate coefficients, k, for ion-molecule reactions. Assuming that an ion, C^+, is formed by

$$A^+ + B \rightarrow C^+ + D \tag{7}$$

and lost preferably by ion-ion recombination, a steady-state treatment yields

$$[B] = [C^+] / [A^+] (k\ t_R).$$

Taking $k = 10^{-9}\ cm^3 s^{-1}$, $t_R = 10^4$ s, and a minimum measurable ion abundance ratio, $[C^+]/[A^+] = 10^{-4}$, one obtains a minimum detectable $[B] = 10\ cm^{-3}$ corresponding to a volume mixing ratio on the order of 10^{-16} in the middle stratosphere.

Various trace gases (Table 3) have already been detected using PACIMS. Particularly interesting ones are H_2SO_4 and HSO_3, as they are involved in stratospheric aerosol formation (Fig. 10).

It seems that H_2SO_4-vapor is supersaturated with respect to H_2SO_4-H_2O solution droplets at altitudes below about 30 - 35 km depending on season. Thus, H_2SO_4-H_2O aerosols can exist only below these heights. It also seems that HSO_3 becomes abundant at the lower heights, possibly suggesting (17) that aerosols are formed from HSO_3- rather than H_2SO_4-vapor. It is conceivable that H_2SO_4 is formed in the aerosols rather than in the gas phase, which would be in contrast to most current models of the stratospheric sulfur chemistry.

This example clearly demonstrates the kind of interesting new information on trace gases which was recently obtained from stratospheric in situ ion composition measurements.

Very recently, active chemical ionization mass spectrometry (ACIMS) was employed for stratospheric trace gas detection (18). Here ions are created in the stratospheric medium by an electron bombardment ion source and sampled by an ion mass

TABLE 3 - Stratospheric trace gases detected by chemical ionization mass spectrometry. Altitude range and positive (PI) or negative (NI) ion composition measurements are indicated. Parentheses (+) denote that identification is uncertain.

Trace Gas	PI	NI	Altitude Range
H_2O	+	+	33 - 42
CH_3CN	+		15 - 42
CH_3OH	+		15 - 34
CH_3NO_2	(+)		15 - 34
HNO_3		+	33 - 42
HCN		(+)	34
HSO_3		+	28 - 34
H_2SO_4		+	15 - 40
HNO_2		(+)	34

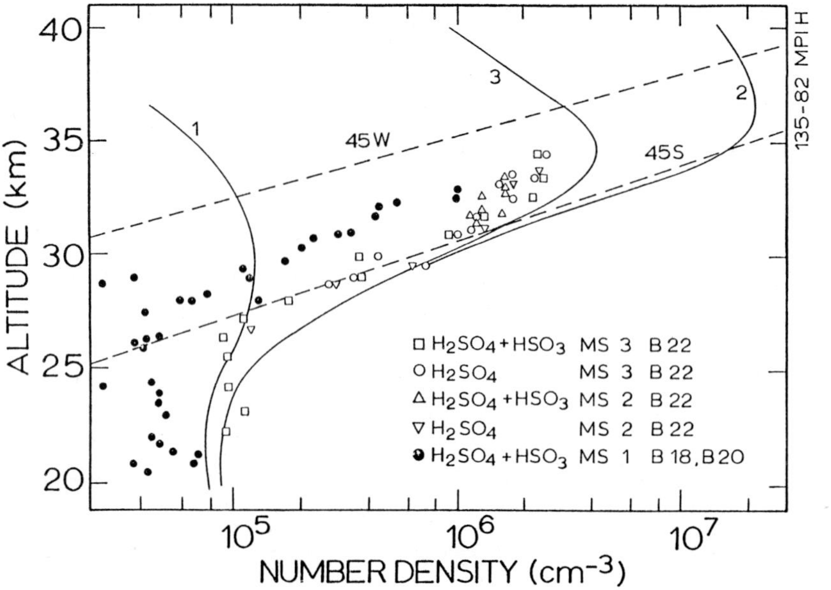

FIG. 10 - Sulfuric acid vapor abundances as obtained by passive chemical ionization mass spectrometry. Curves 1, 2, and 3 are model predictions of Turco et al. (21). Broken curves are equilibrium saturation vapor concentrations for 45° latitude summer (S) and winter (W). (Figure from Arnold and Bührke, submitted for publication).

spectrometer after they have flown with the neutral gas over a distance of about 50 cm. The ions reside for about 20 seconds in the medium before they are sampled. Within this time, they react with trace gases whose concentration can be inferred from the measured abundances of reactant and product ions.

ACIMS has already provided information on HNO_3, H_2SO_4, CH_3CN, and H_2O (7).

Since the large cluster ions interact with condensable vapors similar to aerosols, in situ compositional measurements of large compositional cluster ions also provide interesting information on the composition and thermodynamics of aerosols which exist in the same medium (17). It was found that large negative cluster ions around 30 km altitude are composed mostly of H_2SO_4 and H_2O which is rather similar to the expected aerosol composition. Bond energies for H_2SO_4-ligands were found to be about 20 kcal $mole^{-1}$ which is close to the heat of vaporization for a bulk H_2SO_4-H_2O solution. It was also found that the HSO_3-radical bonds similarly strongly to large negative ion clusters such as H_2SO_4. This finding is of great interest as no laboratory information on HSO_3 is as yet available.

Thus, in situ ion composition measurements can provide new and interesting information on neutral trace gases and aerosols.

Assessment of Tropospheric Ion Chemistry

The nature of tropospheric ions and their possible role in trace gas and aerosol processes is, as already mentioned, largely unknown. Building on recent progress in our understanding of stratospheric ion processes, an assessment of tropospheric ion chemistry will be attempted in the following section.

In the troposphere the most important sources of ionization are radioactivity and galactic cosmic rays (Böhringer, private communication). The former, which is due to α, β, and γ radiation from thoron and radon, is dominant up to about one kilometer altitude. Above this height, as in the stratosphere, galactic cosmic ray ionization is most important.

The total ionization rate, Q, is on the order of 1 - 10 $cm^{-3}s^{-1}$ in the troposphere.

Removal of free ions occurs by two mechanisms: ion-ion recombination (essentially saturated ternary ion-ion recombination; effective binary coefficient $\alpha = 2 \cdot 10^{-6}$ cm^3s^{-1}) and ion-attachment to aerosol particles. The latter process leads to so-called "large ions" which are, in fact, electrically-charged aerosols rather than ions in a strictly physical sense.

Usually, ion-attachment is the most important sink for free ions throughout the troposphere as the tropospheric aerosol content is relatively large. In this respect, the tropospheric ionization-deionization balance differs greatly from the stratospheric one.

On the average, the free ion lifetime in the troposphere is on the order of 10^2 - 10^3 seconds. Taking typical Q-values of 1 - 10 $cm^{-3}a^{-1}$, a steady-state free ion concentration, n, on the order of 10^2 - 10^4 cm^{-3} is obtained. Due to temporal and spatial changes of the aerosol content, tropospheric free ion concentrations may undergo marked changes.

The chemical evolution of tropospheric free ions is rather uncertain. Stages one and two of the positive and negative ion evolutions may be similar to those occurring in the stratosphere. A marked difference, however, may arise from the relatively large tropospheric water vapor abundance possibly leading to marked hydration as early as stages one and two. It is conceivable that in certain cases hydration changes the reactivity of the core ion. For example, reaction of $O_3^-(H_2O)_n$ with CO_2 is much slower or even inefficient compared to that of O_3.

For typical ground level conditions the most prominent positive and negative ions should contain about 10 - 20 water molecules. This has been found from mass spectrometric studies of ions created in ground level air carried out by our group (unpublished data).

Most uncertain, however, are stages three of the positive and negative ion evolutions, particularly as reactant trace gases are not known. Taking the above free ion lifetimes of 10^2 - 10^3 seconds and a maximum rate coefficient for an ion molecule reaction on the order of 10^{-9} cm^3s^{-1}, reactant trace gases having abundances of only 10^6 - 10^7 cm^{-3} (corresponding volume mixing ratio: 10^{-16} - 10^{-15}) can convert a significant fraction of the free ion population.

Potential reactant trace gases may, as in the stratosphere, include molecules possessing large proton affinities or large gas phase acidities. A major difference, however, arises from the fact that trace gases which can be depleted by heterogeneous interaction with aerosols may have small and possibly strongly variable abundances. Such interactions may involve condensation, dissolution, or surface as well as liquid phase chemical reactions.

Since such trace gases seem to be particularly efficient reactants for ions, it is conceivable that the tropospheric ion composition responds sensitively to aerosol conditions and meteorological factors.

This may be illustrated by discussing a possible influence of sulfuric acid vapor, which is a potential reactant trace gas. It may react with negative ions leading to $HSO_4^-(H_2SO_4)_\ell(H_2O)_m$ mixed clusters, and it may also react with large positive water clusters leading to a ligand "shell" composed of a mixture of H_2SO_4 and H_2O.

In the troposphere, sulfuric acid is formed from sulfur-bearing precursor gases and ultimately leads to H_2SO_4-H_2O aerosol solution droplets, as in the stratosphere. However, in contrast to the stratospheric situation, no significant H_2SO_4 supersaturation with respect to the H_2SO_4-H_2O phase should occur. Thus, the atmospheric H_2SO_4-vapor concentration should be roughly equal to the equilibrium saturation concentration over the aerosol which depends critically on the relative humidity and on temperature. It may vary between about 10^5 and 10^{11} cm^{-3} for typical ground level conditions. Thus, the H_2SO_4-vapor concentration may be lower or higher than the critical reactant trace gas abundance level of 10^6 - 10^7 cm^{-3}. Consequently, the influence of tropospheric sulfuric acid vapor on the ground level ion composition may vary from negligible to important, depending on conditions.

This example demonstrates the possible response of tropospheric ions to aerosol and meteorological conditions.

Other potential reactant trace gases besides H_2O and H_2SO_4 are NH_3, CH_3CN, and acids such as HNO_3 and HCl. Ammonia, for example, is highly soluble in water and therefore may become depleted from the gas phase. According to in situ measurements, tropospheric ammonia vapor abundances greatly exceed the critical reactant trace gas level. Consequently, NH_3 may markedly influence the positive ion chemistry. The same may be true for CH_3CN which seems to originate from the troposphere.

Thus, positive mixed cluster ions of the type $NH_4^+(NH_3)_k(CH_3CN)_\ell(H_2O)_m$ may be formed. However, it is conceivable that other trace gases which have the potential to react with positive cluster ions are also present in sufficiently large abundances.

For both NH_3 and CH_3CN it was found in the laboratory that only a few molecules can be incorporated into the cluster, forming an "inner" ligand shell, and that H_2O becomes the preferred ligand in the "outer" ligand shell.

Potential reactant gases for negative ions under conditions of low H_2SO_4-vapor abundances are acids such as HNO_3, HCl, and HNO_2, where mixed cluster ions of the type $NO_3^-(HNO_3)_\ell(H_2O)_m$ may be most prominent.

In order to investigate tropospheric ion chemistry, several so-called "simulation experiments" involving ionization of laboratory air at atmospheric or elevated pressures and subsequent ion detection by differentially-pumped mass spectrometers have been carried out during recent years. However, due to the comparatively small ion residence times encountered in these experiments as a result of ion losses to the walls of the apparatus, at best only stages one and two of the ion evolution can be simulated. The decisive stage three which determines the nature of the terminal ions cannot be simulated as its characteristic time scale of 10^2 - 10^3 s is much larger than the ion residence time. Therefore, these "simulation studies" have to be interpreted with care.

Now a possible role of free ions in tropospheric trace gas and aerosol processes will be discussed.

As far as aerosol formation is concerned, conventional ion nucleation (IN) which requires relatively large supersaturation ratios ($S \gtrsim 4$) is probably inefficient. Due to the relatively large tropospheric aerosol content, large S values can hardly build up but supersaturated vapors should condense on preexisting aerosols. Polyion nucleation may be operative but it is not known whether small polyions can grow to condensation nuclei sizes before they are scavenged by preexisting aerosols.

Concerning a possible role of ions in tropospheric trace gas processes little can be said at present. Besides ion-molecule reactions, ion-ion recombination and ion-catalyzed reactions may in this respect be important. The latter may also include "quasi liquid phase reactions" occurring in relatively large cluster ions or polyions. Maximum rates for trace gas destruction and formation are similar to those estimated for the stratosphere (see section on Potential Role of Ions in Stratospheric Trace Gas and Aerosol Processes).

Finally, analytical and diagnostic applications of tropospheric free ions will be discussed.

As in the stratosphere, in situ ion composition measurements should offer an enormous potential for neutral trace gas detection. Likely candidates for PACIMS trace gas detection are the reactant trace gases discussed above. Diagnostic applications for probing aerosol properties also seem promising. Since tropospheric cluster ions are relatively large, they should resemble aerosol solution droplets.

As already mentioned, the first in situ positive ion composition measurements were recently made in the uppermost part of the troposphere (Arnold et al., unpublished). Prospects for an extension of these measurements towards lower altitudes during the next years are good.

Acknowledgements. Stimulating discussions with E.E. Ferguson, R. Turco, P.J. Crutzen, D.H. Ehhalt, and B. Keesee are acknowledged.

REFERENCES

(1) Albritton, D.L. 1978. Ion-neutral reaction rate constants measured in flow reactors through 1977. In Atomic Data and Nuclear Tables, vol. 22, No. 1. New York: Academic Press.

(2) Arijs, E.; Nevejans, D.; Frederick, P.; and Ingels, J. 1981. Geophys. Res. Lett. 8: 121.

(3) Arijs, E.; Nevejans, D.; and Ingels, J. 1980. Unambiguous mass determination of major stratospheric positive ions. Nature 288: 684.

(4) Arnold, F. 1980. Multi-ion complexes in the stratosphere - Implications for trace gases and aerosol. Nature 284: 610.

(5) Arnold, F. 1982. Ion nucleation a potential source for stratospheric aerosols. Nature 299: 134-137.

(6) Arnold, F.; Böhringer, H.; and Henschen, G. 1978. Composition measurements of stratospheric positive ions. Geophys. Res. Lett. 5: 653.

(7) Arnold, F.; Fabian, R.; Henschen, G.; and Joos, W. 1980. Stratospheric trace gas analysis from ions: H_2O and HNO_3. Planet. Space Sci. 28: 681.

(8) Arnold, F., and Ferguson, E.E. 1980. Ions in the middle atmosphere: Their composition chemistry and role in atmospheric processes. In Proceedings of the VI International Conference on Atmospheric Electricity, Manchester, 28 July - 1 August 1980.

(9) Arnold, F., and Henschen, G. 1978. First mass analysis of stratospheric negative ions. Nature 257: 521.

(10) Arnold, F., and Henschen, G. 1982. Positive ion composition measurements in the upper stratosphere - Evidence for an unknown aerosol component. Planet. Space Sci. 30: 101.

(11) Arnold, F.; Henschen, G.; and Ferguson, E.E. 1981. Mass spectrometric measurements of fractional ion abundances in the stratosphere. I. Positive ions. Planet. Space Sci. 29: 185.

(12) Arnold, F.; Kissel, J.; Wieder, H.; and Zähringer, J. 1971. Negative ions in the lower ionosphere: A mass spectrometric measurement. J. Atmos. Terr. Phys. 33: 1669.

(13) Arnold, F.; Viggiano, A.A.; and Schlager, H. 1982. Detection of large negative cluster ions in the stratosphere - Implications for trace gases and aerosols. Nature 297: 1-5.

(14) Bates, D.R. 1982. Recombination of small ions in the troposphere and lower stratosphere. Planet. Space Sci., in press.

(15) Böhringer, H., and Arnold, F. 1981. Acetonitrile in the stratosphere - Implications from laboratory studies. Nature 290: 321.

(16) Castleman, Jr., A.W. 1979. Nucleation and molecular clustering about ions. In Advanced Colloid and Interface Science, "Nucleation," ed. A. Zettelmoyer, pp. 73-128. Amsterdam: Elsevier Press.

(17) Crutzen, P. 1976. The possible role of CSO for the sulfate layer of the stratosphere. J. Geophys. Res. 3: 73.

(18) Fahey, D.W.; Böhringer, H.; Fehsenfeld, F.C.; and Ferguson, E.E. 1982. Reaction rate constants for $O_2(H_2O)_n$ ions n = 0 to 4, with O_3, NO, SO_2 and CO_2. J. Chem. Phys. 76: 1799.

(19) Fehsenfeld, F.; Dotan, F.C.; Albritton, D.; Howard, C.; and Ferguson, E. 1978. Stratospheric positive ion chemistry of formaldehyde and methanol. J. Geophys. Res. 83: 1333.

(20) Ferguson, E.E. 1978. Sodium hydroxide ions in the stratosphere. Geophys. Res. Lett. 5: 1035.

(21) Ferguson, E.E., and Arnold, F. 1981. Ion chemistry of the stratosphere. Acc. Chem. Res. 14: 327.

(22) Ferguson, E.E.; Fehsenfeld, F.C.; and Albritton, D.L. 1979. Ion chemistry of the earth's atmosphere. In Gas Phase Chemistry, ed. M.T. Bowers. New York: Academic Press.

(23) Harris, G.W.; Kleindienst, T.E.; and Pitts, Jr., J.N. 1981. Rate constants for the reaction of OH radicals with CH_3CN, C_2H_5CN and CH_2=CH-CN in the temperature range 298-424 K. Chem. Phys. Lett. 80: 479.

(24) Henschen, G., and Arnold, F. 1980. Detection of new positive ion species in the stratosphere. Nature 291: 211.

(25) Henschen, G., and Arnold, F. 1981. Extended positive ion composition measurements in the stratosphere - Implications for neutral trace gases. Geophys. Res. Lett. 8: 999.

(26) McCrumb, J.L., and Arnold, F. 1981. High-sensitivity detection of negative ions in the stratosphere. Nature 294: 136.

(27) Mühleisen, R. 1957. Atmosphärische Elektrizität. In Handbuch der Physik, vol. XLVIII, p. 541. Heidelberg: Springer-Verlag.

(28) Narcisi, R.S., and Bailey, A.D. 1965. Mass spectrometric measurements of positive ions at altitudes from 64 to 112 kilometers. J. Geophys. Res. 70: 3687.

(29) Narcisi, R.S.; Bailey, A.D.; Della Luca, L.; Sherman, C.; and Thomas, D.M. 1971. Mass spectrometric measurements of negative ions in the D- and lower E-regions. J. Atmos. Terr. Phys. 33: 1147.

(30) Rowe, B.R.; Viggiano, A.A.; Fehsenfeld, F.C.; Fahey, D.W.; and Ferguson, E.E. 1982. Reactions between neutrals clustered to ions. J. Chem. Phys. 76: 742.

(31) Rosen, J.M., and Hofmann, D.J. 1981. Balloon-borne measurements of the small ion concentration. J. Geophys. Res. 86: 7399.

(32) Smith, D.; Adams, N.G.; and Alge, E. 1981. Ion-ion mutual neutralization and ion-neutral switching reactions of some stratospheric ions. Planet. Space Sci. 29: 449.

(33) Turco, R.P.; Toon, O.B.; Hamill, P.; and Whitten, R.C. 1981. Effects of meteoric debris on stratospheric aerosols and gases. J. Geophys. Res. 86: 1113.

(34) Turco, R.P.; Whitten, R.C.; and Toon, O.B. 1982. Stratospheric aerosols: Observation and theory. Rev. Geophys. Space Phys. 20: 233.

(35) Viggiano, A.A., and Arnold, F. 1981. The first height variation measurements of the negative ion composition of the stratosphere. Planet. Space Sci. 29: 895.

(36) Webber, W. 1962. The production of free electrons in the ionospheric D-layer by solar and galactic cosmic rays and the resultant absorption of radiowaves. J. Geophys. Res. 67: 5091.

Atmospheric Chemistry, ed. E.D. Goldberg, pp. 301-312. Dahlem Konferenzen 1982.
Berlin, Heidelberg, New York: Springer-Verlag.

Homogeneous Gas Phase Oxidation Processes in the Troposphere

H. Niki
Scientific Research Laboratory
Ford Motor Company, Dearborn, MI 48121, USA

Abstract. Principal gas phase oxidation reactions governing in situ formation and removal of various C-, N-, S-, and halogen-containing trace gases in the global troposphere are reviewed briefly. Many important gas phase reactions are common to both the troposphere and the stratosphere. However, because of the substantially different chemical compositions and physical conditions between the two spheres, there exist numerous gas phase oxidation processes which can be considered unique to the troposphere. O_3 plays the key role as the primary oxidizing agent. It reacts with compounds such as NO_x ($NO + NO_2$) and unsaturated hydrocarbons directly, and more importantly, leads to the formation of HO radical upon photodissociation. HO-initiated oxidation provides the major path for transformation of a large variety of tropospheric compounds and determines their chemical lifetimes. In these processes, numerous oxygen containing intermediates including free radicals are produced and subsequently converted to thermodynamically more stable products which are eventually removed from the gas phase by heterogeneous processes. Notably, chemical transformations of various trace gases are closely coupled and mutually interrelated by reactions involving common oxidizing or reducing species. At present, there is considerable uncertainty in the identity and fate of compounds produced in situ in the troposphere. Some of the existing uncertainties concerning the potentially important gas phase oxidation processes are indicated.

INTRODUCTION

Recent field and laboratory studies have provided ample evidence that chemical transformation occurring in the global troposphere constitutes a key component in the biogeochemical cycle of various elements such as carbon, nitrogen, oxygen, sulfur, and halogens (e.g., (3,5-7). Notably, numerous reduced gaseous species introduced into the troposphere are converted to thermodynamically more stable oxidized products. Thus, gas phase oxidation processes provide in situ sources for a variety of oxygen-containing gases such as CO_x, NO_x, and SO_x ($x \geq 1$). Gas phase chemical reactions involved in these processes are generally referred to as oxidation reactions. However, a series of the elementary reaction steps, leading to the final oxidized products, consists of both oxidation and reduction reactions involving intermediate products. Also, some of the oxidized products encountered in the atmosphere are thermodynamically less stable than their precursors, e.g., O_3 vs. O_2 and NO_3 vs. NO_2. Formation of these species, therefore, requires additional chemical energy, the primary source of which is solar radiation. The term "photooxidation" is commonly used to describe these processes, although absorption of solar photons by atmospheric gases and subsequent chemical transformation to produce reactive species are generally involved in only a few critical reaction steps. In this respect, it is important to note that some reactive species, e.g., O_3, present at nighttime can carry "dark" oxidation processes. Also, the majority of tropospheric trace gases are sufficiently inert against direct attack by molecular oxygen in the ground electronic state. Thus, either absorption of active solar radiation or collisional interaction with "reactive species," particularly free radicals, is required to trigger oxidation processes involving many trace gases. Among the "reactive species," hydroxyl (HO) radical produced photochemically from O_3 is unique. It reacts readily with virtually all the important reduced gases and governs their chemical fates in the troposphere (1).

To illustrate this, listed in Table 1 are the representative C-, N-, S-, and halogen-containing tropospheric trace gases and their estimated HO-lifetimes (τ) derived from the relation $\tau = 1/k[HO]$, where k is the rate constant for the corresponding HO reaction and [HO] is the global, seasonally averaged HO concentration of 5×10^5 cm^{-3} derived from model calculations, as

TABLE 1 - HO lifetimes of C-, N-, S-, and halogen-containing compounds.

Gas	Mixing ratio (ppbv)	HO Lifetime
CO	50 - 200	3 mo
CH_4	1500 - 2000	11 yr
C_2H_6	~1	3 mo
C_3H_8	~0.2	13 d
C_5H_8 (isoprene)	<10	8 hr
$C_{10}H_{16}$ (pinenes)	<5	11 hr
NO_x ($NO + NO_2$)	<0.1	2 d
HNO_3	<0.3	10 mo
NH_3	<3	2 mo
SO_2	<0.2	5 d
H_2S	<0.1	3 d
$(CH_3)_2S$	<0.5	3 d
CS_2	<0.4*	>14 d
OCS	0.3 - 0.5	>1 yr
CH_3Cl	~0.8	2 yr
CH_3Br	<0.01	2 yr
$CHCl_3$	~0.02	1 yr

*Some measurements in free troposphere are <0.005.

discussed in detail by Crutzen (this volume). Suffice it to say, HO-radical reactions of compounds listed in this table give rise to a large variety of oxidation intermediates including free radicals. Because of their inherently great chemical reactivity, free radicals are generally very short-lived and are present in the troposphere at concentrations much too low to be detected with the state of the art analytical tools. Thus, fundamental knowledge on potentially important atmospheric reactions involving highly reactive species comes largely from carefully controlled laboratory studies. Some crucial kinetic data on the major tropospheric reactions are becoming available for evaluating the balance of various trace gases (3,6). These pertinent reactions are too numerous to cite individually and are shown here only schematically. For instance, a typical chemical model dealing with atmospheric oxidation of CH_4 alone includes approximately 50 elementary reactions and 30 chemical species. At present, there are numerous experimental difficulties for obtaining needed kinetic data on many important atmospheric reactions (5). Our current knowledge of the tropospheric oxidation processes is highlighted here with particular emphasis on existing uncertainties.

MAJOR PHOTOOXIDATION PROCESSES

Photochemical reactions involving O_3, H_2O, and NO_x ($NO + NO_2$) provide the principal source of "reactive species," particularly the HO radical. Subsequent reactions of HO with O_3, NO_x, and carbon-containing gases CO and CH_4 constitute the tropospheric photooxidation processes (3,6,7). These reactions are discussed briefly in this section. At present, many important elementary gas phase reactions encompassing these chemical systems appear to be sufficiently well characterized and serve as a basis for examining the tropospheric balance of various other trace gases (2). However, it must be emphasized that many important reactions involving reactive species are currently not feasible for laboratory studies because of difficulties associated with selective generation and direct detection of these species under simulated atmospheric conditions. Therefore, the majority of available kinetic information suffers from varying degrees of uncertainties reflecting assumptions and approximations employed in data interpretation.

Reactions Involving the O_3 - NO_x System

The major elementary reactions governing the O_3 - NO_x chemistry are illustrated schematically in Fig. 1. Various O_x and NO_x ($x \geq 1$) species are shown to be closely coupled and interrelated by numerous reactions channels. Namely, the photodissociation of O_3 forms electronically excited oxygen atom O('D) which then reacts, in part, with H_2O to generate two HO radicals. The HO radical is oxidized by O_3 to the HO_2 radical. The HO_2 can be reduced by O_3 to HO. NO is oxidized to NO_2 by either O_3 or HO_2, and NO_2 to NO_3 by O_3. Conversely, NO acts as a reducing

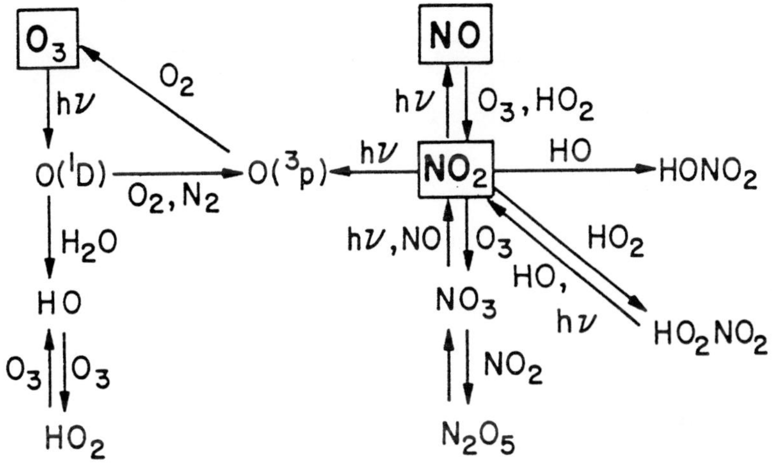

FIG. 1 - Major atmospheric reactions initiated by O_3 and NO_x ($NO + NO_2$).

agent for O_3, HO_2, and NO_3. The NO_2 photolysis plays a unique role in forming both the oxidizing agent, O_3, and the reducing agent, NO, simultaneously. NO_2 subsequently reacts with HO and HO_2 to form nitric acid HNO_3 and peroxynitric acid HO_2NO_2 and also converts NO_3 to N_2O_5. Both HO_2NO_2 and N_2O_5 are thermochemically unstable and can redissociate spontaneously. Their corresponding lifetimes are highly sensitive to both the temperature and pressure encountered in the atmosphere. There exist considerable uncertainties in the kinetics and mechanisms of gas phase reactions involving these nitrogenous products and the intermediate NO_3. Also, these products are likely to be incorporated into condensed phases by heterogeneous processes.

HO Chain Oxidation of CO

Among various reactive species, HO appears to be solely responsible for the conversion of CO to CO_2. The HO-reaction of CO concomitantly produces HO_2 which can, in turn, regenerate HO via the NO reaction. Thus, both CO and NO are oxidized by $HO-HO_2$ radical chain reactions, as illustrated schematically in Fig. 2. The chain efficiency for these processes is highly sensitive to the NO_x concentration [NO_x]. Namely, at low [NO_x],

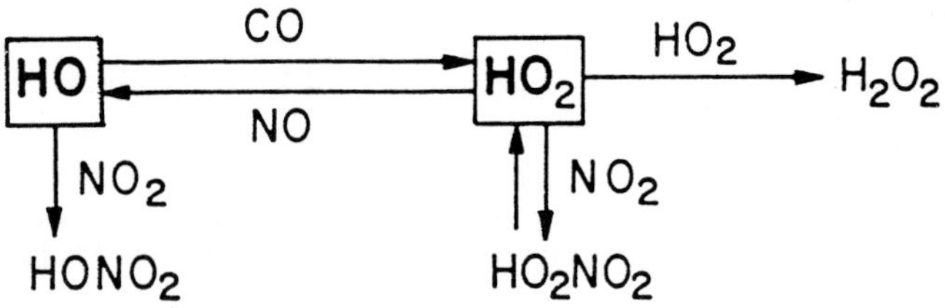

FIG. 2 - HO-HO_2 chain oxidation of CO and NO.

the HO_2 radicals react among themselves to yield hydrogen peroxide H_2O_2 rather than regenerating HO by the NO-HO_2 reaction. Also, NO_2 competes with CO and NO to remove both HO and HO_2 from the chain reactions. In view of the extremely important atmospheric role of these chain reactions, it should be considered crucial to obtain more extensive, reliable data on these elementary reactions than are available at present.

HO-initiated Oxidation of CH_4

One of the major reaction pathways for CH_4 oxidation is the formation of CO via the intermediate product formaldehyde CH_2O, as shown in Fig. 3. All of the free radical species involved, except CH_3OO, i.e., CH_3. CH_3O, and CHO, appear to react almost exclusively with O_2 under the tropospheric conditions. The alkylperoxy radical CH_3OO behaves analogously to the simplest peroxy radical HO_2. It can be either reduced by NO to RO, or converted to hydroperoxide ROOH (R = CH_3) by HO_2. Thus, the relative efficiency of production of CH_2O and CO vs. CH_3OOH is determined by the concentration ratio $[NO]/[HO_2]$. It follows that $[NO]/[HO_2]$ as well as $[NO]/[O_3]$ provides a good indication of the extent to which the atmosphere is a "reducing" or "oxidizing" environment. The atmospheric fate of CH_3OOH is rather uncertain. Its reaction with HO either regenerates CH_3OO or forms CH_2O with HO as a catalyst. Possibly, both reaction channels may be operative in competition with heterogeneous loss processes. In addition to the photochemical and HO-reactions indicated in Fig. 3, formaldehyde CH_2O may partly undergo an addition reaction with HO_2 to form RO_2 (R = CH_2OH), leading to the corresponding ROOH and formic acid HC(O)OH. Also, methyl

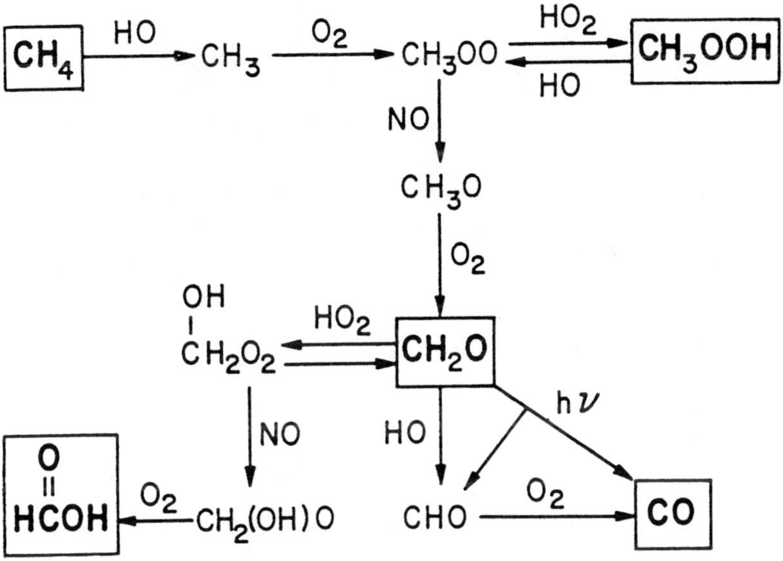

FIG. 3 - Schematics for HO-initiated oxidation of CH_4.

peroxynitrate RO_2NO_2 ($R = CH_3$) should be present in photochemical equilibrium with RO_2 and NO_2 and may play some atmospheric role.

OXIDATION PROCESSES INVOLVING OTHER TRACE GASES

The majority of C-, N-, S-, and halogen-containing compounds listed in Table 1, but not discussed in the preceding section, appear to undergo primarily HO-initiated oxidation processes. However, in general, not much is known about the ensuing reaction products, and it is not possible to decide on an a priori basis which ones among many thermochemically feasible reaction channels are the dominant processes. Moreover, for many potentially important intermediate species, their thermodynamic parameters are not known with sufficient accuracy to determine all the feasible reaction channels. Also, reactions involving reactive species other than HO, i.e., O_3, HO_2, NO_3, and photolysis, may contribute significantly but are largely uncertain at present.

Non-CH_4 Hydrocarbons

Ethane C_2H_6 follows the oxidation paths analogous to those shown for CH_4 in Fig. 3. In this case, the primary radical, R, is C_2H_5, and it forms the corresponding RO_2 and RO. The major departure from the CH_4 oxidation occurs in the subsequent HO-reaction of the carbonyl product, CH_3CHO. Namely, while O_2 removes H from the simplest carbonyl radical, CHO, derived from CH_2O, it adds to CH_3CO to form RO_2 (R = CH_3CO). This RO_2 can then react competitively with NO, NO_2, and HO_2 as indicated below.

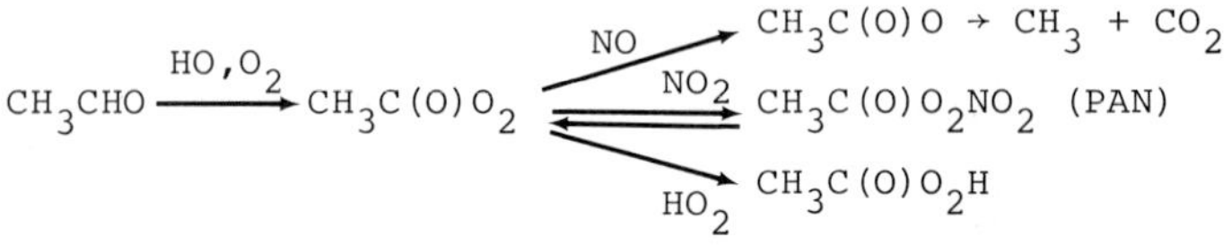

Peroxyacetyl nitrate (PAN) is thermochemically more stable than other types of $ROONO_2$ (R = H or alkyl group C_nH_{2n+1}, $n \geq 1$) and may indeed serve as an important reservoir for NO_x and RO_2 radicals (3). PAN may also decay in the atmosphere via photodissociation and HO-reaction, although these processes are still uncertain. The next larger alkane C_3H_8 leads to the formation of two types of R, $CH_3CH_2CH_2\cdot$ and $CH_3\dot{C}HCH_3$, which produce aldehyde, C_2H_5CHO, and ketone, $(CH_3)_2CO$, respectively. The fate of C_2H_5CHO should be analogous to that of CH_3CHO. The oxidative degradation of $(CH_3)_2CO$ is likely to yield the CH_3CO radical and CH_2O. For saturated hydrocarbons containing more than three carbon atoms, their HO-initiated oxidation processes involve many more varieties of intermediates and become progressively more complex (1,5). In particular, the RO-type radicals can dissociate, isomerize, or react with O_2. Relative importance of these competitive channels varies greatly depending on the type of RO, but reliable, empirical relationships are not as yet firmly established.

For olefinic hydrocarbons, HO generally adds to double-bonded carbons to form HO-substituted alkyl radicals which follow the reaction paths similar to those of alkyl radicals. One of the unique chemical behaviors exhibited by these compounds is their high reactivity towards O_3. As a result, O_3 and HO reactions of some unsaturated hydrocarbons can be of comparable importance in the troposphere. To illustrate this, atmospheric lifetimes of several compounds estimated from their HO and O_3 reactions are listed in Table 2. These compounds may eventually produce CO, via both HO and O_3 reactions, thereby providing an important in situ source in the atmosphere (3). However, in order to quantify this CO source, more experimental work is needed to better characterize the relevant reactions and products. Available laboratory data suggest that the simplest unsaturated compound C_2H_4 reacts with O_3 to form CH_2O and reactive intermediate CH_2OO. The CH_2OO may undergo dissociation to form CO and CO_2 and, alternatively, react with carbonyl compounds, H_2O and SO_2, to form products which are not as yet fully characterized. Similar mechanisms may be operative, at least in part, for large unsaturated compounds listed in Table 2. It should also be noted that many more trace organic compounds other than those listed in this table are present in the troposphere (4). Their fate is highly uncertain at present.

TABLE 2 - HO and O_3 lifetimes of unsaturated hydrocarbons.

Gas		Lifetime (hr)*	
		HO	O_3
isoprene		4.7	17
α-pinene	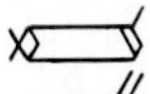	6.3	1.9
β-pinene	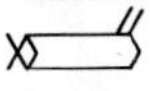	5.5	7.5
d-limonene	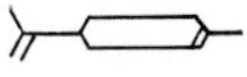	2.5	0.4

*$[O_3]$ = 40 ppbv; [HO] = 8 x 10^5 molecule cm^{-3} assumed.

Reduced Nitrogen Compounds

The HO-initiated reaction of NH_3 may provide either a sink or source of NO_x depending on $[NO_x]$. Namely, the primary radical NH_2 is rather inert towards O_2 but reacts readily with NO, NO_2, and O_3, i.e.,

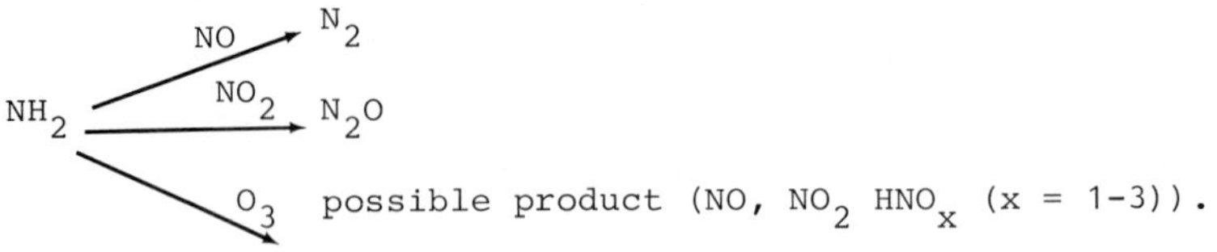

Available laboratory data suggest that the NH_2 can effectively convert NO_x to N_2 and N_2O at $[NO_x]>60$ ppt. At lower $[NO_x]$, the O_3-reaction would be predominant, although the identity of the ensuing nitrogenous product(s) is uncertain. There are no known gas-phase reactions for N_2O in the troposphere. The HO-reaction of HCN may be significant but is largely uncertain at present.

Sulfur Compounds

The reduced S-containing gases H_2S and $(CH_3)_2S$ appear to react mainly with HO. These compounds may also react with HO_2 to some extent. Another related species, CH_3SH, has not been observed to date in the troposphere. Notably, it has an approximately 10 times shorter HO-lifetime as compared with H_2S and $(CH_3)_2S$ and can become important if present even at correspondingly lower concentrations. However, the relevant rate constants are not yet firmly established, particularly with respect to their temperature dependence. In the HO-initiated oxidation (in the presence of NO_x), SO_2 has recently been observed as a product with its yield ranging from ~100% for H_2S to ~25% for $(CH_3)_2S$. The remaining S-containing product(s) of $(CH_3)_2S$ has not been positively identified but does not appear to be either OCS or H_2SO_4. It is not at all clear whether HO adds to S or abstract H from these compounds. On the other hand, the HO-initiated oxidation of SO_2 must involve the formation of an adduct, $HOSO_2$, as the primary radical species. Its eventual fate is probably the production of H_2SO_4 aerosols, although the detailed kinetics and mechanisms are unknown. HO-reactions of both CS_2 and OCS and generally considered too slow to be important in the troposphere. This notion may not be fully warranted at least for CS_2. Several direct measurements indicating very low values for this rate constant have all been made in the presence of inert diluent gases only. There is some evidence that the HO-initiated oxidation of CS_2 leading to OCS and SO_2 is greatly accelerated in the presence of air. CS_2 may also undergo photo-oxidation via electronically excited CS_2^*.

$$CS_2 \xrightarrow{h\nu > 280\ nm} CS_2^* \xrightarrow{O_2} CS + SO_2$$

OCS, SO, and SO_2 are possible S-containing products formed in the subsequent oxidation of CS by either O_2 or O_3. Again, this is one of the many atmospheric reactions which require further studies in order to remove existing kinetic and mechanistic uncertainties.

Halocarbons

As indicated in Table 1, the halogenated methanes CH_3Cl, CH_3Br, and $CHCl_3$ can be removed by HO at significant rates in the troposphere. The oxidation mechanism for these compounds should be somewhat analogous to that of CH_4 (cf. Fig. 3). Namely, the corresponding primary radical species CH_2Cl, CH_2Br, and CCl_3 lead to the formation of either ROOH via RO_2 or carbonyl products via RO. For RO = CH_2ClO, the H-atom transfer reaction with O_2 to form formylchloride, CHClO, and HO_2 radical appears to be dominant over the unimolecular dissociation to form CH_2O and Cl atom. Similarly, the CH_2BrO is expected to form formylbromide, CHBrO, rather than CH_2O. On the other hand, the CCl_3O can only form phosgene, CCl_2O, and a Cl atom. The subsequent fate of these halocarbonyl compounds is likely to be governed by photodissociation. These compounds would have rather low reactivity towards HO, if they behaved similarly to their corresponding Cl-atom reactions. It is possible that the photolysis of these gas compounds yields molecular products rather than halogen atoms exclusively. Halogen atoms possibly produced from these compounds as well as man-made halocarbons, e.g., CH_3CCl_3, in the troposphere would lead to various reactions considered for the modeling of stratospheric chemistry. However, it is not clear that these reactions would play a significant chemical role comparable to those involving HO in the troposphere.

CONCLUSION

At present, a large body of laboratory data already exists on elementary reactions involving relatively simple and readily observable reactive species such as those encountered in the oxidation of CH_4. In contrast, there is a dearth of information concerning the identity and fate of intermediate products produced from the majority of "complex" tropospheric trace compounds, primarily because of numerous experimental difficulties. Throughout this paper, attempts were made to point out existing uncertainties in these oxidation processes.

REFERENCES

(1) Atkinson, R.; Darnall, K.R.; Lloyd, A.C.; Winer, A.M.; and Pitts, Jr., J.N. 1979. Kinetics and mechanism of the gas phase. Adv. Photochem. 11: 375-488.

(2) Baulch, D.L.; Cox, R.A.; Hampson, Jr., R.F.; Kerr, J.A.; Troe, J.; and Watson, R.T. 1980. Evaluated kinetic and photochemical data for atmospheric chemistry. J. Phys. Chem. Ref. Data 9: 295-471.

(3) Crutzen, P.J. 1983. Atmospheric interactions - Homogeneous gas reactions of C, N, and S-containing compounds. In SCOPE 24. New York, London: John Wiley and Sons, in press.

(4) Graedel, T.E. 1978. Chemical Compounds in the Atmosphere. New York: Academic Press.

(5) Herron, J.T.; Huie, R.E.; and Hodgeson, J.A., eds. 1978. Chemical Kinetic Data Needs for Modeling the Lower Troposphere. NBS Special Publication 557.

(6) Logan, J.A.; Prather, M.J.; Wofsy, S.C.; and McElroy, M.B. 1981. Tropospheric chemistry: A global perspective. J. Geophys. Res. 86: 7210-7254.

(7) Seinfeld, J.H. (Chairman). 1981. Report on the NASA Working Group on Tropospheric Program Planning. NASA Reference Publication 1062.

Atmospheric Chemistry, ed. E.D. Goldberg, pp. 313-328. Dahlem Konferenzen 1982.
Berlin, Heidelberg, New York: Springer-Verlag.

The Global Distribution of Hydroxyl

P. J. Crutzen
Max-Planck-Institut für Chemie, Abt. Luftchemie
6500 Mainz, F. R. Germany

Abstract. Two tropospheric, meridional sets of OH distribution are calculated with a two-dimensional model. One of the solutions represents the expected minimum OH concentrations in the absence of radical scavenging by aerosol particles in the atmosphere. The other solution provides excellent descriptions of the observed $CFCl_3$, CF_2Cl_2 distributions, good approximations of the ozone observations, and a very good agreement with the global observations of CH_3CCl_3 by Rasmussen and co-workers (22). The derived distributions of hydroxyl may, therefore, represent reasonable approximations to the average meridional OH distributions in the atmosphere. However, if the methylchloroform observations of other workers are also considered, up to two times larger OH concentrations are possible. Despite rough approximation, rather fortuitously, it may, therefore, be possible to propose OH distributions which may approximate the average meridional OH distributions to within a factor of two. The OH concentrations are most difficult to estimate in the continental boundary layer because of reactions with effluents from biological and anthropogenic processes which are most pronounced over land. For gases with average residence times of less than a few weeks, the removal rates from the atmosphere are determined by photochemical processes taking place over the continents, so that for such gases a three-dimensional treatment is required.

INTRODUCTION

More than a decade has passed since the essential role of the hydroxyl radical in the chemistry of the troposphere was proposed (14), and although the idea is now generally accepted, there is still little direct quantitative knowledge available about the prevailing hydroxyl distribution in the troposphere. It is clear that direct measurements of hydroxyl are exceedingly difficult (2,7,19,20,30,31), and it is even doubtful whether truly average OH distributions in the troposphere can be derived by a global measurement program of hydroxyl alone, even when the existing experimental difficulties with the measurements are resolved. We may expect it to be highly variable in space and time and to be dependent on many changing factors such as global distribution of key chemical species O_3, CO, NO_x, H_2O, CH_4, and non-methane hydrocarbons and the intensity of solar ultraviolet radiation which in turn depends on cloudiness, atmospheric aerosol content, and, of course, the total ozone column (15,17). It will, therefore, become prohibitively expensive to make sufficiently long series of measurements of hydroxyl at enough places in the troposphere to supply the necessary data base to derive global average OH distributions. Instead, we must rely on a combination of observational and theoretical efforts. In the first place, we must test whether our understanding of atmospheric chemistry is adequate enough so that a measured OH concentration can be explained in terms of the postulated photochemical scheme and observations of the chemical factors which are important in the scheme. For instance, because of low concentrations of non-methane hydrocarbons and particulate matter (9,14,24), it is generally expected that in the background marine troposphere the concentration of hydroxyl is mainly governed by the local concentrations of ozone, water vapor, carbon monoxide, methane, nitric oxide, hydrogen peroxide, methyl-hydro-peroxide, formaldehyde, and the intensity of ultraviolet radiation in the 295-325 nm wavelength region. With the ongoing development of measurement techniques for peroxides and hydroxyl, the concentration of all important, photochemically active species can soon be measured (e.g., (3,14,18,25)) so that the existing photochemical theory can then be tested. This was in fact one of the aims of major aircraft research expeditions which were carried out during 1977 and 1978 but did not succeed in this particular task because of unrecognized observational requirements. (Many of the highly interesting results obtained on these research flights of atmospheric trace gases have been presented in (13). It is quite likely that another global measurement program of this type aimed at testing the photochemical theory will be tried in the marine, free troposphere in only a few years' time. As follows from the paper by Niki (this volume), it will be much more difficult, however, to test photochemical theory in the continental boundary layer because additional chemical compounds, especially reactive hydrocarbons, will play a significant role in determining the OH concentrations. For example, during our own work over the forested areas of Brazil, we have observed isoprene volume mixing ratios averaging almost 2 ppbv above the tropical forests, while average carbon monoxide and methane volume mixing ratios were equal to 300 ppbv and 1.6 ppmv, respectively (unpublished data). Becasue the reaction

$$OH + C_5H_8 \rightarrow C_5H_9O \ (\rightarrow \text{products}) \qquad \text{(R1)}$$

has a rate coefficient ($\simeq 8 \times 10^{-11}$ cm^3 $mol.^{-1}$ s^{-1}) about 100 times larger than that of

$$OH + CO \rightarrow H + CO_2 \ (= 2.5 \times 10^{-13} \ cm^3 \ mol.^{-1} \ s^{-1}) \qquad \text{(R2)}$$

and 10^4 times larger than that of

$$OH + CH_4 \rightarrow H_2O + CH_3 \ (= 8 \times 10^{-15} \ cm^3 \ mol.^{-1} \ s^{-1}), \qquad \text{(R3)}$$

it is clear that the atmospheric concentrations of OH in the forest boundary layer will be much more dependent upon the isoprene chain reactions than those involving carbon monoxide and methane. As the breakdown of isoprene can lead to a long chain of reactants which are far from well established, a test of the photochemical theory in the boundary layer above forests is most likely going to be extremely difficult. (In addition, we should be aware that there are substantial practical difficulties involved with measurements in remote, tropical environments which are located in developing nations with limited technical facilities.)

Considering these major difficulties, special emphasis should be assigned to global measurements of atmospheric trace gases with well-defined atmospheric input rates and locations. For this purpose, the atmospheric chemistry community has at its disposal a number of industrial, halogenated organic compounds with reasonably well-defined sources (1,4) and a variety of reaction rate coefficients with hydroxyl. As the most interesting compounds we identify methylchloroform (CH_3CCl_3), tetrachloroethylene (C_2Cl_4), trichloro-ethylene (C_2HCl_3), F-11 ($CFCl_3$) and F-12 (CF_2Cl_2) with widely different rate coefficients with hydroxyl. The latter two compounds, F-11 and F-12, in fact do not react at all with OH and are, therefore, especially well suited to test transport descriptions in advanced global meteorological-photochemical models. The described method to arrive at knowledge of the global OH distribution from methylchloroform observations has been adopted by atmospheric chemists (8,26,27), and the approach has been discussed by Penkett (this volume). Unfortunately, presently available information on industrial production rates and global distribution of the aforementioned gases are not well enough known to derive the global OH distribution better than with a factor of two uncertainty, as we will discuss below.

GLOBAL DISTRIBUTION OF HYDROXYL

It is possible to derive a rough global picture of the minimum concentrations of hydroxyl in large sections of the troposphere by equating the OH production rate by the reactions

$$O_3 + h\nu \rightarrow O(^1D) + O_2 \quad (\lambda \lesssim 310\ nm), \qquad (R4)$$

$$O(^1D) + H_2O \rightarrow 2\ OH,\ \text{and} \qquad (R5)$$

$$O(^1D) + M \rightarrow O(3P) + M, \qquad (R6)$$

with the known OH loss rates through the reactions

$$CO + OH \rightarrow H + CO_2, \qquad (R2)$$

$$CH_4 + OH \rightarrow CH_3 + H_2O,\ \text{and} \qquad (R3)$$

$$O_3 + OH \rightarrow HO_2 + O_2, \qquad (R7)$$

neglecting for the moment the possibility of hydroxyl-recycling or additional production by reactions such as

$$H + O_2 + M \rightarrow HO_2 + M, \qquad (R8)$$

$$HO_2 + O_3 \rightarrow OH + 2\ O_2, \qquad (R9)$$

$$HO_2 + NO \rightarrow OH + NO_2, \tag{R10}$$

$$HO_2 + HO_2 \rightarrow H_2O_2 + O_2, \tag{R11}$$

$$H_2O_2 + hv \rightarrow 2\ OH, \tag{R12}$$

$$CH_3 + O_2 + M \rightarrow CH_3O_2 + M, \tag{R13}$$

$$CH_3O_2 + NO \rightarrow CH_3O + NO_2, \tag{R14}$$

$$CH_3O + O_2 \rightarrow CH_2O + HO_2, \tag{R15}$$

$$CH_2O + hv \rightarrow H + HCO, \text{ and} \tag{R16a}$$

$$HCO + O_2 \rightarrow HO_2 + CO, \tag{R17}$$

so that the minimum concentration of OH can be written as

$$(OH)_{min} = \frac{2\ k_5(O^1D)\ (H_2O)}{k_2(CO) + k_3(CH_4) + k_7(O_3)} ,$$

with

$$(O^1D) = \frac{J_4(O_3)}{k_6(M) + k_5(H_2O)} .$$

Other loss reactions in the troposphere, including reactions on aerosol particles, are probably unimportant (32). There seem to be only minor pathways to associations of OH with H_2O and O_2. Using the recommended reaction kinetic data and observations on the average distributions of ozone, water vapor, carbon monoxide, and methane, the distribution of the minimum hydroxyl concentrations can be calculated relatively easily. The global, tropospheric 24-hour average OH concentration is equal to 1.3×10^5 molecules cm^{-3}.

Not all OH which reacts with CO, CH_4, and O_3 is, however, lost from the atmosphere, because there is at least a transfer back from HO_2 to OH by the reaction R9 with ozone, so that in fact minimum OH concentrations can be determined by assuming that the atmospheric NO_x concentrations are so low (we assume a volume mixing ratio of 10^{-14}) and removal of peroxides so fast that recycling of OH is only due to the reaction R9. For the background marine atmosphere, observations on aerosol concentrations (14) can be used to show that for most of the troposphere removal of HO_2 by aerosol can be important only if the accommodation coefficient of HO_2 to aerosol approaches unity. We neglect this possibility (see (32) for a more thorough discussion). The calculated OH distributions for model I are given in Figs. 1a-1e. On an average the global mean of the minimum tropospheric OH concentrations equals 2.3×10^5 molecules cm^{-3}, and the average column ozone loss rate is equal to 1.5×10^{11} molecules cm^{-2} s^{-1}. This is three times larger than is supplied from the stratosphere (10), and, consequently, ozone production must take place in the troposphere to compensate for this photochemical loss. This requires the presence of NO_x, which in turn leads to larger OH concentrations through reaction R10.

The effect of recycling of OH by reactions involving NO_x was estimated with a two-dimensional, photochemical model in which

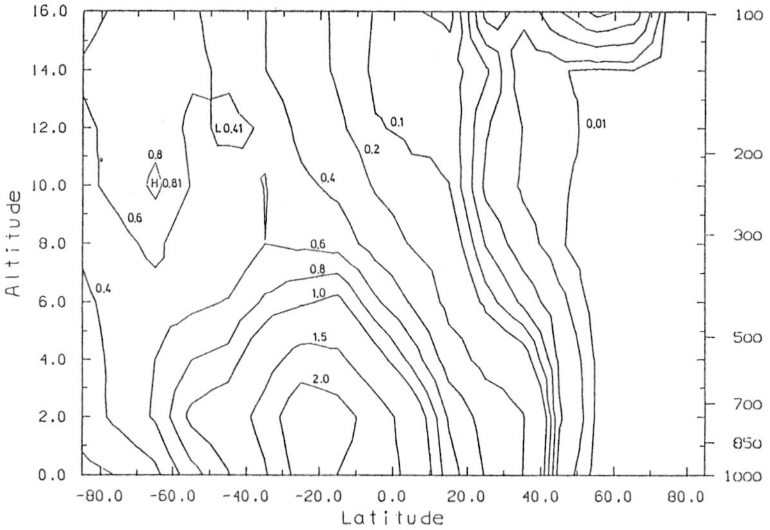

FIG. 1a.

FIGS. 1a-1e - Calculated tropospheric OH concentrations during the sunlit period of the day considering recycling of OH only by the reaction $HO_2 + O_3 \rightarrow OH + 2\ O_2$ for Feb. 1, May 1, July 1, and Oct. 1, and annual average, respectively. Unit: 10^6 molecules OH per cm^3.

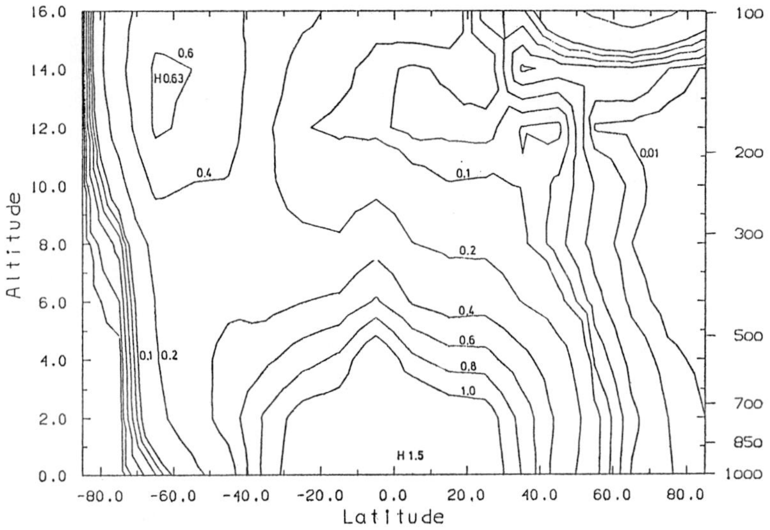
Altitude
16.0
14.0
12.0
10.0
8.0
6.0
4.0
2.0
0.0
-80.0 -60.0 -40.0 -20.0 0.0 20.0 40.0 60.0 80.0
Latitude
100
200
300
500
700
850
1000
0.6
H 0.63
0.4
0.1
0.01
0.2
0.4
0.6
0.8
1.0
0.1 0.2
H 1.5

FIG. 1b.

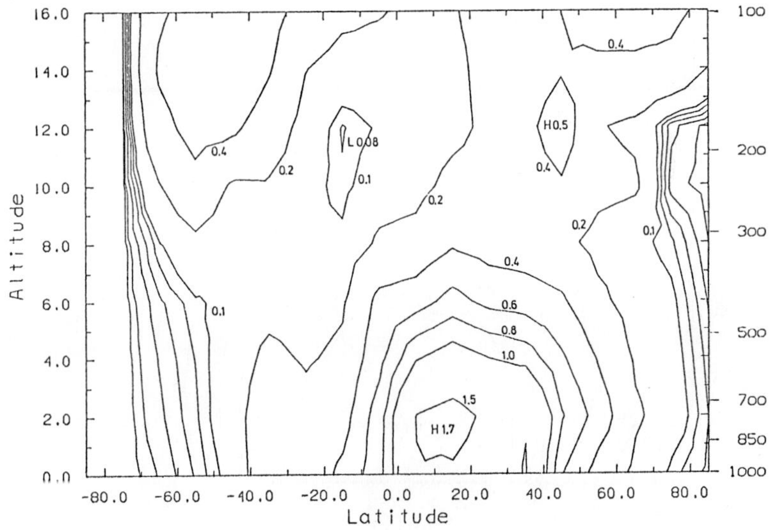
Altitude
16.0
14.0
12.0
10.0
8.0
6.0
4.0
2.0
0.0
-80.0 -60.0 -40.0 -20.0 0.0 20.0 40.0 60.0 80.0
Latitude
100
200
300
500
700
850
1000
0.4
H 0.5
L 0.08
0.4
0.2
0.1
0.4
0.2
0.2
0.1
0.4
0.1
0.6
0.8
1.0
1.5
H 1.7

FIG. 1c.

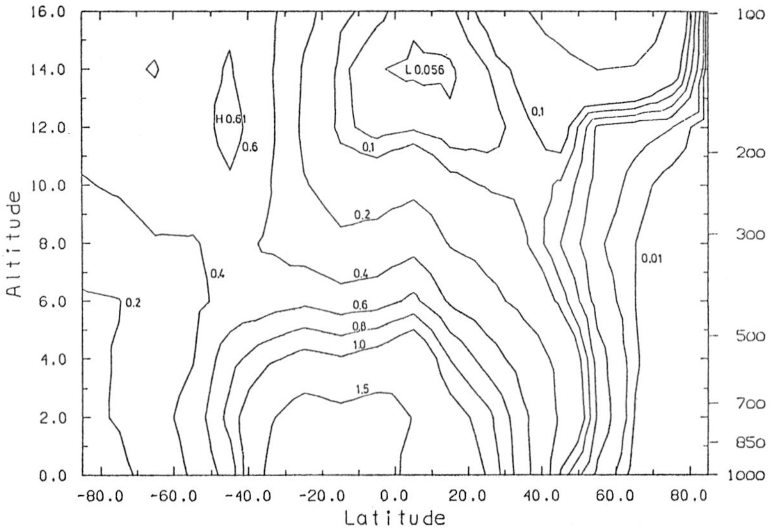
Altitude
Latitude
16.0
14.0
12.0
10.0
8.0
6.0
4.0
2.0
0.0
-80.0
-60.0
-40.0
-20.0
0.0
20.0
40.0
60.0
80.0
100
200
300
500
700
850
1000
H 0.61
0.6
L 0.056
0.1
0.1
0.2
0.4
0.4
0.2
0.6
0.8
1.0
1.5
0.01

FIG. 1d.

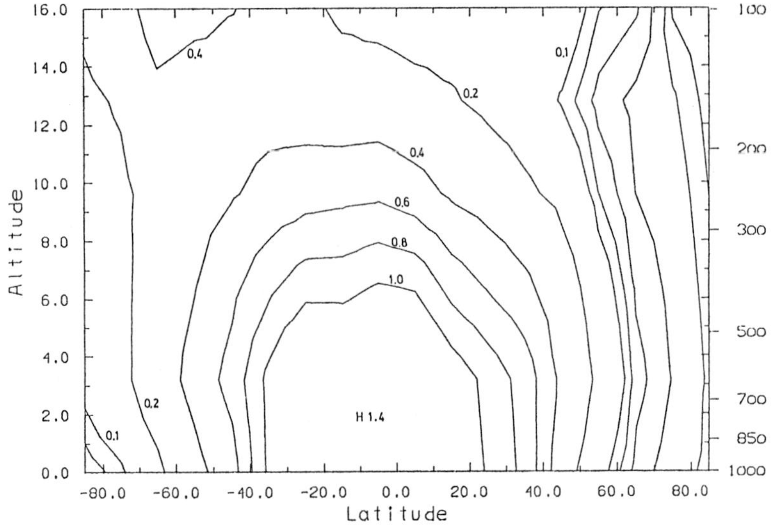
Altitude
Latitude
16.0
14.0
12.0
10.0
8.0
6.0
4.0
2.0
0.0
-80.0
-60.0
-40.0
-20.0
0.0
20.0
40.0
60.0
80.0
100
200
300
500
700
850
1000
0.4
0.1
0.2
0.4
0.6
0.8
1.0
0.2
0.1
H 1.4

FIG. 1e.

hydroxyl, ozone, and NO_x fields were calculated, using inputs of NO_x from industrial processes, lightning, and stratosphere (11,18,28). The calculated distributions of OH are shown in Figs. 2a-2e. According to this model, annual average OH concentrations during a 24-hour period in the troposphere were equal to 5×10^5 and 6×10^5 molecules cm^{-3} in the northern and southern hemispheres, respectively. We notice, therefore, that the recycling of OH from HO_2 and H_2O_2 yielded about two times more OH than the minimum concentrations shown in Figs. 1a-1e. The ozone distributions agree fairly well with the observations (5).

In Table 1, we compare the calculated distributions of the industrial trace gases $CFCl_3$, CF_2Cl_2, and CH_3CCl_3 on the same model with observations (33). We notice a strikingly good agreement for each compound between the observations of

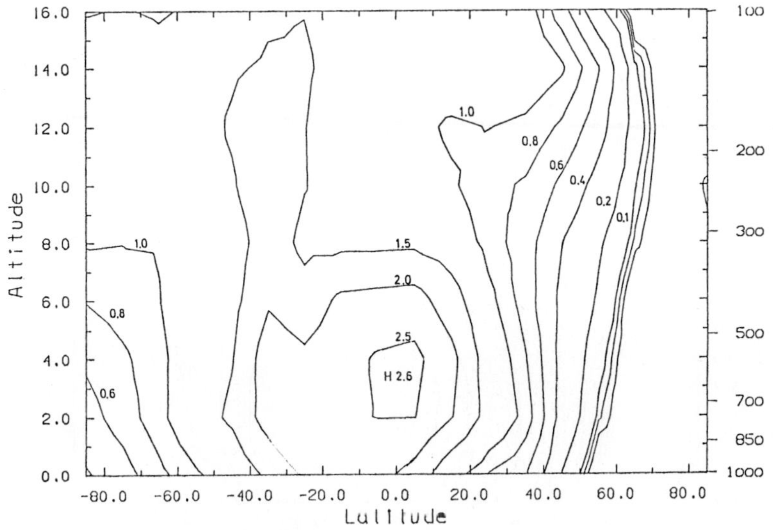

FIG. 2a.

FIGS. 2a-2e - Calculated OH concentrations also considering recycling of OH by the reaction $HO_2 + NO \rightarrow OH + NO_2$. These distributions provide rough agreement with observed tropospheric ozone concentrations.

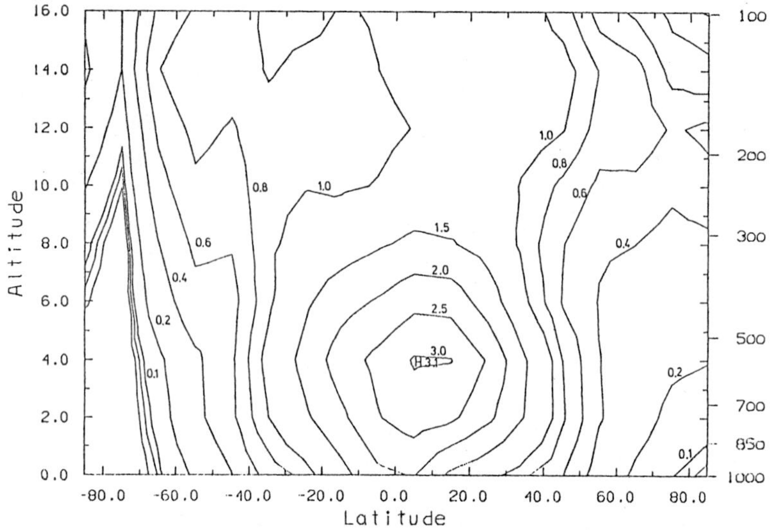
Altitude
Latitude
16.0
14.0
12.0
10.0
8.0
6.0
4.0
2.0
0.0
-80.0
-60.0
-40.0
-20.0
0.0
20.0
40.0
60.0
80.0
100
200
300
500
700
850
1000
0.1
0.2
0.4
0.6
0.8
1.0
1.5
2.0
2.5
3.0
H3.1

FIG. 2b.

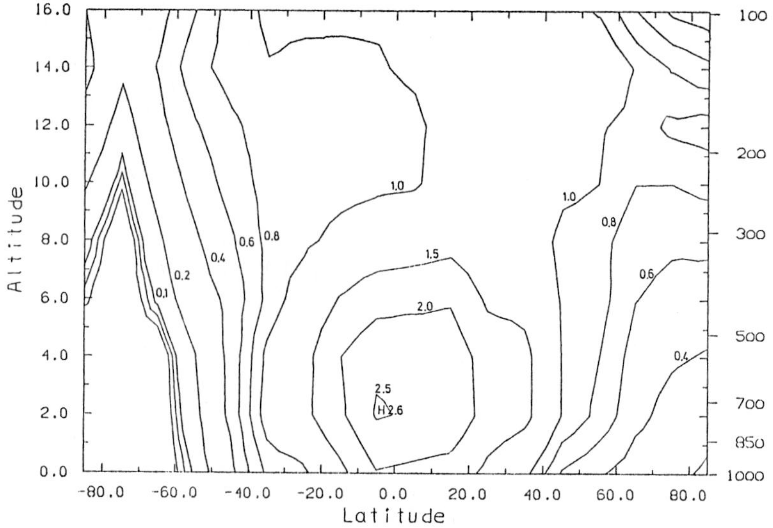
Altitude
Latitude
16.0
14.0
12.0
10.0
8.0
6.0
4.0
2.0
0.0
-80.0
-60.0
-40.0
-20.0
0.0
20.0
40.0
60.0
80.0
100
200
300
500
700
850
1000
0.1
0.2
0.4
0.6
0.8
1.0
1.5
2.0
2.5
H2.6

FIG. 2c.

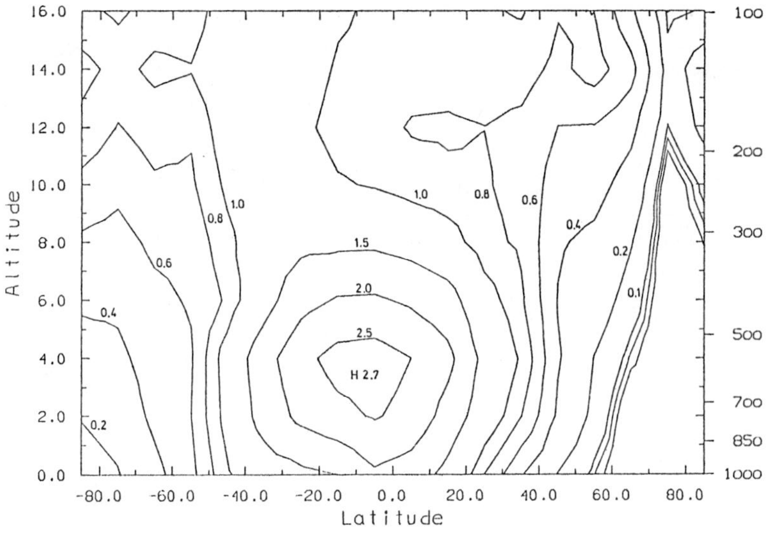
16.0
14.0
12.0
10.0
8.0
6.0
4.0
2.0
0.0
Altitude
-80.0
-60.0
-40.0
-20.0
0.0
20.0
40.0
60.0
80.0
Latitude
100
200
300
500
700
850
1000
1.0
0.8
0.6
0.4
0.2
1.5
2.0
2.5
H 2.7
0.1

FIG. 2d.

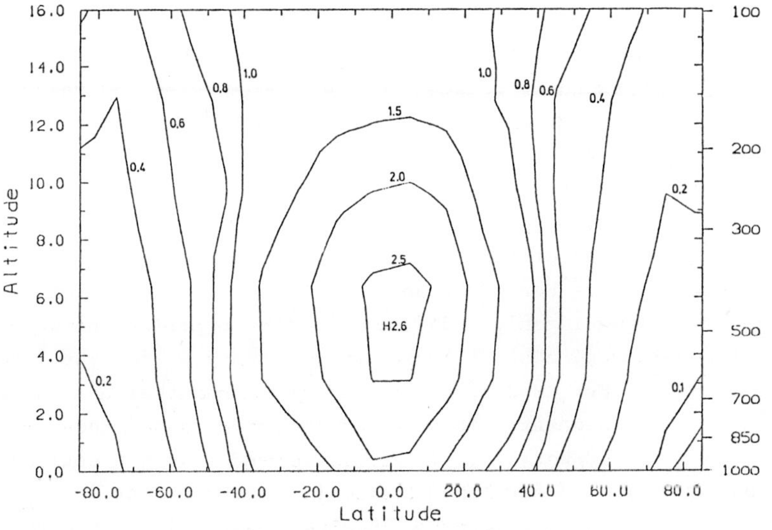
16.0
14.0
12.0
10.0
8.0
6.0
4.0
2.0
0.0
Altitude
-80.0
-60.0
-40.0
-20.0
0.0
20.0
40.0
60.0
80.0
Latitude
100
200
300
500
700
850
1000
1.0
0.8
0.6
0.4
0.2
1.5
2.0
2.5
H2.6
0.1

FIG. 2e.

TABLE 1 - Observed and calculated average volume mixing ratios for $CFCl_3$, CF_2Cl_2 and CH_3CCl_3, mostly for late 1979. The observational data have been compiled from a recent WMO Document (33).

Gas (pptv)	Northern Hemisphere	Southern Hemisphere	Comment
CF_2Cl_2	315±6	282±6	WMO, late 1979
	300	277	Rasmussen et al., late 1979
	299	276	Model, late 1979
$CFCl_3$	184±5	164±5	WMO, late 1979
	174	155	Rasmussen et al., late 1979
	178	161	Model, late 1979
CH_3CCl_3	125±20	94±12	WMO, late 1979
		103	Rasmussen et al., late 1979
	145	107	Model, late 1979
	136		Rasmussen et al., late 1978
	132	96	Model, late 1978

Rasmussen and co-workers and calculations which can only be true if both exchange between the two hemispheres and calculated OH concentrations are roughly correct. However, as shown by laboratory intercomparison studies (21,23), various laboratories report somewhat different values for the concentrations of chlorocarbon gases for the same samples, so that there is a spread of reported values which is likewise shown in Table 1 (33). For $CFCl_3$ and CF_2Cl_2 the relative spread in reported values is only 2-3%, so that the comparison between model results and observations remains good even for the larger set of data. For CH_3CCl_3 the calculated concentrations agree only with the highest values in the reported range, those by Rasmussen and co-workers. If it is assumed that the estimated global CH_3CCl_3 industrial emission rates are correct, the mixing ratios of CH_3CCl_3 without chemical loss would have been calculated to be equal to 185 pptv in the northern and 170 pptv in the southern hemisphere (33). These compare with the reported ranges of 105-145 and 82-106 pptv, respectively. The range of reported CH_3CCl_3 volume mixing ratios would, therefore, allow OH concentrations to be at most two times larger than those shown in Figs. 2a-2e. Existing uncertainties in the statistics of global CH_3CCl_3 emission rates which are based on incomplete data from nations outside the United States affect especially the OH estimate range on the low side. At the minimum allowed OH concentrations of Figs. 1a-1e are about a factor of two smaller than those presented in Figs. 2a-2e, we may conclude that the average OH concentrations are roughly within a factor of two from those presented in Figs. 2a-2e.

DISCUSSION AND CONCLUSIONS

In this study we have discussed some of the factors which determine the concentrations of hydroxyl in the troposphere. We have calculated the minimum concentrations of hydroxyl which are possible in the troposphere and which lead to a global 24-hour average of 2.3×10^5 molecules cm^{-3} in the troposphere. Calculations with a two-dimensional model which were in excellent agreement with the observations for $CFCl_3$ and CF_2Cl_2 and good agreement for O_3 yielded OH distributions with an average OH concentration of 5.5×10^5 molecules cm^{-3}. With the same model, methylchloroform concentrations are calculated which agree remarkably well with the observations by Rasmussen and co-workers (33). If all reported concentrations of CH_3CCl_3 are, however, taken into account, OH concentrations which are larger by a factor of two than those calculated may still be possible. It seems, therefore, justifiable to state that the average global hydroxyl concentration distributions are roughly as shown in Figs. 2a-2e with an uncertainty of at most a factor of two. The global average, tropospheric OH concentration is, therefore, equal to about 5×10^5 molecules cm^{-3} with a most likely range of $3 - 10 \times 10^5$ molecules cm^{-3}. This range agrees very well with the range of values $4.5 - 9.5 \times 10^5$ molecules cm^{-3} based on global production rates and observations of ^{14}CO (29).

Adopting the average distribution of OH as presented in Figs. 2a-2e and estimates of global distributions of chemical constituents which react mainly with OH in the troposphere, it is next possible to determine their atmospheric loss rates to better than a factor of two. For instance, for CH_4, CO, and CH_3Cl, the calculated, annual, global loss rates are 3.2 x 10^{14}g CH_4, 2 x 10^{15}g CO, and 1.9 x 10^{12}g CH_3Cl (5). Most of the destruction of those compounds occurs in the tropics requiring large input rates at tropical latitudes. The sources can be provided by biomass burning (6), emissions from the forests (34), and marine emissions (16). Finally we must emphasize that the OH distributions presented in this study are probably more representative for oceanic than for continental regions, e.g., because of the neglect of aerosol scavenging of HO_2 and H_2O_2 and reactions of reactive hydrocarbons. This implies that these OH distributions cannot be used for such gases with mainly continental sources and shorter atmospheric residence times than a few weeks. Such gases will be much more abundant over land than over ocean.

REFERENCES

(1) Bauer, E. 1979. A catalogue of perturbing influences on stratospheric ozone 1955-1975. J. Geophys. Res. 84: 6929-6940.

(2) Campbell, M.J.; Sheppard, J.C.; and Au, B.F. 1979. Measurements of hydroxyl concentration in boundary layer air by monitoring CO oxidation. Geophys. Res. Lett. 6: 175-178.

(3) Chatfield, R.B., and Harrison, H. 1977. Tropospheric ozone: variations along a meridional band. J. Geophys. Res. 82: 5969-5976.

(4) Chemical Manufacturers Association. 1979. World Production and Release of Chlorofluorocarbons 11 and 12 through 1978. Washington, DC.

(5) Crutzen, P.J., and Gidel, L.T. 1982. A two-dimensional photochemical model of the atmosphere. J. Geophys. Res., accepted for publication.

(6) Crutzen, P.J.; Heidt, L.E.; Krasnec, J.P.; Pollock, W.H.; and Seiler, W. 1979. Biomass burning as a source of atmospheric gases CO, H_2, N_2O, NO, CH_3Cl and COS. Nature 282: 235-256.

(7) Davis, D.D.; Heaps, W.; and McGee, T. 1976. Direct measurement of natural tropospheric levels of OH via an airborne tunable dye laser. Geophys. Res. Lett. 3: 331-333.

(8) Derwent, R.G., and Eggleton, A.E.J. 1981. Two-dimensional model studies of methyl chloroform in the troposphere. Q. J. R. Meteorol. Soc. 107: 231-242.

(9) Eichmann, R.; Ketseridis, G.; Schebeske, G.; Jaenicke, R.; Hahn, J.; Warneck, P.; and Junge, C. 1980. N-alkane studies in the troposphere - II. Gas and particulate concentrations in Indian Ocean air. Atmos. Env. 14: 695-703.

(10) Fabian, P., and Pruchniewicz, P.G. 1977. Meridional distribution of ozone in the troposphere and its seasonal variations. J. Geophys. Res. 85: 7546-7552.

(11) Fishman, J., and Crutzen, P.J. 1978. The origin of ozone in the troposphere. Nature 274: 855-858.

(12) Heidt, L.E.; Krasnec, J.P.; Lueb, R.A.; Pollock, W.H.; Henry, B.E.; and Crutzen, P.J. 1980. Latitudinal distribution of CO and CH_4 over the Pacific. J. Geophys. Res. 85: 7329-7336.

(13) Imhof, W.L.; Reagan, J.B.; and Gaines, E.E. 1980. Measurements of innerzone electron precipitation. J. Geophys. Res. 85: 9-16.

(14) Jaenicke, R. 1981. Atmospheric and global climate. In Climatic Variations and Variability: Facts and Theories, ed. A. Berger, pp. 577-597. Dordrecht: Reidel.

(15) Levy II, H. 1971. Normal atmosphere: Large radical and formaldehyde concentrations predicted. Science 173: 141-143.

(16) Lovelock, J.E. 1975. Natural halocarbons in the air and in the sea. Nature 256: 193-197.

(17) McConnell, J.C.; McElroy, M.B.; and Wofsy, S.C. 1971. Natural sources of atmospheric CO. Nature 233: 187-188.

(18) Noxon, J.F. 1978. Tropospheric NO_2. J. Geophys. Res. 83: 3051-3057.

(19) Perner, D.; Ehhalt, D.H.; Pätz, H.W.; Platt, U.; Röth, E.P.; and Volz, A. 1976. OH radicals in the lower troposphere. Geophys. Res. Lett. 3: 466-468.

(20) Perner, D.; Ehhalt, D.H.; Pätz, H.W.; Platt, U.; Röth, E.P.; and Volz, A. 1981. Comment on "Improved Airborne Measurements of OH in the Atmosphere Using the Technique of Laser-Induced Fluorescence." J. Geophys. Res. 86: 12155.

(21) Rasmussen, R.A. 1978. Interlaboratory comparison of fluorocarbon measurements. Atmos. Env. 12: 2505-2508.

(22) Rasmussen, R.A., and Khalil, M.A.K. 1981. Global atmospheric distribution and trend of CH_3CCl_3. Geophys. Res. Lett. 8: 1005-1007.

(23) Rasmussen, R.A., and Khalil, M.A.K. 1981. Interlaboratory comparison of fluorocarbons-11, -12, methylchloroform and nitrous oxide measurements. Atmos. Env. 15: 1559-1568.

(24) Rudolph, J., and Ehhalt, D.H. 1981. Measurements of C_2 and C_5 hydrocarbons over the North Atlantic. J. Geophys. Res. 86: 11959-11964.

(25) Seiler, W. 1974. The cycle of atmospheric CO. Tellus 24: 117-135.

(26) Singh, H.B. 1977. Atmospheric halocarbons: evidence in favor of reduced hydroxyl radical concentration in the troposphere. Geophys. Res. Lett 4: 101-104.

(27) Singh, H.B. 1977. Preliminary estimation of average tropospheric OH concentrations in the northern and southern hemisphere. Geophys. Res. Lett. 4: 453-456.

(28) Turman, B.N., and Edgar, B.C. 1982. Global lightning distributions at dawn and dusk. J. Geophys. Res. 87: 1191-1206.

(29) Volz, A.; Ehhalt, D.H.; and Derwent, R.G. 1981. Seasonal and latitudinal variation of ^{14}CO and the tropospheric concentration of OH radicals. J. Geophys. Res. 86: 5163-5171.

(30) Wang, C.C.; Davis, Jr., L.I.; Selzer, P.M.; and Munoz, R. 1981. Improved airborne measurements of OH in the atmosphere using the technique of laser-induced fluorescence. J. Geophys. Res. 86: 1181-1186.

(31) Wang, C.C.; Davis, Jr., L.I.; and Selzer, M. 1981. Reply. J. Geophys. Res. 86: 12156.

(32) Warneck, P. 1974. On the role of OH and HO_2 radicals in the troposphere. Tellus 26: 39-46.

(33) WMO (World Meteorological Organization). 1981. The Stratosphere 1981, Theory and Measurements. WMO Global Ozone Research and Monitoring Project Report No. 11, Geneva, Switzerland.

(34) Zimmerman, P.R.; Chatfield, R.B.; Fishman, J.; Crutzen, P.J.; and Hanst, P.L. 1978. Estimates on the production of CO and H_2 from the oxidation of hydrocarbon emissions from vegetation. Geophys. Res. Lett. 5: 679-682.

Atmospheric Chemistry, ed. E.D. Goldberg, pp. 329-355. Dahlem Konferenzen 1982.
Berlin, Heidelberg, New York: Springer-Verlag.

Non-methane Organics in the Remote Troposphere

S. A. Penkett
Environmental and Medical Sciences Division
AERE, Harwell, Oxfordshire OX11 ORA, England

Abstract. The atmosphere is now understood to contain many gaseous species at concentrations varying by many orders of magnitude. At one end of the concentration scale, gases such as oxygen are vital to sustaining animal respiration, whilst at the other end, trace species such as nitrogen oxides control fundamental geochemical phenomena such as the earth's ozone shield and the level of hydroxyl radicals present in the troposphere. The most numerous trace gases present are organic molecules which have a variety of sources including combustion processes, biogenic decay on land and in the ocean, solvent usage, natural gas leakage, etc. The measurement of these many compounds at concentration levels down to parts in 10^{12} and below has presented atmospheric scientists with severe analytical problems. These are now being solved by using a variety of modern techniques based upon spectroscopy and gas chromatography. Data on the distribution of many compounds is still exceedingly scarce. This is urgently needed to quantify major sources and to test current theories of atmospheric oxidation leading to the creation of soluble molecules capable of being removed by rain. Some of the limited amount of concentration data available is discussed by compound type: namely, halocarbons, hydrocarbons, oxygenated compounds, and sulfur compounds. The main object of this discussion is an attempt to find some consistent indicators for the degree of chemical reaction which can be attributed to attack by hydroxyl radicals in the troposphere. It appears that the level of oxidation indicated by the latitudinal distribution of molecules such as perchloroethylene and acetylene, and possibly benzene, toluene, and some other compounds, is considerably less than that predicted from a study of the carbon monoxide cycle. However, much more data on the distribution of these molecules throughout the troposphere are needed before final conclusions can be drawn. A low oxidation rate for many molecules tends to lower the amounts that can be removed each year by natural processes occurring in the atmosphere. This makes pollution more of a problem, of course, and at the same time it affects the current assessments of the efficiency of the biosphere for producing trace gases in the atmosphere, including those involved in the creation of condensation nuclei by direct gas to particle conversion processes. Finally, there are indications from a comparison of the concentration distribution of acetylene and sulfur dioxide, both of which are largely pollution derived, that oxidation of soluble species in atmospheric droplets may be a very efficient process, possibly by reaction with hydrogen peroxide.

INTRODUCTION

Study of the composition of atmospheric air has been a respected branch of science for some hundreds of years. The Greeks considered it to be immutable and classified it as one of the four elements: air, earth, fire, and water. In the seventeenth century, however, Robert Boyle wrote, "The air is a confused aggregate of effluviums from such differing bodies, that, though they all agree in constituting by their minuteness and various motions one great mass of matter, yet perhaps there is scarcely a more heterogeneous body in the world." The comment in magnificent English of a very farsighted man.

Elucidating the composition of air has led to major advances in the overall science of chemistry. It was recognized in the seventeenth century that the substance that supported combustion and allowed respiration was but a constituent of air. Priestly prepared this substance, which he called "dephlogesticated air," by heating mercury oxide and showed its unique power of supporting combustion and respiration. Lavoisier on hearing of Priestly's findings repeated them in a quantitative manner. He showed that 1/6 of the air over hot mercury was converted into mercuric oxide, which could then be reconverted into mercury and the original volume of "air" that was consumed in the initial experiment. In this way he showed that approximately 15% of the air was made up of a substance he called oxygen and that chemistry, like physics, was a quantitative science. This allowed the development of the subject to proceed in a much more logical manner. Lavoisier, in fact, with his simple experiments on the take-up and release of oxygen did for chemistry what Newton had done for physics.

Carbon dioxide was discovered before oxygen by Joseph Black in Edinburgh in about 1755, and the next finding, if one excludes water, was of argon in the 1890s by Lord Rayleigh and W. Ramsey who both received the Nobel prize in 1904. Ramsey proposed that argon was part of another period of the periodic table and went on to discover the presence of krypton, xenon, helium, and neon in the residue from the evaporation of liquid air.

The terminology at this time referred to permanent gases, which could not be liquified by pressure alone, and rare gases, which were a function of their concentration in relation to oxygen and nitrogen in the air. By the 1950s a number of other atmospheric species had been identified at concentrations similar to those of the least abundant rare gases (i.e., 1 ppm and less). These are currently called trace gases. Glueckauf (13), in a penetrating review of the composition of atmospheric air written in 1951, referred to gases of constant percentage, including oxygen, nitrogen, carbon dioxide, rare gases, nitrous oxide, methane, and hydrogen. The presence of nitrous oxide and methane was established by Adel (1) and Migeotte (35) using spectroscopic methods, and also hydrogen was found in the neon fraction of liquid air. Its concentration (0.51 ppm) was accurately established at Harwell by Glueckauf and Kitt (14) in 1954, again by separation from liquid air. (Glueckauf also made an estimate of the methane concentration in the early 1950s using methods similar to those used for the hydrogen measurements. He obtained a value of 1.1 ppm at this time.)

Gases which indicated the presence of chemical reactivity in the atmosphere and which were classified by Glueckauf as constituents of variable constituency include ozone, sulfur dioxide, nitrogen dioxide, formaldehyde, iodine, ammonia, and carbon monoxide. The theory of ozone formation in the stratosphere by ultraviolet photolysis of oxygen was proposed by Chapman in 1930 (6). At that time chemical processes in the troposphere were mainly thought of in terms of oxidation of sulfur dioxide in fogs giving rise to the well-known phenomenon of the London smoke fogs or smogs, a very bad example of which was responsible for the deaths of some 3,000 people in 1952. The phenomenon of Los Angeles smog first noted around 1950, however, led Haagen-Smit (15) to the realization that chemical processes in the troposphere could also produce ozone from a complex series of photochemical reactions involving hydrocarbons and oxides of nitrogen. Attempts to understand this chemistry acted as a catalyst in promoting ideas about wider chemical phenomena occurring throughout the troposphere. In this respect carbon monoxide proved to be a very important species since a study of its C^{14} content by Weinstock and Niki showed that it must have a short atmospheric lifetime (50) - which could only be the case if it were removed by oxidation by hydroxyl radicals. Free radical chemistry has only very recently been accepted as a normal phenomenon in the troposphere (24), and its presence has concentrated efforts to detect both radical species and reactive gases at levels of less than one millionth of the gases previously thought of as rare gases. Over the last thirty years there has therefore been a large increase in the number of trace gases, particularly organic gases, that have been observed in the atmosphere.

The paper will first consider the most likely sources of many of these compounds and will then give some examples of the capabilities of modern analytical methods for obtaining data on the concentrations of species at levels down to parts in 10^{12} by volume. The main body of the paper will subsequently concern itself with a discussion of data by compound type. It will consider the limited amount of information available on the concentration distribution of many compounds in the clean troposphere against a background of published information on the rate of reaction of these compounds with hydroxyl radicals.

An attempt will be made to determine whether any consistent pattern emerges as to the extent of free radical chemistry in the lower atmosphere. It must be emphasized at the outset that data are only just beginning to be produced for the concentration distribution of many species, and the conclusions presented may ultimately be subject to revision in the light of new information.

It is hoped, however, to indicate the value of this basically experimental approach, using existing analytical techniques, to solving problems currently exercising the thoughts of many atmospheric chemists.

SOURCES OF ORGANIC TRACE GASES

A number of sources of trace gases have been identified at the earth's surface; some of these are natural and some are not. These include volcanoes, which were the ultimate source of all material in the atmosphere, the hydrosphere, and the lithosphere. Today, volcanoes are thought to represent a relatively minor source of trace gases in the troposphere, but explosive events, such as the recent Mount St. Helens' eruption, will undoubtedly inject substantial quantities of sulfur compounds into the stratosphere.

Perhaps the best way to delineate and quantify present sources of gases containing organic carbon (which is the main subject of this review) is to consider their contribution to the production of carbon monoxide in the atmosphere. This is because most gaseous carbon compounds are commonly oxidized to carbon monoxide by the hydroxyl radical chemistry (26,43,49). A major source of carbon monoxide is combustion, which is roughly evenly divided between fossil fuels and biomass burning (11). Inefficient combustion, of course, will also give rise to direct emissions of hydrocarbons and partially oxidized species such as aldehydes and ketones. Leaving aside CH_4 oxidation, the next largest source of atmospheric CO comes from emissions of natural hydrocarbons from the biosphere. These include molecules such as C_2H_4 and also C_5H_8 and terpenes (54). The ocean is a relatively small source of CO, but as will be shown later, it is an important source for organic species of other elements such as sulfur (31) and the halogens (29). Industrial emission of hydrocarbons generally comes from motor vehicle exhausts and from fuel and solvent evaporation; however, there may also be a substantial source from natural gas leakage, which has not been properly quantified so far. According to most assessments the biosphere is still the largest source of CO, but the level of anthropogenic disturbance is not insignificant; it is responsible for at least a third of the total amount of CO in the atmosphere at present (26).

The emission ratio of organic compounds from natural and anthropogenic sources is less easy to assess than CO and requires measurements to be made of individual types of carbon compounds throughout the atmosphere. The types include hydrocarbons, oxidation products of hydrocarbons, halocarbons, sulfur compounds, nitrogen compounds, etc. Modern methods of gas analysis are now capable of measuring most of these species at atmospheric levels.

ANALYTICAL METHODS

It is not an exaggeration to say that much of the increase in knowledge of atmospheric processes over the last decade has been brought about by the application of increasingly sophisticated analytical techniques. In the field of reaction kinetics, spectroscopic methods have been devised to measure very fast reaction rates and low concentrations of transient species such as hydroxyl radicals. In the field of general gas analysis, improvements in spectroscopy and in detectors used in gas chromatography have revolutionized the capabilities. A review article describing the progress in this field has recently been published (37). It was pointed out here that spectroscopic methods are most usefully applied to unstable species such as free radicals and reactive gases. It was also suggested that chromatographic methods are more suitable for more stable molecules that will pass through a chromatographic column. This provides a very efficient way of separating the individual components of a complex mixture such as air. At the same time, it concentrates the component into a pulse which can be fed into a convenient detector system. Further concentration techniques such as cryogenic trapping of liquid air can be applied without difficulty to extend the range of measurement down to very low levels (below 1 part in 10^{12} in many cases).

One of the most sensitive detector systems ever devised is probably Lovelock's electron capture detector (27) which is also rather specific for certain types of molecules containing chlorine (such as chlorofluorocarbons, CFCs) and nitro-groups (such as PAN). Lovelock's own measurements of the CFCs led Molina and Rowland (36) to propose that these molecules could deplete the ozone concentration in the stratosphere.

A more universal detector which also combines high specificity with high sensitivity is the mass spectrometer (MS). When combined with a gas chromatograph into a GC/MS combination it represents a very powerful analytical tool, which in principle should be capable of defining and measuring the total organic content of atmospheric air.

Data on the composition of atmospheric air in more remote parts of the troposphere is slowly becoming available. At present there are only very small quantities obtained by a handful of researchers. It is therefore not an easy task to write a definitive review on this subject. The discussion which follows is largely based upon data obtained by Rasmussen on the GAMETAG experiment, which involved sample collection by aircraft flying from the United States across the Pacific ocean to latitudes in the range of $50^{o}S$. This experiment, organized by Davis, Crutzen, Duce, and Kiang and involving many US scientists, was one of the largest yet undertaken in the field of tropospheric chemistry. Other data have been obtained on the German research vessel, Meteor, which has been used extensively for many years in experiments designed to probe the chemistry of the remote marine atmosphere over the Atlantic ocean.

GC/MS ANALYSIS OF THE TOTAL ORGANIC CONTENT OF ATLANTIC AIR

All compounds that will pass through a gas chromatograph (GC) will be detected by a mass spectrometer (MS). When used in combination with a cryogenic concentrator, which removes the major constituents oxygen and nitrogen, the output from this instrument gives a good indication of the total trace gas content of the atmosphere. Compounds with boiling points from -50°C, such as COS, to over 150°C, such as long-chain hydrocarbons, are observed. Those which are not observed include CH_4, C_2H_6, C_2H_4, C_2H_2, C_3H_8, and C_3H_6 which do not have mass peaks in the range scanned. Other molecules not observed include aldehydes and labile species which do not enter the mass spectrometer from the gas chromatograph.

Values for the concentration of many species obtained from the GC/MS analysis are shown in Table 1. These numbers were derived from a comparison of peak areas at the masses shown with that for fluorocarbon 11 in the chromatogram at mass 101, constructed from the mass spectrometer data system. The values

TABLE 1 - Concentrations of organic trace species in Atlantic air derived from the mass spectrometer scan analysis.

Compound	Mass No.	Estimated Conc. (pptv)	Independent Assessment (pptv)
CCl_3F	101	184	184
CCl_2F_2	85	288	300
$CFCl_2CF_2Cl$	101	25	19
CF_2ClCF_2Cl	135	19	9
CHF_2Cl	51	45	50
CH_3CCl_3	97	84	115
CCl_4	117	87	102
CH_3Cl	50	341	627
CH_2Cl_2	84	60	37
$CHCl_3$	83	27	29
CH_3Br	96	11	12
CH_3I	142	2.4	2.0
C_2HCl_3	130	6.3	-
C_2Cl_4	129	35	56
C_6H_6	78	95	-
C_7H_8	91	7	-
$(CH_3)_2CO$	58	480	-
COS	60	400	430
CS_2	76	150	150

are compared with values obtained independently on other analytical systems either by Rasmussen (39), Singh (46), Lovelock (29), or by the author. There is sufficient agreement between the two sets of data to suggest that (a) this is a valuable technique for estimating the concentration of a wide range of species from a single analysis, and (b) the sample analyzed was indeed representative of a clean air sample with little sign of contamination. This is an important point since the concentrations of a number of species previously not determined, or not known well, are shown. These species include C_6H_6, C_7H_8, C_2Cl_4, C_2HCl_3, CH_3Br, CH_2Cl_2, and $(CH_3)_2CO$.

In Table 2 estimates of the concentrations of ten interesting species are shown for each of five air samples that were collected at a similar latitude (35^oN). Using fluorocarbon 11 as an internal standard, concentrations of the other species were estimated as above. The data are within the limits of error expected for this rather simple approach to analysis, certainly for the species which are not expected to vary much, CH_3CCl_3 and CCl_4. All the other species vary more than these two with C_7H_8 and C_2HCl_3 showing the highest variance, closely followed by C_6H_6 then by C_2Cl_4, $(CH_3)_2CO$, $CHCl_3$, and CH_2Cl_2.

This situation would probably be expected if the compounds above were being removed at different rates by hydroxyl radical

TABLE 2 - Average concentrations (pptv) of organic species in Atlantic air.

	CH_3CCl_3	CCl_4	C_2HCl_3	C_2Cl_4	$CHCl_3$	CH_2Cl_2	C_6H_6	C_7H_8	$(CH_3)_2CO$
	84	87	6.3	35	26	60	95	7	480
		83	4.2	39	34	53	60	17	627
	70	74	2.5	33	26	64	97	11	412
	68	80	14	21	37	41	19	15	421
	72	86	3.1	43	24	60	61	34	408
$\bar{x}$	73.5	82	6.0	34.2	31.4	55.6	66.4	16.8	470
σ	7.2	5.2	4.7	8.3	6.0	9.1	31.9	10.4	93

chemistry of the type which is believed to prevail in the troposphere (10,24,53). The most reactive compounds, C_7H_8 and C_2HCl_3, react considerably faster than $CHCl_3$ or CH_2Cl_2. The main removal mechanism for $(CH_3)_2CO$ is probably direct photolysis, and the data in Table 2 suggest its lifetime in the atmosphere lies somewhere between C_2Cl_4 and $CHCl_3$. All these arguments linking reactivity with variance, of course, are very empirical.

DISCUSSION OF SOME RECENT DATA RELATING TO THE ORGANIC COMPOSITION OF THE REMOTE TROPOSPHERE WITH PARTICULAR REFERENCE TO ATMOSPHERIC REACTIVITY

Halocarbons

Halocarbons make up a large fraction of the organic carbon content of the atmosphere, other important compounds being sulfur compounds, hydrocarbons, and oxidized hydrocarbons. The halocarbons positively identified as well mixed components of the atmosphere are listed in Table 3, along with their average concentrations, their predominant source, and their rate coefficient for reaction with hydroxyl radicals.

Only three of the compounds listed are definitely known to have natural sources. CH_3Cl, CH_3Br, and CH_3I (29) have been measured at elevated concentrations in ocean surface waters; also GC/MS analysis of coastal seawater has shown the presence of $CHCl_3$ and CH_2Cl_2 (Penkett and Rasmussen, unpublished data). The equal distribution of $CHCl_3$ in both hemispheres, even though it should be removed by atmospheric oxidation, argues in favor of its being a natural compound. Also its direct anthropogenic release is quite small (15,000 tons/year in 1977). It is possible that this compound may have larger indirect anthropogenic sources, and some reports suggest that its concentration in the southern hemisphere is much lower than that shown in Table 3 (46). There are significant anthropogenic releases of CH_2Cl_2 (about 430,000 tons/year in 1977), but it may also be produced naturally. There is no data available on its southern hemisphere concentration at present.

TABLE 3 - Halocarbons in clean air.

Compound	Hydroxyl rate coef. x 10^{12} cm^3 $molec^{-1}$ s^{-1}	Concentration pptv N.H.	S.H.	Source	Investigator
CH_3Cl	0.036	626	625	n (a)	Rasmussen (39)
CH_2Cl_2	0.16	60		(n) a	
$CHCl_3$	0.10	29	29	n a	Rasmussen (39)
CCl_4	small	102		a	
$CFCl_3$	small	161	148	a	Rasmussen (39)
CF_2Cl_2	small	274	255	a	Rasmussen (39)
CHF_2Cl	0.005	42	36	a	Rasmussen (39)
$CHFCl_2$	0.026	0.1	0.1	a	
CH_3Cl	small	3.4	3.4	a	
CF_4	negligible	70	70	a	
CF_2ClCF_2Cl	small	12	10	a	Singh (46)
$CFCl_2CF_2Cl$	small	13	12	a	Singh (46)
C_2F_6	small	3.8	3.8	a	
C_2F_5Cl	small	4		a	
CH_3CCl_3	0.022	117	90	a	Rasmussen (39)
C_2HCl_3	2.0	6		a	
C_2Cl_4	0.17	56	14	a	Rasmussen (39)
$C_2H_2Cl_2$		11	11	n?	
CH_3Br	0.035	13		n	
CF_3Br	small	0.7		a	
$C_2H_4Br_2$	0.25	< 5		a	Singh (46)
CH_3I	-	3.1	2.2	n	Rasmussen (39)

All rate coefficients derived from Atkinson, Darnall, Lloyd, Winer, and Pitts (3). n = natural; a = anthropogenic; ( ) indicates a minor contribution; N.H. and S.H. indicate northern and southern hemispheres, respectively.

The small number of natural compounds compares with sixteen com- compounds of anthropogenic origin, many of which contain the element fluorine. All of the natural compounds have major sinks in the troposphere either by reaction with hydroxyl radicals (CH_3Cl, CH_2Cl_2, $CHCl_3$, and CH_3Br) or by photolytic degradation of CH_3I. In contrast, most of the anthropogenic compounds are unreactive in the troposphere, certainly with respect to hydroxyl radicals. Multifluorinated compounds such as CF_4 appear to have no stratospheric sink either (7).

Anthropogenic halogen compounds which are almost entirely removed in the troposphere include C_2HCl_3 and C_2Cl_4. Larger quantities of these chlorine compounds have been released than either fluorocarbon 11 or 12, but their average concentrations are much lower, indicating some considerable atmospheric reactivity. The release of C_2Cl_4 into the northern hemisphere has been continuing at an approximately even rate (600,000 tons/year) for over a decade. Because of its short lifetime, a steady state situation has now been reached whereby the difference in concentrations in the northern and southern hemisphere (see Fig. 1) represents the amount removed during the time for interhemispheric transfer (about 400 days). This simple notion can be used to derive a value for the global average hydroxyl radical concentration [OH], provided of course that C_2Cl_4 is removed entirely by reaction with this species. In this case [OH] is approximately 2.5 x 10^5 molecules/cm^3, which is significantly lower than estimates derived from CO data (6 x 10^5 molecules/cm^3) (26,49). The lower estimate for [OH] would lead to a steady state value of only 25 ppt for C_2Cl_4 in the northern hemisphere, which is in reasonable agreement with the number shown in Table 2 (34 ppt). In the case of C_2HCl_3, the two lowest values shown in Table 2 would also be close to the expected equilibrium concentration for this compound. The major flaw in this argument is associated with the variation of reaction rate with temperature. If this is large, there is still room for more OH radicals, although it is very unlikely that an average value of 6 x 10^5 molecules/cm^3 would be attained.

Only two organic bromine compounds have been positively identified so far in the remote troposphere. These are CH_3Br and CF_3Br. Nine separate measurements of the CH_3Br concentration have been made using the GC/MS in the single ion mode on Atlantic air collected in an area from latitude 30-35°N longitude

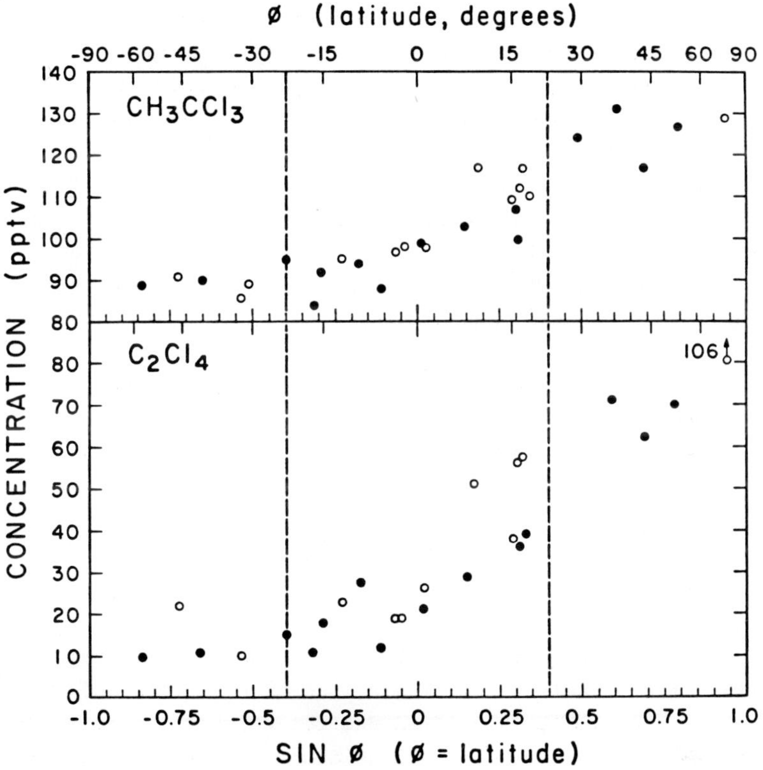

FIG. 1 - Latitudinal transepts of CH_3CCl_3 and C_2Cl_4 obtained by Rasmussen in the GAMETAG experiments (39).

15-25°W. These gave a mean of 12.9 pptv and a standard deviation of only 1.8 pptv. Other values determined on air samples from the upper troposphere at 50°N from the continental USA and from southern England gave a similar average concentration with a similar small spread of values (Penkett and Rasmussen, unpublished data). Other possibilities for organic bromine compounds are CH_2Br_2 and $CHBr_3$ which have both been detected in high concentrations by GC/MS analysis of seawater (Penkett and Rasmussen, unpublished data). The absorption spectra of both these compounds should overlap into the range of the solar emission spectrum at the base of the troposphere; they should therefore be photolyzed giving rise to bromine atoms, which could take part in chemical reactions with many trace gases.

CH_3I is the only organic iodine species that has been positively identified in the atmosphere so far ((31), and Penkett and Rasmussen, unpublished data). Its average concentration over the ocean lies in the range of 1.5 to 3 pptv. Other organic iodine species may also exist. Substantial levels of C_2H_5I, n-C_3H_7I, and i-C_3H_7I have been found in seawater in equilibrium with different types of seaweed (Penkett and Rasmussen, unpublished data), and evidence for the presence of these higher iodides was found in the air samples collected on the 1981 Meteor cruise. These compounds will probably be photolyzed more rapidly than CH_3I, thus increasing the concentration of iodine atoms and related species in the atmosphere. It has recently been speculated that reactions involving photolysis fragments from organic iodine compounds could provide a removal process for ozone in the troposphere (5). Perhaps a more likely fate, though, for oxidized iodine fragments is the formation of an aerosol (Cox, personal communication).

Hydrocarbons

Hydrocarbons represent the most numerous species of compound in clean air. Table 4 contains a list of nine compounds that have been positively identified and for which there is good quantitative data. Many others no doubt exist. All of the compounds listed in Table 4 react quite rapidly with hydroxyl radicals. The reaction rates cover an interesting range of values which makes a study of their latitudinal transepts and vertical profiles of some significance to understanding the general level of chemical activity in the troposphere. Probably the most extensive data set of hydrocarbon concentrations which has been built up over many years belongs to Rasmussen (39). His most recent data on the latitudinal variation of C_2H_2, C_2H_4, and C_2H_6 concentrations, obtained on the GAMETAG flights, is shown in Fig. 2. The concentrations differ in magnitude from some other values published recently (41) but agree with the author's own data.

C_2H_2 is a very interesting compound in that it has no known natural sources. The anthropogenic source of acetylene is

TABLE 4 - Hydrocarbons in clean air.

Compound	Hydroxyl rate coef. x 10^{12} cm^3 $molec.^{-1}$ s^{-1}	Concentration pptv N.H.	S.H.	Source	Investigator
CH_4	0.0095	1575 x 10^3	1510 x 10^3	n (a)	Rasmussen (39)
C_2H_6	0.29	1030	240	n a	Rasmussen (39)
C_2H_4	7.8	80	100	n	Rasmussen (39)
C_2H_2	0.68	160	20	a	Rasmussen (39)
C_3H_8	1.6	85		n a	
C_3H_6	25.1	53		n	
C_4H_{10}	2.7	10		a	Rudolph (41)
C_6H_6	1.2	64		a	
C_7H_8	6.4	17		a	

All rate coefficients derived from Atkinson, Darnall, Lloyd, Winer, and Pitts (3). n = natural; a = anthropogenic; ( ) indicates a minor contribution; N.H. and S.H. indicate northern and southern hemispheres, respectively.

approximately 1 million tons (MT) per year (Derwent, personal communication), and the sink can be calculated using known rate constants and estimated [OH]. Using a value of 6.8 x 10^{-13} cm^3 $molecule^{-1}$ s^{-1} for the rate coefficient gives rise to a global average figure of 2.5 x 10^5 molecules/cm^3 for [OH], which is identical to the number derived earlier using the C_2Cl_4 data.

Turning to olefins, it is clear that the ocean must be a major source for both C_2H_4 and C_3H_6. The source strength deduced from its annual turnover caused by reaction with hydroxyl radicals is about 20 MT for C_2H_4 and over 50 MT for C_3H_6. These numbers can probably be doubled when the amount of olefin destroyed by the ozone-olefin reaction is taken into account. It is an interesting point to bear in mind that for substances containing a doubly-bonded carbon moity, as the value of [OH] drops, so the relative importance of the ozone-olefin reaction increases.

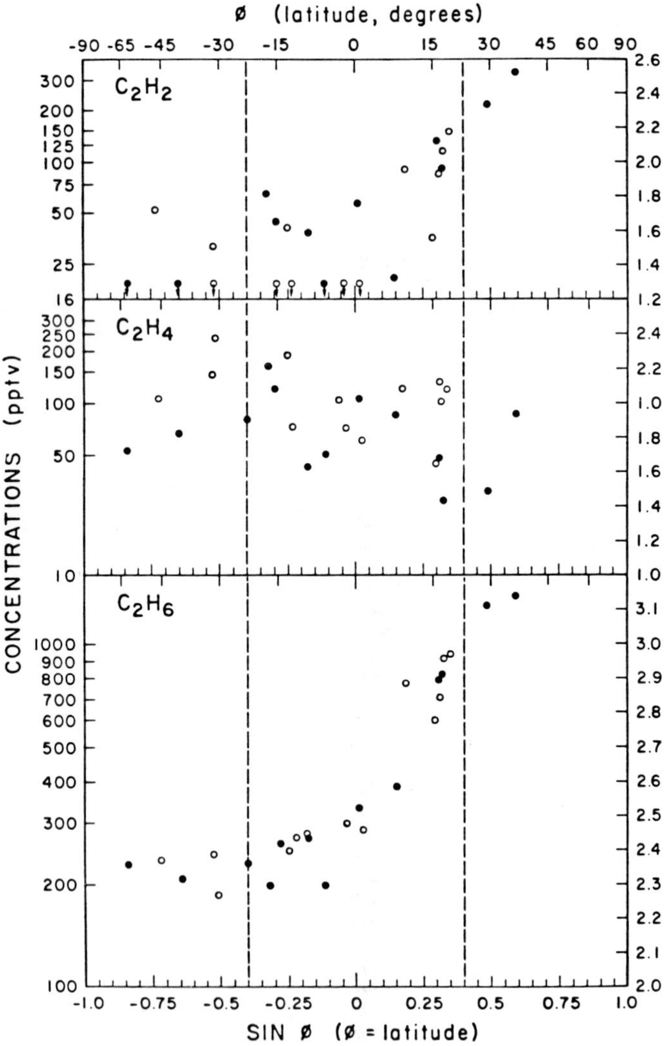

FIG. 2 - Latitudinal transepts of C_2H_2, C_2H_4, and C_2H_6 obtained by Rasmussen in the GAMETAG experiments (39).

The concentration data in Table 4 for paraffins suggest that about 5 MT of C_2H_6 is emitted per year, which is approximately twice that estimated to be released from leakage of natural gas (Derwent, personal communication). The C_3H_8 concentration is about twice that expected if natural gas were the only source, and this may be interpreted either in terms of a lower [OH] than 2.5 x 10^5 molecules/cm^3 or in terms of a natural source for paraffins.

There is some data to suggest that the ocean is a source of C_2H_6 and C_3H_8 (23), but generally speaking the biosphere acts as a source of unsaturated rather than saturated hydrocarbons.

Recent measurements of the concentration of C_4 and C_5 hydrocarbons over the north Atlantic suggest values in the 10 pptv range (41). The only other hydrocarbons specifically identified in clean maritime air are C_6H_6 and C_7H_8, which are probably produced entirely from anthropogenic emissions. This being the case, the average concentrations of C_6H_6 (64 pptv) and C_7H_8 (17 pptv) again argue for an [OH] value lower than 6 x 10^5 molecules/cm^3.

Measurements of other hydrocarbons in clean air are rather sparse. This comment was made in 1977 (12), and things have not changed a great deal since then. There has been considerable speculation concerning the emissions of terpenoid hydrocarbons from growing plants, and indeed measurements have been made of the release rate of isoprene and α-pinene amongst others (17,38). However, there is very little data on the concentrations of these species in the free atmosphere away from plant canopies; they would be expected to be very low because of their reactivity with ozone and hydroxyl radicals.

It has been estimated that 350 MT/year of isoprene and 500 MT/year of terpenes are emitted (54). Isoprene is a good source of CO, and its contribution to the global CO budget is to be compared with possible oceanic emissions of C_2H_4 and C_3H_8 in the range of 150 to 250 MT/year. Terpenes will also give rise to some CO when oxidized, but their main importance in the atmosphere is probably as a source of organic aerosol material.

According to some estimates, perhaps 100 to 200 MT/year of carbonaceous aerosol is formed (12,19) from the oxidation of large olefinic molecules, most of which must be in the form of terpenes.

Nitrogen Compounds, Oxygen Compounds, and Nitro Compounds

Atmospheric composition continues to expand to include more and more molecules. Just recently the presence of HCN has been detected in the stratosphere at a concentration close to 170 pptv, using sophisticated infrared spectroscopic methods (8). This is not a reactive compound and its source is unknown; it is, however, produced during combustion (11) and thus could be mainly of anthropogenic origin.

A series of measurements of CH_2O over the Atlantic ocean has just been reported (33). These are of considerable significance since they provide insight into the workings of the atmospheric oxidation process in very remote areas. The concentration appears to be relatively constant at about 0.2 ppbv between 50°N and 40°S. There is evidence of a diurnal variation indicating the presence of chemical change in the natural atmosphere, but the extent is quite small as shown in Fig. 3. The midday maximum concentration of hydroxyl radicals required to

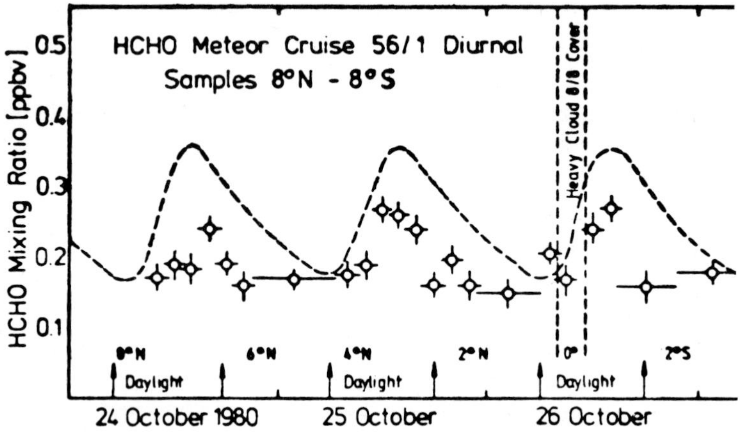

FIG. 3 - Measurements of the diurnal variation of CH_2O over the equatorial Atlantic obtained by Lowe and Schmidt on the research vessel Meteor in 1980 (33).

account for the observations would be about 1 x 10^6 molecules/ cm^3, assuming, of course, that other oxidation processes are absent. The source of the CH_2O is probably CH_4 oxidation, but some could also be produced both from C_2H_4 and C_3H_6 oxidation. C_2H_6 and C_3H_6 oxidation will yield CH_3CHO, and C_3H_8 oxidation will yield $(CH_3)_2CO$, which can also be produced from higher aliphatic hydrocarbons. Both $(CH_3)_2CO$ and CH_3CHO can be photolyzed and react with hydroxyl radicals to produce peroxy acetyl nitrate (PAN) in the presence of NO_2. Recent calculations based only upon C_2H_6 and C_3H_8 precursors suggest that the lower troposphere would contain 22 pptv of CH_3CHO, 111 pptv of $(CH_3)_2CO$, and 17 pptv of PAN (45).

Measurements of these components in the clean atmosphere are sparse indeed. There is no data on CH_3CHO, a species that would provide invaluable information on the workings of oxidation processes in the clean atmosphere. Some $(CH_3)_2CO$ data were obtained on the 1981 Meteor cruise, and these are shown in Table 2. The value, 470 ± 93 pptv, is higher than the calculated value, but it is more than likely that the photolytic lifetime of $(CH_3)_2CO$ used in the calculations is seriously underestimated. (The degree of variation observed in Table 2 would argue for quite a long life.) The data on these oxygen compounds are extremely preliminary and need to be repeated before any firm conclusions can be drawn. The first measurements of PAN in remote areas were made in 1974 (32). These indicated that long-range transport was possible over the oceans and suggested that PAN could be made in situ under favorable conditions, perhaps with C_3H_6 as its major precursor. PAN measurements made at Harwell in extremely clean Atlantic air suggest that levels between 50 and 100 pptv are typical in mid-latitudes (Brice and Penkett, unpublished data). There is an urgent need for coincidental PAN, aldehyde, ketone, and hydrocarbon data to be determined to try to assess the level of photochemical activity in the background atmosphere. If hydroxyl radical concentrations average out globally to about 2 x 10^5 molecules/ cm^3 and do not exceed $10^6/cm^3$, they are going to be very difficult to measure directly.

Organic Sulfur Compounds

Atmospheric sulfur chemistry used to be thought of in purely inorganic terms with emissions either of H_2S from natural sources or SO_2 from combustion sources. The proposal that natural sulfur was mainly emitted in the form of $(CH_3)_2S$ changed all that (31). Other organic forms of sulfur (COS and CS_2) have since been discovered in the atmosphere (16,42). The chemistry of these sulfur compounds is very interesting. H_2S is oxidized to give SO_2, $(CH_3)_2S$ gives SO_2 and a sulfur-containing aerosol, and CS_2 is oxidized to give an equimolar mixture of SO_2 and COS. SO_2 also gives rise to a sulfuric acid aerosol on oxidation. The rates of all of these processes have been studied recently under conditions likely to prevail in the lower troposphere (9,22), and they are summarized in Table 5. It is also known that CS_2 can be photolyzed in the troposphere ($j = 4.5 \times 10^{-5}$ s^{-1}, $\phi = 0.012$) (22,51).

The relative importance of one of these compounds, CS_2, in the overall sulfur cycle, is best seen through its conversion to

TABLE 5 - Volatile sulfur compounds in the atmosphere.

Compound	Hydroxyl rate coef.* x 10^{12} cm^3 $molec^{-1}$ s^{-1}	Concentration pptv northern hemisphere	southern hemisphere	Investigator
H_2S	5.0	7		Natusch (47)
CH_3SH	90	-		
$(CH_3)_2S$	9.1	4.3		Andreae (2)
$(CH_3)_2S_2$	223	-		
COS	0.005	523	498	Bandy (48)
CS_2	1.7	33		Bandy (4)
SO_2	0.72	89	57	Bandy (34)

*All rates from Cox and Shepherd (9) except for CS_2 which came from Jones, Burrows, Cox, and Penkett (21). This differs from much lower values published recently by Iyer and Rowland (18), and Wine, Shah, and Ravishankara (52), being obtained in the presence of oxygen.

COS. This is widely spread throughout the lower atmosphere at a concentration close to 500 pptv (48), and it is the most abundant form of gaseous sulfur in remote regions. There are direct emissions from biomass burning (0.1 MT), from volcanoes, from some industrial processes and possibly from the ocean. However, this is far from certain and indeed the ocean has been suggested as a sink (40). The chemical lifetime of COS has been estimated recently to be about 4 to 7 years (20), which would require a maximum input of 0.5 MT of sulfur to sustain its concentration at the 500 pptv level. This would be provided by emission of sulfur, in the form of CS_2, of 0.15 MT from the ocean surface (25,28) and up to 0.4 MT from solvent evaporation. The CS_2 concentration in the remote troposphere which would be required to sustain the level and extent of COS observed is only 3 pptv. Recent measurements in clean air suggest values of about 30 pptv in the boundary layer and less than 3 pptv in the free troposphere over the North American continent (4).

Transferring the argument to SO_2 formation suggests that between 0.15 and 0.5 MT of sulfur per year is available from CS_2. Oceanic input from CH_3SCH_3 has been calculated from flux measurements to vary between 3.7 and 36 MT of sulfur (2,25). However, only about 20% of the sulfur would be converted to SO_2, the rest going into an aerosol form (22). A contribution of only 1 MT is suggested, though, from the average CH_3SCH_3 concentration over the open ocean (4.3 pptv) with a low [OH] concentration (see above) (2). The estimated contribution from H_2S has been shrinking for many years. If the average oceanic concentration is 7 pptv (47), then this would give rise to the annual production of 1.2 MT of sulfur in the form of SO_2. Many measurements of the concentration of SO_2 in the background atmosphere have recently become available (34). They suggest that the average is close to 100 pptv, which would require a sulfur input of about 2.1 MT. This would limit the size of the sulfur input in the form of CH_3SCH_3 to approximately 0.7 MT, or about 3.5 MT of CH_3SCH_3 emitted with a 20% conversion factor. Assuming that all compounds emitted ultimately become sulfate aerosols, this adds up to a yearly production of close to 5 MT, expressed as sulfur, from natural processes. This number is far lower than that derived from early attempts based on budgetary methods. Even if the average [OH] approaches 6×10^5 molecules/cm^3, the yearly natural production of sulfur will probably not exceed 15 MT, which is much lower than the 70 MT emitted as a pollutant.

One of the problems that these recent data highlight is the missing SO_2 emitted from combustion sources. Roughly one hundred times more SO_2 is emitted than C_2H_2, but their concentrations in clean air are rather similar. Since the gas-phase oxidation rates by hydroxyl radicals are also similar, other very efficient removal processes must exist for SO_2. These include surface deposition and most probably wet-phase oxidation involving hydrogen peroxide.

CONCLUSION

Data on the concentration distribution of many organic compounds over the ocean provide clear evidence for chemical change occurring in the remote atmosphere. This is observed in the removal of reactive halogen compounds and hydrocarbons, and also in the formation of oxygen and nitrocompounds. Using admittedly empirical arguments, the extent of chemical change of these species through reaction with hydroxyl radicals is somewhat smaller than would be predicted from assessments based solely on the CO budget. This situation was first alluded to in connection with the CH_3CCl_3 budget (30,44), and it may suggest that some of the reasoning behind the cycle of CO in the atmosphere is incorrect.

The amounts of sulfur compounds emitted from the biosphere appears to be small compared to amounts emitted from combustion processes. In this case it is possible that volcanic activity can make a substantial contribution to the natural sulfur budget. There is a strong indication that much of the combustion-produced SO_2 is removed before it is oxidized to a H_2SO_4 aerosol in the gas phase by hydroxyl radicals.

Assuming a low sulfur budget, then much of the mass of aerosol formed by gas to particle processes in the atmosphere is in the form of carbonaceous matter. Perhaps 20 MT of sulfate are produced per year by natural processes compared with estimates of 200 MT of carbonaceous aerosol. This latter estimate, however, is very uncertain and may well be too high. The mass of secondary aerosol material is far exceeded by a primary aerosol which is believed to consist of 1000 MT of sea salt and 500 MT of wind-raised soil. The smaller particle size of the secondary aerosol will allow it to act more efficiently in forming condensation nuclei. Also, the secondary aerosol is probably more widely spread throughout the troposphere, but it is not clear whether it controls cloud formation.

Acknowledgements. The author wishes to thank R.A. Rasmussen, D. Lowe, and M.O. Andreae who graciously provided preprints of their work for inclusion in this review. Work at AERE Harwell was funded by the Department of the Environment.

REFERENCES

(1) Adel, A. 1949. Selected topics in the infrared spectroscopy of the solar system. In The Atmospheres of the Earth and Planets, ed. G.P. Kuiper, p. 269. Chicago: University of Chicago Press.

(2) Andreae, M.O.; Bernard, W.R.; and Ammons, J.M. 1981. The Biological Production of Dimethylsulfide in the Ocean and its Role in the Global Sulfur Budget. 5th International Symposium on Environmental Biogeochemistry, Stockholm.

(3) Atkinson, R.; Darnall, K.R.; Lloyd, A.C.; Winer, A.M.; and Pitts, J.N. 1979. Kinetics and mechanisms of the reactions of the hydroxyl radical with organic compounds in the gas phase. Adv. Photochem. 11: 1975.

(4) Bandy, A.R.; Maroulis, P.J.; Shalaby, L.; and Wilner, L.A. 1981. Evidence for a short tropospheric residence time for carbon disulfide. Geophys. Res. Lett. 8: 1180.

(5) Chameides, W.L., and Davis, D.D. 1980. Iodine: its possible role in tropospheric photochemistry. J. Geophys. Res. 85: 7383.

(6) Chapman, S. 1930. A theory of upper atmospheric ozone. Mem. Roy. Meteor. Soc. 3: 103.

(7) Cicerone, R.J. 1979. Atmospheric carbon tetrafluoride: A nearly inert gas. Science 206: 59.

(8) Coffey, M.J.; Mankin, W.G.; and Cicerone, R.J. 1981. Spectroscopic detection of stratospheric hydrogen cyanide. Science 214: 333.

(9) Cox, R.A., and Sheppard, D. 1980. Reactions of OH radicals with gaseous sulfur compounds. Nature 284: 330.

(10) Crutzen, P.J. 1979. The role of NO and NO_2 in the chemistry of the troposphere and the stratosphere. Ann. Rev. Earth Planet Sci. 7: 443.

(11) Crutzen, P.J.; Heidt, L.E.; Krasnec, J.P.; Pollock, W.H.; and Seiler, W. 1979. Biomass burning as a source of atmospheric gases CO, H_2, N_2O, NO, CH_3Cl and COS. Nature 282: 253.

(12) Duce, R.A. 1978. Speculations on the budget of particulate and vapor phase non-methane organic carbon in the global troposphere. In Influence of the Biosphere on the Atmosphere, ed. H.E. Dutsch, p. 244. Basel and Stuttgart: Birkhauser Verlag.

(13) Glueckauf, E. 1951. The Composition of Atmospheric Air. Compendium of Meteorology, p. 3. Boston, MA: American Meteorology Society.

(14) Glueckauf, E., and Kitt, G.P. 1957. The hydrogen content of atmospheric air at ground level. Quart. J. Roy. Meteor. Soc. 83: 522.

(15) Haagen-Smit, A.J. 1952. Chemistry and physiology of Los Angeles smog. Ind. Eng. Chem. 44: 1342.

(16) Hanst, P.C.; Spiller, L.L.; Watts, D.M.; Spence, J.W.; and Miller, M.F. 1975. Infrared measurements of fluorocarbons, carbon tetrachloride, carbonyl sulfide and other trace gases. J. Air Polln. Contr. Ass. 25: 1220!

(17) Holdren, M.W.; Westberg, H.H.; and Zimmerman, P.R. 1979. Analysis of monoterpene hydrocarbons in rural atmospheres. J. Geophys. Res. 84: 5083.

(18) Iyer, R.S., and Rowland, F.S. 1980. A significant upper limit for the rate of formation of OCS from the reaction of OH with CS_2. Geophys. Res. Lett. 7: 797.

(19) Jaenicke, R. 1978. The role of organic material in atmospheric aerosols. In Influence of the Biosphere on the Atmosphere, ed. H.U. Dutsch, p. 283. Basel and Stuttgart: Birkhauser Verlag.

(20) Johnson, J.E. 1981. The lifetime of carbonyl sulfide in the troposphere. Geophys. Res. Lett. 8: 938.

(21) Jones, B.M.R.; Burrows, J.; Cox, R.A.; and Penkett, S.A. 1982. OCS formation in the reaction of OH with CS_2. Chem. Phys. Lett. 88: 372.

(22) Jones, B.M.R.; Burrows, J.; Cox, R.A.; and Penkett, S.A. 1983. The role of CS_2 in atmospheric sulfur chemistry. Atmos. Env., in press.

(23) Lamontagne, R.A.; Swinnerton, J.W.; and Linnenbom, V.J. 1974. C_1-C_4 hydrocarbons in the North and South Pacific. Tellus 26: 72.

(24) Levy, H. 1971. Normal atmosphere: large radical and formaldehyde concentrations predicted. Science 173: 141.

(25) Liss, P.S., and Slater, P.G. 1974. Flux of gases across the air-sea interface. Nature 247: 181.

(26) Logan, J.A.; Prather, M.J.; Wofsy, S.C.; and McElroy, M.B. 1981. Tropospheric chemistry: a global perspective. J. Geophys. Res. 86: 7210.

(27) Lovelock, J.E. 1961. Ionization methods for the analysis of gases and vapors. Anal. Chem. 33: 162.

(28) Lovelock, J.E. 1974. CS_2 and the natural sulfur cycle. Nature 248: 625.

(29) Lovelock, J.E. 1975. Natural halocarbons in the air and in the sea. Nature 256: 193.

(30) Lovelock, J.E. 1977. Methyl chloroform in the troposphere as an indicator of the OH radical abundance. Nature 267: 32.

(31) Lovelock, J.E.; Maggs, R.J.; and Rasmussen, R.A. 1972. Atmospheric dimethyl sulfide and the natural sulfur cycle. Nature 237: 452.

(32) Lovelock, J.E., and Penkett, S.A. 1974. PAN over the Atlantic and the smell of clean linen. Nature 249: 434.

(33) Lowe, D.C.; Schmidt, U.; and Ehhalt, D.H. 1981. The tropospheric distribution of formaldehyde. Berichte der Kernforschungsanlage, Nr. 1756, Jülich, FRG.

(34) Maroulis, P.J.; Torres, A.L.; Goldberg, A.B.; and Bandy, A.R. 1980. Atmospheric SO_2 measurements on project GAMETAG. J. Geophys. Res. 85: 7345.

(35) Migeotte, M.V. 1949. On the presence of CH_4, N_2O, and NH_3 in the earth's atmosphere. In The Atmospheres of the Earth and Planets, ed. G.P. Kuiper, p. 284. Chicago: University of Chicago Press.

(36) Molina, M.J., and Rowland, F.S. 1974. Stratospheric sink for chlorofluoromethanes: chlorine atom catalyzed destruction of ozone. Nature 249: 810.

(37) Penkett, S.A. 1981. The application of analytical techniques to the understanding of chemical processes occurring in the atmosphere. Toxicol. Env. Chem. 3: 291.

(38) Rasmussen, R.A. 1972. What do hydrocarbons from trees contribute to air pollution? J. Air Polln. Contr. Ass. 22: 537.

(39) Rasmussen, R.A., and Khalil, M.A.K. 1982. Latitudinal distributions of trace gases in and above the boundary layer. J. Geophys. Res., in press.

(40) Rowland, F.S. 1979. The atmospheric and oceanic sinks for carbonyl sulfide. 4th International Conference of the Commission on Atmospheric Chemistry and Global Pollution, Boulder, Colorado.

(41) Rudolph, J., and Ehhalt, D.H. 1981. Measurements of C_2-C_5 hydrocarbons over the north Atlantic. J. Geophys. Res. 86: 11959.

(42) Sandalls, F.J., and Penkett, S.A. 1977. Measurements of carbonyl sulfide and carbon disulfide in the atmosphere. Atmos. Env. 11: 197.

(43) Seiler, W. 1976. The cycle of atmospheric CO. Tellus 26: 46.

(44) Singh, H.B. 1977. Atmospheric halocarbons: evidence in favor of a reduced average hydroxyl radical concentration in the troposphere. Geophys. Res. Lett. 4: 101.

(45) Singh, H.B., and Hanst, P.L. 1981. Peroxyacetyl nitrate (PAN) in the unpolluted atmosphere: an important reservoir for nitrogen oxides. Geophys. Res. Lett. 8: 941.

(46) Singh, H.B.; Salas, L.J.; Shigeishi, H.; Smith, A.J.; Scribner, E.; and Cavanagh, L.A. 1979. Atmospheric distributions, sources and sinks of selected halocarbons, hydrocarbons, SF_6 and N_2O. Stanford Research Institute Project Report No. 4487.

(47) Slatt, B.J.; Natusch, D.F.S.; Prospero, J.M.; and Savoie, D.L. 1978. Hydorgen sulfide in the atmosphere of the northern equatorial Atlantic ocean and its relation to the global sulfur cycle. Atmos. Env. 12: 981.

(48) Torres, A.L.; Maroulis, P.J.; Goldberg, A.B.; and Bandy, A.R. 1980. Atmospheric OCS measurements on project GAMETAG. J. Geophys. Res. 85: 7357.

(49) Volz, A.; Ehhalt, D.H.; and Derwent, R.G. 1981. Seasonal and latitudinal variation of ^{14}CO and the tropospheric concentration of OH radicals. J. Geophys. Res. 86: 5163.

(50) Weinstock, B., and Niki, H. 1972. Carbon monoxide balance in nature. Science 176: 290.

(51) Wine, P.H.; Chameides, W.L.; and Ravishankara, A.R. 1980. Potential role of CS_2 photooxidation in tropospheric sulfur chemistry. J. Phys. Chem. 84: 2499.

(52) Wine, P.H.; Shah, R.C.; and Ravishankara, A.R. 1981. Rate of reaction of OH with CS_2. Geophys. Res. Lett. 8: 543.

(53) Wofsy, S.C.; McConnell, J.C.; and McElroy, M.B. 1972. Atmospheric CH_4, CO and CO_2. J. Geophys. Res. 77: 4477.

(54) Zimmerman, P.R.; Chatfield, R.B.; Fishman, J.; Crutzen, P.J.; and Hanst, P.C. 1978. Estimates on the production of CO and H_2 from the oxidation of hydrocarbon emissions from vegetation. Geophys. Res. Lett. 5: 679.

Group on Tropospheric Gases, Aerosols and Photochemical Reactions

Standing, left to right:
Reinhard Zellner, Ruprecht Jaenicke, Stu Penkett, Frank Arnold, Pat Zimmerman, Hiromi Niki, and Franz Meixner.

Seated, left to right:
Ralph Cicerone, Paul Crutzen, Sherry Rowland, Mario Molina, and Donald Hornig.

Atmospheric Chemistry, ed. E.D. Goldberg, pp. 357-372. Dahlem Konferenzen 1982.
Berlin, Heidelberg, New York: Springer-Verlag.

Tropospheric Gases, Aerosols and Photochemical Reactions Group Report

R. J. Cicerone, Rapporteur
F. Arnold, P. J. Crutzen, D. F. Hornig, R. Jaenicke, F. X. Meixner,
M. J. Molina, H. Niki, S. A. Penkett, F. S. Rowland, R. E. Zellner,
P. R. Zimmerman

INTRODUCTION

The exploration of chemical reactions in the atmosphere near the Earth became a subject of intense interest only one decade ago. Earlier investigations were limited to components of stratospheric ozone photochemistry or of specific portions of urban air-pollution problems. It is indeed difficult for one to comprehend today how little was known conceptually two or three decades ago, unless one reviews how little was known of the composition of the air at that time. Quantitatively, one was sure only of the atmospheric amounts of N_2, O_2, the noble gases, CO_2, H_2O below the tropopause, and O_3 in the stratosphere. By 1950, CH_4, N_2O, and CO had been detected but measured only to about 50% accuracy. The presence of tropospheric ozone was attributed to the subsidence of air from the stratosphere. The existence of the airborne particulate was known, but little information, even on bulk properties, was available. Extant data on the visible- and ultraviolet-light spectrum of the sun allowed the speculation that stratospheric ozone was important as an ultraviolet shield, but other possible absorbers were unexplored. The roles of CO_2, H_2O, and O_3 in climate and atmospheric dynamics were identified qualitatively but were not well understood.

In brief summary, the atmosphere near the Earth was viewed as the fluid that moved moisture and heat. It also transported pollutants away from cities, factories, and fires. The chemical species of air were regarded as essentially inert and for fairly good reason - those that were known to exist were mostly inert gases. A fair amount was known of the isotopic composition of atmospheric chemicals, indeed the use of radiochemical techniques was more prevalent in 1960 than in 1980. Tropospheric ozone was a curiosity.

Since those early days progress has been rapid and it is accelerating. The realization that chemical reactions can occur and do proceed in the troposphere, although late in its advent, is now complete. This realization and our growing understanding, both conceptual and detailed, are the result of new data from chemical measurements and new theoretical insights. It has also become clear to the atmospheric chemist and to mankind that humans are increasingly capable of perturbing the atmosphere; direct perturbations and more subtle yet pervasive influences are possible. These direct and sometimes subtle perturbations, while often inadvertent and unforeseen, can extend to the Earth's soils, waters, fauna, flora, and climate. They also can alter the electromagnetic spectrum of the Earth as seen from space.

Anthropogenic perturbations and influences on the chemistry of the atmosphere have been identified so rapidly in the past decade or so that our knowledge of the system has not kept pace - even though our knowledge has grown explosively. Far from accidental, the identification of potential atmospheric effects of man's activities has resulted from our growing understanding of the pathways of atmospheric chemical transformations and dynamics. As the chemical composition of the air has become better known and these latter pathways have become defined, we have been able to draw from laws of chemistry and physics and from laboratory experiments to deduce what other chemical substances are in the air and how they must behave. Even though the identification of an anthropogenic effect on the atmosphere represents progress in itself, it is also necessary to obtain more quantitative understanding of the dynamics of the perturbation. Effective control strategies require this. Because of the rapid increase in man's population, technology, and consumption, it has not been possible for the relatively small community of atmospheric chemists to keep pace with, or even foresee, the need for more and more quantitative information on air chemistry.

RECENT PROGRESS AND PRESENT KNOWLEDGE

The principal gaseous species of the atmosphere have been measured accurately at least in some locations and many other trace gases have been identified since about 1965. Detailed data are now available on CO_2 concentrations, latitudinal and seasonal variations, and secular trends. For CH_4, N_2O, and selected chlorofluorocarbons, corresponding but fewer data are in hand. Our data base from measurements of a number of other organic gases in the atmosphere is summarized by Penkett (this volume). Many advances in analytical chemistry have permitted (and often have been stimulated by) the assay of atmospheric gases at levels of one part in 10^{12} (1 ppt), well below the part per million or so threshold for analysis of 1955-60. Data on the present trace-gas and aerosol composition of the atmosphere, including spatial variations, are essential prerequisites to understanding the system.

A class of chemical species that are critical in controlling atmospheric composition is that of free radicals. Just over ten years ago, Levy (6) proposed that the well-known ultra-violet photolysis of ozone,

$$O_3 + h\nu \rightarrow O(^1D) + O_2, \quad (1)$$

followed by a small fraction of the product $O(^1D)$ (metastable, electronically excited oxygen atoms) reacting with H_2O vapor would produce hydroxyl radicals:

$$O(^1D) + H_2O \rightarrow OH + OH. \quad (2)$$

It was proposed that reactions 1 and 2 would result in the presence of significant background levels of gaseous OH radicals in the free troposphere where UV light (in particular, 290 nm < λ < 310 nm), H_2O, and O_3 are available. Further work by Crutzen (3) showed that interesting possibilities for tropospheric photochemical reactions followed from reactions 1 and 2. Indeed the presence of ozone guaranteed that ozone-producing and ozone-destroying reactions would occur. It soon became clear that tropospheric OH is a critical constituent (see discussion in Logan et al. (8)).

Irreversible processes such as the oxidation of methane, carbon monoxide, certain chlorocarbons, organic sulfur species, and many hydrocarbons are initiated by OH. The oxidation of SO_2 and of NO_2 to sulfuric and nitric acid, respectively, can be initiated by reaction with OH. Furthermore, OH and other radicals can initiate and propagate chain reactions such as:

$$CO + OH \rightarrow H + CO_2 \quad (3)$$

$$H + O_2 + M \rightarrow HO_2 + M \quad (4)$$

$$HO_2 + O_3 \rightarrow OH + 2\,O_2 \quad (5)$$

net: $CO + O_3 \rightarrow CO_2 + O_2$, and

$$CO + OH \rightarrow H + CO_2 \quad (6)$$

$$H + O_2 + M \rightarrow HO_2 + M \quad (7)$$

$$HO_2 + NO \rightarrow OH + NO_2 \quad (8)$$

$$NO_2 + h\nu \rightarrow NO + O \quad (\lambda \lesssim 430 \text{ nm}) \quad (9)$$

$$O + O_2 + M \rightarrow O_3 + M \quad (10)$$

net: $CO + 2O_2 \rightarrow CO_2 + O_3$.

The reaction chain 3-5 destroys ozone and thus decreases the free radical source (reaction 1) while the chain 6-10 produces ozone, and more radicals follow. In a given region the balance between ozone destruction and production depends on the competition between the overall rates of reactions 5 and 8, effectively the ratio of the concentrations of O_3 and another radical, NO.

The gaseous reactions listed above also serve to illustrate an important concept of chemistry that in turn shows how very low concentrations of gases can affect important chemical balances in the atmosphere - the concept of chemical catalysis. Consider, for example, the behavior of the stratospheric ozone system. Although present at concentrations of 10 ppm and less, stratospheric ozone is the only atmospheric gas capable of significantly absorbing the biologically damaging sunlight between 240 nm and 320 nm wavelength; these relatively small amounts of ozone in the stratosphere absorb nearly all this light before it reaches the Earth's surface. Catalytic chemical reactions enable chemical species present in amounts far less than 10 ppm to regulate stratospheric O_3 levels. The chlorine-ozone reactions,

$$Cl + O_3 \rightarrow ClO + O_2 \quad (11)$$

$$O_3 + h\nu \rightarrow O + O_2 \quad (\lambda < 1000\ nm) \quad (12)$$

$$ClO + O \rightarrow Cl + O_2 \quad (13)$$

$$\text{net: } 2O_3 + h\nu \rightarrow 3O_2$$

are chain reactions propagated by the Cl and the ClO radicals. Present knowledge demonstrates that the reaction chain length (or catalytic efficiency) of chlorine in destroying ozone is as high as 10^5 in parts of the stratosphere. As catalytic efficiencies of 10^6 are not uncommon in chemical manufacturing processes, one can see how stratospheric ozone (1-10 ppm) can be controlled by substances present at concentrations of, e.g., 100 ppt. This was proposed by Molina and Rowland (9) as the mechanism by which anthropogenic chlorofluoromethanes could perturb the global ozone layer. Similar reactions involving nitrogen oxides (NO_x) and ozone are also very important in the natural stratosphere and in an NO_x-polluted atmosphere (3). Other cases illustrate the potential importance, even critical role, of trace gases and radicals at very low concentrations, e.g., OH in one part in 10^{13} or 10^{14} in the troposphere. For example, theory now indicates that concentrations of NO and NO_2 between 10 and 30 ppt are sufficient to allow reaction 8 to dominate reaction 5 in ground-level air, thus producing O_3 in the background troposphere (4).

The above reasoning about gas-phase chemical catalysis can be augmented by a similar potential for catalysis of aqueous reactions by metals in cloud, raindrops, or aerosols. Furthermore, important infrared radiative effects can arise from gases that absorb in the atmospheric window region of 8-12 µm wavelength. The considerable potential for a global warming due to increasing usage of man-made chlorofluorocarbons was recognized by Ramanathan (10). Lacis et al. (5) have shown that the calculated warming (0.14°K) due to global CO_2 increases between 1970 and 1980 is less than twice as large as that (~0.10°K) due to the measured increases in CCl_2F_2 and CCl_3F and the apparent increases in N_2O and CH_4 in the same period. It is now clear that very small concentrations of gases or of solid catalysts can have dramatic effects on atmospheric chemical and thermal balances, effects far larger in proportion than the concentrations of the relevant trace substances.

Research on atmospheric particulate material, while also stimulated by the need to assess man's impact on the environment, has seen more gradual progress. Physical as well as chemical properties of the atmospheric aerosol particle have received (and require) a great deal of attention; their static physical properties, e.g., size, mass, and index of refraction, and their role as tracers of atmospheric motion have been studied more than their chemical kinetics. Field measurements and examination of basic aerosol particle properties have established that particle residence times in the atmosphere depend on their size: the physical processes of sedimentation, impaction, wet removal, and coagulation lead to a maximum residence time for particles with radii near 0.3 µm. The elemental composition of aerosol particles has been measured well if not for all particle sizes in all regions; data exist for 40 or more elements. In Aitken particles (radius $\lesssim$ 0.1 µm), the most abundant elements are S and C. Free acids, principally H_2SO_4, are frequently found in background aerosols. The presence of organic matter in background aerosol particles is firmly established; typical background aerosols carry 1 $\mu g/m^3$ of organic matter. Minerals and inorganic materials dominate the sea-salt aerosol, e.g., Na^+, Cl^-, Br^-, K^+, SO_4^{2-}, NO_3^-, and PO_4^{-3}.

Total number densities (cm^{-3}) of airborne particles are typically: 300-600 over ocean surface (and above the boundary layer), 300,000 in cities, and in very remote areas can occasionally be as low as 5-10. Over continents, number densities range from several hundred per cm^3 to about 15,000, depending on proximity to population centers and on meteorological and surface conditions. When produced, aerosol particles often exhibit a multi-modal size distribution due to different sources, but the distribution relaxes to a unimodal shape as the aerosols age. The various sources of aerosol particles include gas-phase nucleation reactions, mechanical action such as wind action on soils, waves and shears on ocean waters, and possibly gas-phase ionic reactions, both ion-molecule and ion-ion reactions. Mechanical action produces mostly large particles ($r \gtrsim 1$ µm), while gas-to-particle conversion builds smaller particles ($r \lesssim .1$ µm).

The various potential influences of aerosol particles on the natural and perturbed atmosphere include radiative and chemical roles as well as impacts on global and regional biology. Possible climatic effects of aerosols have long been suspected, especially after clusters of large volcanic eruptions that give rise to long-lasting stratosphere aerosol layers (either from direct ash injections or gas-to-particle conversion or both) or after large amounts of dust are lofted by meteoritic impact. Proper assessment of such effects requires specification of several physical properties: mass loading, size distribution, particle shapes, and the imaginary part of the index of refraction. Aerosol transport carries bio-active materials such as NH_4^+, NO_3^-, and PO_4^{-3} (nutrients), sulfates, toxins, viruses, and spores and is capable of important material transfer over long distances. Long-range transport (> 1000 km) of aerosol particles has been demonstrated repeatedly, and man's activities are capable of influencing the Arctic and possibly the Antarctic regions.

Regional-scale phenomena due to aerosol particles, such as decreased visibility, are fairly widely recognized if not well understood. The possibility that the surface energy budget of a polluted region can be affected by the particulate loading of that region's atmosphere (2) needs serious exploration. Ball and Robinson (1) estimate that the solar energy received at the Earth's surface is 7.5% less (annually averaged in the east-central US) than if aerosols were absent. The origin of these airborne particles must be determined. Over the northern Atlantic Ocean, the dust cover from the Saharan desert has been estimated to be of equal importance as the cloud cover.

While field monitoring and characterizations of remote background aerosols must continue, many aspects of aerosol behavior require theoretical investigation and quantitative modeling. For example, the possible mechanisms of tropospheric and stratospheric aerosol formation need investigation. Ions are present in the Earth's troposphere and stratosphere, but the degree of ionization is very much smaller than in the higher-lying ionosphere where as many as one atom or molecule out of every 10^5 in the background atmosphere is ionized. In the stratosphere and troposphere, ionization Debye lengths are much larger than collision mean-free paths, and where H_2O is present (as in the troposphere) large clusters are formed, so ionic mobilities are very low. In the troposphere ion-cluster densities have been measured and data on ion mobilities has allowed estimations of ion-mass ranges, but the chemical nature of the ions is unknown. With so few data it is premature to assess possible atmospheric chemical or health effects.

In the stratosphere, the data base is more substantial, and it offers hope for future extensions to the troposphere. Stratospheric ion number densities and the chemical identities of the major positive and negative ions have been characterized by in situ mass spectrometer measurements. Such measurements are beginning to permit the indirect deduction of ambient concentrations of H_2O, HNO_3, and ultra-trace neutral species, e.g., CH_3CN, CH_3OH, HCN, H_2SO_4, and HSO_3. Results to date indicate that ion nucleation in the stratosphere can produce small particles.

The roles of laboratory photochemistry and of laboratory chemical kinetics have been crucial in the progress of the last ten years of atmospheric chemistry. Modern state-of-the-art laboratory techniques have generated fundamental and reliable data on photoabsorption cross sections, quantum yields for photodissociation, and on reaction-rate constants for elementary processes, especially for small molecules, of, e.g., six atoms or less. Gas-handling procedures have been advanced, especially for radicals, and more importantly, high sensitivity, high specificity spectroscopic techniques have been developed and applied to many important atmospheric reactions, both bimolecular and termolecular. Key techniques include a) flash photolysis resonance fluorescence and absorption, and b) discharge-flow followed by mass spectroscopy, resonance fluorescence, or laser magnetic resonance for radical detection. While Niki (this volume) has summarized briefly where we stand in this field, it is useful to note that these laboratory kinetic studies have allowed the quantitative assessment of the rates of many chemical reaction pathways. For example, the rates of reactions 1-13 here are known to be within 50%; indeed, most to within 20% for tropospheric and stratospheric pressures and temperatures. Outstanding examples of advances in chemical kinetics are those of OH and HO_2 reactions with atmospheric species. For many of these reactions, directly measured rate constants are either available now or can often be reasonably extrapolated empirically from data on other species from related chemical groups. Theoretical calculations have provided useful information on molecular and electronic structures of the ensuing transient products as well as theoretical guidance in extrapolating laboratory data to pressures and temperatures not investigated in certain experiments.

It should be emphasized that progress in chemical kinetics in the past decade has been greatly stimulated by the needs and interests of atmospheric chemistry. In 1970 the chemical reactions of OH radicals with stable molecules were usually interpreted in terms of simple bimolecular abstraction reactions, expected to be independent of pressure and to decrease in rate with decreasing temperature in response to the activation energy for the bimolecular process. Reactions of OH with other radicals were for the most part not investigated at all. The advances of the past decade have shown that many OH reactions of interest in the atmosphere are much more complex than given by the simple bimolecular picture and involve additional reactions to form intermediate complexes (e.g., OH reactions with CO, HNO_3, and possibly with CS_2 and HNO_4) with subsequent further reactions. Frequently, the rates of such chemical reactions have been found to be faster at low temperatures and to exhibit dependence on pressure. Precision measurements have been required over the entire range of temperatures and pressures of atmospheric interest and for appropriate levels of O_2, H_2O, and N_2. The considerable complexity of these supposedly simple OH reactions illustrates the formidable tasks ahead in unraveling the details of gas-to-particle conversions, e.g., those initiated by the reactions of OH with SO_2 or NO_2.

The combined progress of the extended analytical techniques for field measurements of atmospheric chemical composition and of photochemical data on key molecules has permitted the refinement of our conceptual understanding and mathematical models of atmospheric chemistry. This has allowed examination of the effects of man's increasing activities. To assess properly these effects it has proven desirable and necessary to map out the pathways and relationships between atmospheric chemistry and the biosphere, hydrosphere, and soils; the atmospheric portions of the elemental cycles have become more clearly defined conceptually and in some cases more quantitatively. From this set of refined conceptualizations of the cycles, the greatly improved laboratory data (and methods) for gaseous kinetics, and the rapidly expanded data base on atmospheric chemical composition, we can try now to move forward with the assessment of the atmospheric impact of man's present and planned activities.

Challenging questions to be answered include: why are methane and nitrous oxide concentrations increasing globally? Will the trends continue and what will be the effects? Will the observed (and predicted to continue) increases in global concentrations of CCl_2F_2, CCl_3F, CH_3CCl_3 and similar compounds result in major perturbations to the global stratosphere as now predicted? Will the infrared greenhouse effects of each of these trace gases and CO_2 combine to produce a significant global warming? Is ozone increasing in the upper troposphere as is predicted (7) due to commercial aircraft NO_x emissions? Have ozone concentrations near ground level increased in regionally (or N. hemisphere) polluted air? What amounts of food-crop loss and property damage are due to incidents of high ozone and oxidant levels in high air pollution episodes? How large of a surface heating (through IR greenhouse effect plus direct absorption of visible and UV sunlight) should be expected from ozone increases in the upper troposphere? Will there ensue an IR greenhouse effect of such an O_3 increase? If tropospheric CH_4, CO, and hydrocarbons increase, what will tropospheric OH do? Has the atmospheric burden of airborne particulates and/or atmospheric turbidity increased or will it do so? What are appropriate control strategies to alleviate acid-deposition problems?

RESEARCH GOALS AND STRATEGIES

The atmospheric reaction rates and pathways of gases, particles, and ions must be assessed in much more detail before we can obtain an adequate understanding of atmospheric chemical composition and its changes, of the atmospheric portions of global element cycles and the effects of man's various industrial, agricultural, and aerospace activities and the effects of nature's own perturbations, e.g., volcanic activity, dust storms. While mankind has little if any control over nature's geophysical perturbations, there is a need and some hope that human activities can be controlled better. It seems clear that human population will continue to increase in the next 30 to 50 years and that an ever higher standard of living will be sought. Consequently, many industrial activities will become more intense and higher technology will probably be employed, e.g., in energy conversion, mining, combustion, chemical manufacturing, etc. The atmospheric cycles of many chemical species and of aerosol particles will be affected by these intensified human activities. To be able to conduct these activities safely and efficiently will require advanced technical planning and processes; advanced knowledge of many atmospheric chemical processes and pathways is needed first. At present we can set some goals in our efforts to characterize the chemistry of the atmosphere, its dynamical processes, and temporal trends in the system. Principal goals and strategies for the attainment of these goals are discussed briefly in the following paragraphs.

1. Detect and measure key reactive tropospheric gases, including OH, HO_2, NO, and NO_2. This will require continued development of advanced instruments with exquisite sensitivity and specificity. In situ measurements will be required with substantial altitude and latitude coverage to enable meaningful surveys of tropospheric behavior and tests of photochemical theories.

2. Extend and refine data base on aerosol chemical composition and size and shapes. Organic constituents need to be identified and individual particle surface areas require determination. Technique development is needed first. Vertical profiles are needed.

3. Determine temporal trends in tropospheric ozone concentrations. For an adequate assessment of temporal trends in the O_3 vertical profiles and upper tropospheric column amounts vs. latitude (especially in the polluted northern hemisphere), more accurate and stable ozone-measurement instruments are needed. Proper experimental network design for meteorological and statistical factors must be employed.

4. Explore mechanisms for in situ aerosol formation through theory and laboratory experiments. Gas-phase reactions that produce particles as well as possible ion-molecule and ion-ion routes need investigation to determine regimes of variables where each mechanism is significant. Thermodynamic and kinetic properties of ion clusters need to be determined.

5. Obtain field data to characterize the tropospheric background levels of NO_x species to include NO, NO_2, NO_3, HNO_3, HNO_4, and PAN, so that the tropospheric sources, sinks, and transport of these species can be quantified.

6. Develop a new generation of mathematical models of tropospheric photochemistry to couple more fully meteorological motions with chemical reactions and to treat cloud transport, cloud chemistry, and surface deposition more realistically.

7. Extend and improve gas-phase kinetic and photochemical data base and our understanding of reaction mechanisms. Goals include determination of reaction-branching pathways, products and intermediate species, and refinement of reaction-rate theories, even if semi-empirical. This will require the extension of direct detection methods for radicals to higher pressures.

Performing intercomparisons of direct and indirect methods will allow one to extrapolate data more reliably from laboratory to atmospheric conditions.

8. Assess importance of reactions between radicals and aerosol and droplet surfaces both theoretically and experimentally.

9. Determine atmospheric distributions and fates of metal-containing volatile substances, similarly for relatively low volatility organic species, especially for toxic substances, e.g., insecticides, polycyclic aromatics, and urea. Data on the chemical phases of these latter species are also needed. Field measurements and the laboratory determination of photochemical parameters are both required.

10. Measure and estimate emissions of gases and aerosols from biosphere sources and from nonindustrial fires (forest fires, brush fires). The tropics will require special emphasis.

11. Refine our conceptual understanding and quantitative knowledge of biosphere-atmosphere coupling and of global biogeochemical cycles. This will require empirical modeling and multidisciplinary research to understand and account for fundamental mechanisms.

12. Determine the chemical nature of tropospheric ions and attempt to deduce concentrations of related neutral species through equilibrium and steady state arguments.

13. Extend data base on stratospheric ions and related ligands to examine behavior of ions including possibilities for nucleation and to allow the indirect measurement of water-soluble species of tropospheric origin (e.g., NH_3, CH_3CN) for possible use as indicators of tropospheric cloud-removal processes.

14. Determine temporal trends of key stable trace gases and of total particulate loading. This can be achieved by continuing present monitoring of CH_4, CO, CCl_2F_2, CCl_3F, N_2O,

CH_3CCl_3, CCl_4, and $C_2Cl_3F_3$ with some improvements by using available data bases and by enhanced monitoring of particles and turbidity at remote sites.

15. Determine the atmospheric fate of organic gas emissions (e.g., terpenes) from plants. Theoretical and empirical oxidation mechanisms need developing to determine relative yields of gaseous CO and organic aerosols. Laboratory kinetic studies and field measurements of oxidation intermediates such as aldehydes are needed. Field measurements to determine the total organic content and principal organic species of precipitation are needed soon.

16. Test available photochemical theories by detecting and measuring reaction intermediates and products such as PAN, H_2O_2, CH_3OOH, and CH_2O in the atmosphere and comparing with chemical model results.

17. Determine mechanisms of chlorine loss and iodine enrichment in marine aerosols. Identify inorganic gaseous halogen species in the troposphere by experiment, and test theories for same.

REFERENCES

(1) Ball, R.J., and Robinson, G.D. 1982. The origin of haze in the central United States and its effect on solar radiation. J. Appl. Meteor. 21: 171-188.

(2) Bolin, B., and Charlson, R.J. 1976. On the role of the tropospheric sulfur cycle in the shortwave radiative climate of the Earth. Ambio 5: 47-54.

(3) Crutzen, P.J. 1973. A discussion of the chemistry of some minor constituents in the stratosphere and troposphere. Pure Appl. Geophys. 106-108: 1385-1399.

(4) Fishman, J.; Solomon, S.; and Crutzen, P.J. 1979. Observational and theoretical evidence in support of a significant in-situ photochemical source of tropospheric ozone. Tellus 31: 432-446.

(5) Lacis, A.; Hansen, J.; Lee, P.; Mitchell, T.; and Lebedeff, S. 1981. Greenhouse effect of trace gases. Geophys. Res. Lett. 8: 1035-1038.

(6) Levy, H. 1971. Normal atmosphere: Large radical and formaldehyde concentrations predicted. Science 173: 141-143.

(7) Liu, S.C.; Kley, D.; McFarland, M.; Mahlman, J.D.; and Levy, H. 1980. On the origin of tropospheric ozone. J. Geophys. Res. 85: 7546-7552.

(8) Logan, J.A.; Prather, M.J.; Wofsy, S.C.; and McElroy, M.B. 1981. Tropospheric chemistry: A global perspective. J. Geophys. Res. 86: 7210-7254.

(9) Molina, M.J., and Rowland, F.S. 1974. Stratospheric sink for chlorofluoromethanes: Chlorine-atom catalysed destruction of ozone. Nature 249: 810-812.

(10) Ramanathan, V. 1975. Greenhouse effect due to chlorofluorocarbons: Climate implications. Science 190: 50-52.

List of Participants

ANDERSEN, N.R.
Intergovernmental Oceanographic
Commission, UNESCO
75700 Paris, France

Field of research: Marine pollution research and monitoring

ANDREAE, M.O.
Dept. of Oceanography
Florida State University
Tallahassee, FL 32306, USA

Field of research: Air-sea exchange of organosulfur compounds; atmospheric chemistry of sulfur; marine chemistry of tin, germanium, antimony, and arsenic

ARNOLD, F.
Max-Planck-Institut für Kernphysik
6900 Heidelberg, F.R. Germany

Field of research: Ion chemistry - atmospheric ions and trace gases

AYERS, G.P.
Division of Cloud Physics, CSIRO
Sydney, Australia

Field of research: The physics and chemistry of clouds, atmospheric aerosols, and gases

BALZER, W.
Institut für Meereskunde
der Universität Kiel
2300 Kiel 1, F.R. Germany

Field of research: Redox-dependent processes at the sediment/water interface; calcium carbonate saturation state of seawater, uptake of CO_2

BINGEMER, H.G.
Universitätsinstitut für
Meteorologie und Geophysik
6000 Frankfurt am Main 1
F.R. Germany

Field of research: Emissions of reduced biogenic S-gases from oceans and continents, role of natural S-emissions in global S-cycle

BRINCKMAN, F.E.
Chemical and Biodegradation Processes
Group, National Bureau of Standards
Washington, DC 20234, USA

Field of research: Molecular characterization of biogenic organometal(loid)s and their fate in the environment

BRUNER, F.
University of Urbino
Istituto di Scienze Chimiche
61029 Urbino, Italy

Field of research: Assessment of quantitative methodologies for the determination of fluorochlorocarbons in the atmosphere with particular concern for possible tropospheric sinks for halocarbons

CICERONE, R.J.
National Center for Atmospheric
Research
Boulder, CO 80307, USA

Field of research: Atmospheric chemistry - modeling and measurements of trace gases

CRUTZEN, P.J.
Max-Planck-Institut für Chemie
Abt. Luftchemie
6500 Mainz, F.R. Germany

Field of research: Atmospheric chemistry

DUCE, R.A.
Center for Atmospheric Chemistry Studies, Graduate School of Oceanography
University of Rhode Island
Kingston, RI 02881, USA

Field of research: Global cycles of trace substances in the troposphere, with emphasis on the role of the ocean in these cycles

EHHALT, D.H.
Institut für Atmosphärische Chemie
Kernforschungsanlage Jülich
5170 Jülich, F.R. Germany

Field of research: Atmospheric chemistry - measurement of free radicals and light hydrocarbons in the atmosphere

GARRELS, R.M.
Dept. of Marine Science
University of South Florida
St. Petersburg, FL 33701, USA

Field of research: Global biogeochemical cycles

GEORGII, H.W.
Universitätsinstitut für Meteorologie und Geophysik
6000 Frankfurt am Main 1
F.R. Germany

Field of research: Atmospheric chemistry, microphysics of clouds, aerosol physics

GOLDBERG, E.D.
Geological Research Division
Scripps Institution of Oceanography
La Jolla, CA 92093, USA

Field of research: Environmental chemistry

GRAEDEL, T.E.
Bell Laboratories
Murray Hill, NJ 07974, USA

Field of research: Atmospheric chemistry, effects of atmospheric gases on materials

HAHN, J.H.
Max-Planck-Institut für Chemie
Abt. Luftchemie
6500 Mainz, F.R. Germany

Field of research: Tropospheric chemistry, organic compounds

HALLBERG, R.O.
Dept. of Geology
University of Stockholm
113 86 Stockholm, Sweden

Field of research: biogeochemistry of sediments, sedimentary sulfide formation, global sulfur budget, cycling of heavy metals at the sediment/water interface

HAMMER, C.U.
Geophysical Isotope Laboratory
2200 Copenhagen N, Denmark

Field of research: Glaciology and atmospheric chemistry

HORNIG, D.F.
Interdisciplinary Programs in Health
Harvard School of Public Health
Boston, MA 02115, USA

Field of research: Science and public policy

JAENICKE, R.
Institut für Meteorologie
der Universität Mainz
6500 Mainz, F.R. Germany

Field of research: Physical properties of atmospheric aerosols

JØRGENSEN, B.B.
Institute of Ecology and Genetics
University of Aarhus
8000 Aarhus C, Denmark

Field of research: Microbial processes in marine sediments

KLOCKOW, D.G.A.
Universität Dortmund, Abt. Chemie
4600 Dortmund 50, F.R. Germany

Field of research: Analytical chemistry of atmospheric trace compounds, especially strong acids and their ammonium salts

LISS, P.S.
School of Environmental Sciences
University of East Anglia
Norwich NR4 7TJ, England

Field of research: Environmental chemistry - gas transfer at natural air-water interfaces, estuarine chemistry, environmental redox chemistry

LORIUS, C.
Laboratoire de Glaciologie et
Géophysique de l'Environnement
38031 Grenoble Cedex, France

Field of research: Climate and atmospheric chemistry from ice core studies

LOVELOCK, J.E.
Coombe Mill Experimental Station
St. Giles on the Heath, Launceston
Cornwall PL15 9RY, England

Field of research: Planetary biology

MEIXNER, F.X.
Institut für Atmosphärische Chemie
Kernforschungsanlage Jülich
5170 Jülich, F.R. Germany

Field of research: Global atmospheric sulfur cycle, atmospheric reactions and transport of trace gases and aerosols, control and measurement of airborne pollutants

MOLINA, M.J.
Dept. of Chemistry
University of California
Irvine, CA 92717, USA

Field of research: Atmospheric chemistry, gas phase chemical kinetics, and photochemistry

MORGAN, J.J.
Environmental Engineering Science
California Institute of Technology
Pasadena, CA 91125, USA

Field of research: Aquatic chemistry - equilibria and kinetics of natural water systems

NIKI, H.
Scientific Research Laboratory
Ford Motor Company
Dearborn, MI 48121, USA

Field of research: atmospheric chemistry, gas phase chemical kinetics, and photochemistry

PENKETT, S.A.
Environmental and Medical Sciences
Division, AERE, Harwell
Oxfordshire OX11 ORA, England

Field of research: Atmospheric chemistry - chemical composition of the atmosphere

RODHE, H.
Dept. of Meteorology
University of Stockholm
106 91 Stockholm, Sweden

Field of research: Meteorological aspects of tropospheric chemistry, cycles of sulfur and nitrogen, precipitation scavenging

ROWLAND, F.S.
Dept. of Chemistry
University of California
Irvine, CA 92717, USA

Field of research: Atmospheric chemistry and chemical kinetics

RUDOLPH, J.
Institut für Atmosphärische Chemie
Kernforschungsanlage Jülich
5170 Jülich, F.R. Germany

Field of research: Measurements of stable trace gases in the remote troposphere, especially non-methane hydrocarbons

SCHMIDT, U.
Institut für Atmosphärische Chemie
Kernforschungsanlage Jülich
5170 Jülich, F.R. Germany

Field of research: Atmospheric cycles of trace gases, tropospheric and stratospheric measurements

SCHNEIDER, B.
Institut für Meereskunde
der Universität Kiel, Abt. Chemie
2300 Kiel 1, F.R. Germany

Field of research: Transport of pollutants into the Baltic Sea via the atmosphere (mainly trace metals)

SCHNEIDER, S.H.
National Center for Atmospheric Research
Boulder, CO 80307, USA

Field of research: Description, causes, and implications of climate and climatic change

SCHÜTZ, L.
Institut für Meteorologie
der Universität Mainz
6500 Mainz, F.R. Germany

Field of research: Physicochemical properties of atmospheric aerosols and long-range transport of particulate matter

SEILER, W.
Max-Planck-Institut für Chemie
Abt. Luftchemie
6500 Mainz, F.R. Germany

Field of research: Atmospheric chemistry - biosphere-atmosphere interaction

SLINN, W.G.N.
2215 Benton Avenue
Richland, WA 99352, USA

Field of research: Precipitation scavenging, dry deposition, and resuspension

STUIVER, M.
Quaternary Isotope Laboratory
University of Washington
Seattle, WA 98195, USA

Field of research: Geochemistry of the carbon cycle

WAGENBACH, D.
Institut für Umweltphysik
der Universität Heidelberg
6900 Heidelberg, F.R. Germany

Field of research: The chemical composition of atmospheric background-aerosol, including snow and ice-core studies

WALKER, J.C.G.
Space Physics Research Laboratory
Dept. of Atmospheric and Oceanic Science, University of Michigan
Ann Arbor, MI 48109, USA

Field of research: Aeronomy, atmospheric evolution

WIGLEY, T.M.L.
Climatic Research Unit
University of East Anglia
Norwich NR4 7TJ, England

Field of research: Climatology

WRIGHT, Jr., H.E.
Limnological Research Center
University of Minnesota
Minneapolis, MN 55455, USA

Field of research: Quaternary paleoecology

ZAFIRIOU, O.C.
Dept. of Chemistry
Woods Hole Oceanographic Institution
Woods Hole, MA 02543, USA

Field of research: Marine chemistry and marine atmospheric chemistry

ZELLNER, R.E.
Institut für Physikalische Chemie
der Universität Göttingen
3400 Göttingen, F.R. Germany

Field of research: Kinetics of elementary atmospheric reactions, theory of thermal rate constants

ZIMMERMAN, P.R.
National Center for Atmospheric Research
Boulder, CO 80307, USA

Field of research: Biosphere-atmosphere interactions

Subject Index

Author Index

Author Index